# IT MANAGEMENT TITLES
# FROM AUERBACH PUBLICATIONS AND CRC PRESS

**The Executive MBA in Information Security**
John J. Trinckes, Jr
ISBN: 978-1-4398-1007-1

**The Decision Model: A Business Logic Framework Linking Business and Technology**
Barbara von Halle and Larry Goldberg
ISBN: 978-1-4200-8281-4

**The SIM Guide to Enterprise Architecture**
Leon Kappelman, ed.
ISBN: 978-1-4398-1113-9

**Lean Six Sigma Secrets for the CIO**
William Bentley and Peter T. Davis
ISBN: 978-1-4398-0379-0

**Building an Enterprise-Wide Business Continuity Program**
Kelley Okolita
ISBN: 978-1-4200-8864-9

**Marketing IT Products and Services**
Jessica Keyes
ISBN: 978-1-4398-0319-6

**Cloud Computing: Implementation, Management, and Security**
John W. Rittinghouse and
James F. Ransome
ISBN: 978-1-4398-0680-7

**Data Protection: Governance, Risk Management, and Compliance**
David G. Hill
ISBN: 978-1-4398-0692-0

**Strategic Data Warehousing: Achieving Alignment with Business**
Neera Bhansali
ISBN: 978-1-4200-8394-1

**Mobile Enterprise Transition and Management**
Bhuvan Unhelkar
ISBN: 978-1-4200-7827-5

**The Green and Virtual Data Center**
Greg Schulz
ISBN: 978-1-4200-8666-9

**The Effective CIO**
Eric J. Brown, Jr. and William A. Yarberry
ISBN: 978-1-4200-6460-5

**Business Resumption Planning, Second Edition**
Leo A. Wrobel
ISBN: 978-0-8493-1459-9

**IT Auditing and Sarbanes-Oxley Compliance: Key Strategies for Business Improvement**
Dimitris N. Chorafas
ISBN: 978-1-4200-8617-1

**Best Practices in Business Technology Management**
Stephen J. Andriole
ISBN: 978-1-4200-6333-2

**Leading IT Projects: The IT Manager's Guide**
Jessica Keyes
ISBN: 978-1-4200-7082-8

**Knowledge Retention: Strategies and Solutions**
Jay Liebowitz
ISBN: 978-1-4200-6465-0

**The Business Value of IT**
Michael D. S. Harris, David Herron,
and Stasia Iwanicki
ISBN: 978-1-4200-6474-2

**Service-Oriented Architecture: SOA Strategy, Methodology, and Technology**
James P. Lawler and H. Howell-Barber
ISBN: 978-1-4200-4500-0

**Service Oriented Enterprises**
Setrag Khoshafian
ISBN: 978-0-8493-5360-4

# The Supply Chain in Manufacturing, Distribution, and Transportation

## *Modeling, Optimization, and Applications*

# The Supply Chain in Manufacturing, Distribution, and Transportation

## *Modeling, Optimization, and Applications*



Edited by
Kenneth D. Lawrence,
Ronald K. Klimberg,
and Virginia M. Miori

CRC Press
Taylor & Francis Group
Boca Raton London New York

CRC Press is an imprint of the
Taylor & Francis Group, an **informa** business
AN AUERBACH BOOK

Auerbach Publications
Taylor & Francis Group
6000 Broken Sound Parkway NW, Suite 300
Boca Raton, FL 33487-2742





Printed in the United States of America on acid-free paper
10 9 8 7 6 5 4 3 2 1

International Standard Book Number: 978-1-4200-7945-6 (Hardback)



**Library of Congress Cataloging-in-Publication Data**

The supply chain in manufacturing, distribution, and transportation : modeling, optimization, and applications / editors, Kenneth D. Lawrence, Ronald K. Klimberg, Virginia Miori.
p. cm.
Includes bibliographical references and index.
ISBN 978-1-4200-7945-6 (hardcover : alk. paper)
1. Inventory control. 2. Business logistics. 3. Production scheduling. 4. Physical distribution of goods. I. Lawrence, Kenneth D. II. Klimberg, Ronald K. III. Miori, Virginia.

TS160.S86 2011
658.7'87--dc22 2010032487

**Visit the Taylor & Francis Web site at**
**http://www.taylorandfrancis.com**

**and the Auerbach Web site at**
**http://www.auerbach-publications.com**

# Contents

# Preface

This volume is a blind-refereed, multi-authored volume. The objective of this volume is to present state-of-the-art studies in the areas of manufacturing, distribution, and transportation to solve significant problems within the supply chain integration process. This volume focuses on research that integrates the problems of production, distribution, and transportation.

Tactical models support the mid-level decision-making processes that typically extend into a planning horizon of 6 to 18 months. The models featured address a number of areas. High-level production schedules describe the equipment to be used and the hours that a production plant will operate. Product sourcing models assign customers to the most cost-efficient production plant or distribution center as a source of their orders. Network alignment models assist in determining the products to be produced in each production plant, and stored in each distribution center.

Additional tactical models focus on transportation operations with consistent demand. These operations will create static shipment schedules designed to be followed week after week. The physical layout of distribution centers is also a tactical decision. The product lines stored may change by season, requiring reexamination of product storage locations. The goal is the minimization of total distance traveled within the distribution center.

The area of inventory planning is a tactical area that has been the subject of substantial research. Inventory strategies begin with the determination of how much inventory to carry and at what inventory level to reorder the products. At their most complex level, inventory strategies address the possible postponement of final production processes in order to reduce costs. This is most common when standard subassemblies are used for many specialized final products. The subassemblies are lower valued and therefore less expansive to carry. The final production is postponed until an order has been placed for specific products.

Furthermore, operational models involve detailed or day-to-day operations and scheduling processes. The planning horizon for these models ranges from a week to several months. Manufacturing operations cannot effectively run without detailed planning models that schedule the raw material and intermediate product shipments.

These schedules feed into another class of models that schedules the production at these manufacturing facilities. Production schedules include the changeover between products and maintenance of equipment. In cases where multiple production lines are employed, the scheduling task grows very quickly in complexity. Operational planning in the transportation and distribution areas of the supply chain examine the consolidation of small shipments and the breaking down of larger shipments with the goal of cost-efficiency in transportation. In addition, models are addressed that create low-cost truck routes and assignment of various capacity vehicles to these routes. They also meet delivery time windows required by customers.

Both tactical and operational models rely on good quality forecasts of demand. Stochastic customer demand, coordination of supply chain functions, and solution algorithms are of a critical nature and are highlighted in this volume.

The overall integration of transportation, distribution, and production involves the following crossovers:

- Production and outbound transportation
- Inbound transportation and production
- Inbound transportation, production, and outbound transportation

This book is a compilation of scholarly research work involving the utilization of the discussed supply chain concepts, which address a wide variety of organizational issues. It is comprised of a variety of noteworthy works emanating not only from the academic spectrum, but also from business practitioners on a more limited basis.

The book is divided into three sections:

Section I: "Industrial and Service Applications of the Supply Chain"
Section II: "Analytic Probabilistic Models of Supply Chain Problems"
Section III: "Optimization Models of Supply Chain Problems"

# Introduction

## Section I: Industrial and Service Applications of the Supply Chain

Chapter 1, "Multicriteria Decision Making in Ethanol Production Problems: A Fuzzy Goal Programming Approach," applies a multiple objective approach to the optimization of the supply chain over the echelon of material sources, processing mills, and customers for the production of ethanol and associated by-products. The multiple objectives include cost minimization at all levels, as well as the minimization of environmental impact resulting from ethanol production.

Chapter 2, "From Push to Pull: The Automation and Heuristic Optimization of a Caseless Filler Line in the Dairy Industry," applies cost minimization in the transition of a dairy supply chain. The move from the manual "push" supply chain to automated "pull" supply chain not only provided greatly improved internal efficiencies, but also facilitated the opening of new customer channels for the dairy. Despite the fact that supply chain management has been applied extensively across manufacturing and production sectors, more traditional industries, such as the dairy industry, have lagged.

Chapter 3, "Optimization of Medical Services: The Supply Chain and Ethical Implications," provides a positive basis for the resolution of ethical questions while building an alternative "production style" supply chain for the health-care industry. This new supply chain is simulated and validated for continuing work in optimization. Manufacturing and service supply chain optimization models have always run parallel paths with little investigation into the benefits of applying the opposing techniques. While true in all sectors, it is especially the case in the health-care sector where issues of ethical treatment of patients are paramount.

Chapter 4, "Using Hierarchical Planning to Exploit Supply Chain Flexibility: An Example from the Norwegian Meat Industry," provides a supply chain optimization example from the Norwegian meat industry. One of the greatest challenges in developing optimal solutions within the supply chain setting is the existence of stochastic elements in the supply chain. Whether changes occur on a daily basis or less frequently over time, a hierarchical approach to optimal supply chains offers sufficient flexibility to manage these changes.

Chapter 5, "Transforming U.S. Army Supply Chains: An Analytical Architecture for Management Innovation," addresses the need for the U.S. Army to transform from the existing logistics approach to the challenges faced in the world today. It examines the current structure, proposes alternative models, and highlights challenges that will be faced in the ultimate transformation of the U.S. Army supply chains. It is widely understood that the most challenging logistics operations exist within the armed forces. Since World War II, operations research has played a significant role in these operations.

## Section II: Analytic Probabilistic Models of Supply Chain Problems

Chapter 6, "A Determination of the Optimal Level of Collaboration between a Contractor and Its Suppliers under Demand Uncertainty," focuses on the analysis of the collaboration level connected to demand uncertainty and its associated economic costs, based on the number of suppliers utilized by a contractor to maximize the supply chain profile.

Chapter 7, "Online Auction Models and Their Impact on Sourcing and Supply Management," concerns how business relies on online auctions to enhance efficiency and reduce costs within a supply chain. The chapter focuses on the product sourcing problems, and details existing bidding models and organizational dynamics that may influence or be employed to improve bidding strategy.

Chapter 8, "Analytical Models for Integrating Supplier Selection and Inventory Decisions," focuses on analytical models for integrating supplier selection and inventory decisions. The models involve such factors as long-term relationships, quality, delivery performance, quantity discounts, replenishment quantity and timing, and procurement and contractual costs.

Chapter 9, "Inventory Optimization of Small Business Supply Chains with Stochastic Demand," examines the supply chain in a small seasonal business and inventory optimization with stochastic demand.

## Section III: Optimization Models of Supply Chain Problems

Chapter 10, "A Dynamic Programming Approach to the Stochastic Truckload Routing Problem," addresses a continuously changing and challenging problem faced by industry. In addition to traditional out-and-back routing, trucks are permitted to extend their routes until restrictions are reached. This chapter develops and presents the results of applying a dynamic programming solution approach to solve an innovative triplet formulation to solve the stochastic truckload routing problem with time windows.

Chapter 11, "Modeling Data Envelopment Analysis (DEA) Efficient Location/Allocation Decisions II," focuses on the multi-objective nature of the optimal location of facilities. The results of these models have a significant impact on a company's operations and costs. This chapter extends the authors' previous pioneering work of solving the location model and an efficiency model simultaneously by permitting the outputs to be variable. The model is now nonlinear. The results of applying the nonlinear model are presented.

Chapter 12, "Sourcing Models for End-of-Use Products in a Closed-Loop Supply Chain," addresses how a green company places the closed-loop supply chain, which is a combination of the traditional and reverse supply chains, as an integral part of environmentally conscious manufacturing companies. Critical to these environmentally conscious manufacturers is the identification of appropriate end-of-use products from appropriate suppliers. To address these issues, a linear physical programming model is developed to address the desirability selection of a product to be reprocessed and a model that is a combination of analytic network process and goal programming is developed to select suitable suppliers.

Chapter 13, "A Bi-Objective Supply Chain Scheduling," focuses on the coordination of manufacturing and supply with the production and distribution of products as one of the key issues in supply chain management. Integrated and hierarchical approaches are presented and compared to solve the bi-objective, maximize customer service and minimize inventory holding cost, the problem of determining a customer-driven supply chain (i.e., a coordinated schedule for the manufacture of parts by each supplier), for the delivery of parts from each supplier to the producers, and for the assignment of orders to planning periods at the producer. The results from some computational examples are presented.

Chapter 14, "Applying Data Envelopment Analysis and Multiple Objective Data Envelopment Analysis to Identify Successful Pharmaceutical Companies," addresses the development of superior forecasts as a key to a successful supply chain. The chapter presents an innovative approach that incorporates multidimensional performance variables into the regression forecasting model. Results from applying this methodology to a real data set of fifty pharmaceutical companies are presented.

# About the Editors

**Kenneth D. Lawrence, Ph.D.**, is a professor of management and marketing science and decision support systems in the School of Management at the New Jersey Institute of Technology. His professional employment includes over 20 years of technical management experience with AT&T as Director, Decision Support Systems and Marketing Demand Analysis; Hoffmann-La Roche, Inc.; Prudential Insurance; and the U.S. Army in forecasting, marketing planning and research, statistical analysis, and operations research. He is a full member of the graduate doctoral faculty of management at Rutgers, the State University of New Jersey, in the Department of Management Science and Information Systems. Dr. Lawrence has served as the doctoral chairman and thesis advisor for four Rutgers doctoral students. He is a member of the graduate faculty at NJIT in management, transportation, statistics, and industrial engineering. He is also an active participant in professional associations, including the Decision Sciences Institute, Institute of Management Science, Institute of Industrial Engineers, American Statistical Association, and Institute of Forecasters. He has conducted significant funded research projects in health care and transportation.

Professor Lawrence is the associate editor of the *Journal of Statistical Computation and Simulation* and the *Review of Quantitative Finance and Accounting*, and also serves on the editorial boards of *Computers and Operations Research* and the *Journal of Operations Management.* His research work has been cited hundreds of times in 74 different journals, including *Computers and Operations Research, International Journal of Forecasting, Journal of Marketing, Sloan Management Review, Management Science, Sloan Management Review, Technometrics, Applied Statistics, Interfaces, International Journal of Physical Distribution and Logistics,* and the *Journal of the Academy of Marketing Science.* Some articles were published decades ago are often cited. He has 275 publications in the areas of multicriteria decision analysis, management science, statistics, and forecasting, and his articles have appeared in more than 25 journals, including the *European Journal of Operational Research, Computers and Operations Research, Operational Research Quarterly,* and *International Journal of Forecasting and Technometrics.*

Dr. Lawrence is the 1989 recipient of the Institute of Industrial Engineers Award for significant accomplishments in the theory and applications of operations research.

He was recognized in the February 1993 issue of the *Journal of Marketing* for his significant contribution in developing a method of guessing in the no data case, for diffusion of new products, and for forecasting the timing and the magnitude of the peak in the adaption rate. Lawrence is also a member of the following honorary societies: Alpha Iota Delta (Decision Sciences Institute) and Beta Gamma Sigma (Schools of Management). He is the recipient of the 2002 Bright Ideas Award in the New Jersey Policy Research Organization and the New Jersey Business and Industry Associates for his work in auditing, for his use of a goal programming model to improve the efficiency of audit sampling.

In February 2004, Dean Howard Tuckman of Rutgers University appointed Lawrence as an Academic Research Fellow to the Center for Supply Chain Management, because "his reputation and strong body of research is quite impressive." The Center's corporate sponsors include Bayer HealthCare, Hoffmann-LaRoche, IBM, Johnson & Johnson, Merck, Novartis, PeopleSoft, Pfizer, PSE&G, Schering-Plough, and UPS.

Recognition of Professor Lawrence's research work is found in its broad citation, in various sources, in publishing in the finest research publication outlets, and in the recognition of his research abilities and skills by publications in companies and journal editors who continually seek him as a referee and editor. Lawrence's own editorial works are characterized by a thorough blend of the refereed process and contributions by highly recognized scholars, including Nobel Prize winners and chaired professors from both domestic and international universities who are considered worldwide experts in their fields. A majority of these publications are in blind-refereed publications. The Institute of Industrial Engineers honored Lawrence for his "significant accomplishments in the field of operations research, both in application and theory."

Lawrence's research has been listed as breakthrough research by the *Journal of the American Marketing Association* over a period of 35 years. His research in the Rutgers doctoral program has resulted in the awarding of multiple doctoral degrees under his direction. Furthermore, his research expertise and skills have led to his frequent participation on other doctoral dissertation committees.

Another measure of the quality of his research work is a record of multiple fundings of his research work. Furthermore, his work with various high-quality publishing firms in the development of educational material for textbooks also signifies the high quality of his work.

**Ronald K. Klimberg, Ph.D.,** is a professor in the Decision and System Sciences Department of the Haub School of Business at Saint Joseph's University, Philadelphia, Pennsylvania. He received his B.S. in Information Systems from the University of Maryland, his M.S. in operations research from George Washington University, and his Ph.D. in systems analysis and economics for Public Decision Making from the Johns Hopkins University. Before joining the faculty of Saint Joseph's University in 1997, he was a professor at Boston University (10 years), an operations research analyst for the Food and Drug Administration (FDA) (10 years), and a consultant

(7 years). Klimberg was the 2007 recipient of the Tengelmann Award for his excellence in scholarship, teaching, and research.

Dr. Klimberg's research has been directed toward the development and application of quantitative methods, for example, statistics, forecasting, data mining, and management science techniques, such that the results add value to the organization and are effectively communicated. He has published over 30 articles and made over 30 presentations at national and international conferences in the areas of management science, information systems, statistics, and operations management. His current major interests include multiple criteria decision making (MCDM), multiple objective linear programming (MOLP), data envelopment analysis (DEA), facility location, data visualization, data mining, risk analysis, workforce scheduling, and modeling in general. He is currently a member of INFORMS, DSI, and MCDM.

**Virginia M. Miori, Ph.D.,** is an assistant professor in the Decision and System Sciences Department of the Erivan J. Haub School of Business at Saint Joseph's University in Philadelphia, Pennsylvania. She currently has several research streams, the most prolific being in the areas of supply chain modeling, production scheduling optimization, and transportation optimization. She is also working in the areas of health-care supply chain modeling and optimization, EMBA team evaluations using the analytic hierarchy process, and statistical examination of ethical behavior and expectations among high school and college students.

Miori has 12 years of teaching experience and has accumulated more than 15 years of experience in developing and implementing operations research models. These models are applied to problems in the chemical industry, manufacturing industries, logistics, transportation, and supply chain management. She has published a number of articles and presented at numerous conferences. The presentations have been offered in both refereed and invited capacities.

An award for outstanding dissertation was presented to Miori by Drexel University for her work in stochastic truckload transportation optimization. An outstanding research award was also presented by Saint Joseph's University for adapting her transportation scheduling model for use in the dairy production scheduling arena. This work eventually led to the development of a commercial scheduling optimization software product.

As for her extensive educational background, in 2006 Miori earned a doctorate in operations research from the Bennett S. LeBow College of Business at Drexel University. She also holds a Master of Science degree in transportation from the School of Engineering and Applied Sciences at the University of Pennsylvania and a Master of Science degree in operations research from Case Western Reserve University.

Her teaching activities at Saint Joseph's University include both undergraduate and graduate courses in business statistics, quantitative methods, research skills, foundations and applications of Six Sigma for manufacturing, foundations and applications of Six Sigma for service industries, developing decision-making competencies, foundations for business intelligence, and applications of business intelligence.

# Contributors

**Gerard Campagna**
Haub School of Business
Saint Joseph's University
Philadelphia, Pennsylvania

**Kathleen Campbell**
Haub School of Business
Saint Joseph's University
Philadelphia, Pennsylvania

**Anthony Costanzo**
Haub School of Business
Saint Joseph's University
Philadelphia, Pennsylvania

**Surendra M. Gupta**
Laboratory for Responsible Manufacturing
Department of Mechanical and Industrial Engineering
Northeastern University
Boston, Massachusetts

**Christopher M. Keller**
College of Business
East Carolina University
Greenville, North Carolina

**Burcu B. Keskin**
Department of Information Systems, Statistics, and Management Science
University of Alabama
Tuscaloosa, Alabama

**Ronald K. Klimberg**
Haub School of Business
Saint Joseph's University
Philadelphia, Pennsylvania

**John F. Kros**
College of Business
East Carolina University
Greenville, North Carolina

**Hisashi Kurata**
Graduate School of International Management
International University of Japan
Niigata, Japan

**Kenneth D. Lawrence**
School of Management
New Jersey Institute of Technology
Newark, New Jersey

**Sheila M. Lawrence**
Department of MSIS
Rutgers University
Piscataway, New Jersey

**Christopher Matthews**
Haub School of Business
Saint Joseph's University
Philadelphia, Pennsylvania

**Daniel J. Miori**
Millard Fillmore Gates Circle Hospital
Buffalo, New York

**Virginia M. Miori**
Haub School of Business
Saint Joseph's University
Philadelphia, Pennsylvania

**Daniel Mrazik**
Haub School of Business
Saint Joseph's University
Philadelphia, Pennsylvania

**Seong-Hyun Nam**
Department of Management
College of Business and Public Administration
University of North Dakota
Grand Forks, North Dakota

**Dinesh R. Pai**
Department of Business Administration
The Pennsylvania State University
Center Valley, Pennsylvania

**Greg H. Parlier**
Senior Systems Analysis, SAIC
Engineering and Analysis Operation
Institute for Defense Analysis
Huntsville, Alabama

**Kishore K. Pochampally**
Department of Quantitative Studies, Operations and Project Management
School of Business
Southern New Hampshire University
Manchester, New Hampshire

**Harold Rahmlow**
Haub School of Business
Saint Joseph's University
Philadelphia, Pennsylvania

**Samuel J. Ratick**
The George Perkins Marsh Research Institute
The Graduate School of Geography
Clark University
Worcester, Massachusetts

**Tadeusz Sawik**
Department of Operations Research and Information Technology
AGH University of Science & Technology
Krakow, Poland

**Peter Schütz**
Department of Applied Economics
SINTEF Technology & Society
Trondheim, Norway

**Brian W. Segulin**
RoviSys
Aurora, Ohio

**George P. Sillup**
Haub School of Business
Saint Joseph's University
Philadelphia, Pennsylvania

**Vinay Tavva**
Haub School of Business
Saint Joseph's University
Philadelphia, Pennsylvania

**Asgeir Tomasgard**
Department of Industrial Economics and Technology Management
Norwegian University of Science and Technology
and
Department of Applied Economics
SINTEF Technology & Society
Trondheim, Norway

**Kristin Tolstad Uggen**
Department of Applied Economics
SINTEF Technology & Society
Trondheim, Norway

**John Vitton**
Department of Management
College of Business and Public Administration
University of North Dakota
Grand Forks, North Dakota

**Sasanka Vuyyuru**
Haub School of Business
Saint Joseph's University
Philadelphia, Pennsylvania

**George Webster**
Haub School of Business
Saint Joseph's University
Philadelphia, Pennsylvania

# Review Board

# I

# INDUSTRIAL AND SERVICE APPLICATIONS OF THE SUPPLY CHAIN

# Chapter 1

# Multicriteria Decision Making in Ethanol Production Problems: A Fuzzy Goal Programming Approach

Kenneth D. Lawrence, Dinesh R. Pai,
Ronald K. Klimberg, and Sheila M. Lawrence

## Contents

## 1.1 Introduction

Rising fuel prices and policy initiatives have continued to stimulate renewable fuels, including ethanol. Since the mid-1990s the number of ethanol facilities and the plant size have increased gradually. By mid-2006, nearly a hundred ethanol facilities in the United States were producing more than 4 billion gallons of ethanol annually, with 50 to 100 million gallon-per-year plants as standard size (EAA 2006). This chapter deals with a supply chain problem involving the production of ethanol and various by-products. The process includes material sources, the processing mills, and the customers. The primary objectives are to minimize the cost of source materials, production (wet or dry milling), and transportation of final products. Additionally, the minimization of pollution at milling sites is another important management objective.

The goal programming (GP) technique provides an analytical framework that a decision maker can use to provide optimal solutions to multicriteria and conflicting objectives. The GP and its variants have been applied to a wide variety of problems (Ignizio 1976, Romero 1991). The use of GP in process industry problems is not new. Krajnc et al. (2007) have investigated the possibilities of attaining zero-waste emissions in the case of sugar production. Arthur and Lawrence (1982) designed a GP model to develop production and shipping patterns for the chemical and pharmaceutical industries.

The model presented in this chapter is designed to illustrate how preemptive GP can be used as an aid in solving multicriteria production-related problems. Our ultimate goal is to develop a fuzzy goal programming (FGP) model with appropriate tolerance limits. Zimmermann (1978) posed the first method for solving fuzzy linear programming (FLP) problems. Fuzzy optimization focuses first on solving models that reflect real-life uncertainty, and second on transforming them into equivalent crisp problems that benefit from efficient existing solving algorithms. Fuzzy decision is a combination of goals and constraints because it considers that the best fuzzy decision is the union of the aggregated intersections of goal and constraints (Bellman and Zadeh 1970).

## 1.2 Formulation

To formulate the model, we define the following:

### 1.2.1 Notations

Indices:

- $i$: Index of the sources of corn
- $j$: Index of the production process
- $k$: Index of the product (ethanol, corn oil, dry meal, corn gluten, livestock feeding, waste)
- $l$: Index of the customer groups

### 1.2.2 Sets

$I$: Set of the sources of corn
$J$: Set of the production process
$K$: Set of the products
$L$: Set of the customer groups

### 1.2.3 Parameters

$d_{kl}$ = Monthly demand for product type $k$ for customer group $l$
$C_{ijkl}$ = Total cost of producing and shipping the $k$-th product type (including raw materials) from the $I$-th corn source through the $j$-th milling location to the $l$-th customer group
$t_{jk}$ = Unit time to produce a unit of he $k$-th output type at the $j$-th mill
$b_j$ = Production capacity at the $j$-th mill
$p_{jk}$ = Pollution level at the $j$-th mill for the production of a unit of the $k$-th output type (gallons of water)
$H_j$ = Number of hours available on a yearly basis for the $j$-th mill
$TC$ = Total cost
$PG_j$ = Pollution limit for the $j$-th mill
$TW$ = Total waste generated
$P_q$ = Priority labels, where $q$ = 1, 2, 3

### 1.2.4 Variables

$X_{ijkl}$ = Amount of product type $k$ produced from corn from source $i$ in mill type $j$ for customer group $l$
$d_1^+$ = Deviation variable of overachievement of Goal 1
$d_1^-$ = Deviation variable of underachievement of Goal 1
$d_2^+$ = Deviation variable of overachievement of Goal 2
$d_2^-$ = Deviation variable of underachievement of Goal 2
$d_3^+$ = Deviation variable of overachievement of Goal 3
$d_3^-$ = Deviation variable of underachievement of Goal 3

### 1.2.5 Goal Constraints and Objective Functions

Goal 1: Minimize total cost:

$$Z_1 = \sum_{l=1}^{L}\sum_{k=1}^{K}\sum_{j=1}^{J}\sum_{i=1}^{I} C_{ijkl} \cdot X_{ijkl} - d_1^+ + d_1^- \leq TC, \quad X_{ijkl} \geq 0 \tag{1.1}$$

Goal 2: Reduction in level of pollution for the $j$-th mill:

$$Z_2 = \sum_{k=1}^{K}\sum_{j=1}^{J} p_{jk} \cdot X_{ijkl} - d_2^+ + d_2^- \leq PG_j \quad \forall i = 1, ..., I, \quad \forall j = 1, ..., J \tag{1.2}$$

Goal 3: Reduction in level of waste:

$$Z_3 = \sum_{i=1}^{I} \sum_{j=1}^{J} \sum_{l=1}^{L} X_{ijkl} - d_3^+ + d_3^- \leq TW \quad \text{for} \quad k = 6 \tag{1.3}$$

The objective function of the model is to minimize the deviation variables corresponding to various goals. Highest priority is assigned to the total cost goal—to minimize the total cost. Therefore, the undesirable deviational variable in the first priority is overachievement of Goal 1; that is, $d_1^+$ should be minimized. Then, the first priority function in the objective function is $P_1 d_1^+$. In the second priority of pollution level, we wish to minimize the pollution level from each mill to a predetermined safe limit. Therefore, the undesirable deviational variable in the second priority goal is overachievement of Goal 2; that is, $d_2^+$ should be minimized. Then, the second priority function in the objective function is $P_2 d_2^+$. Finally, in the third priority of waste level, we wish to restrict the waste produced within a predetermined limit. Therefore, the undesirable deviational variable in the third priority goal is overachievement of Goal 3; that is, $d_3^+$ should be minimized. Then, the third priority function in the objective function is $P_3 d_3^+$.

Hence, the objective function for the goal programming model is

$$\text{Minimize: } Z = P_1 d_1^+ + P_2 d_2^+ + P_3 d_3^+ \tag{1.4}$$

### 1.2.6 Constraints

The objective functions formulated in the previous section are restricted by two sets of constraints. They are the demand constraints and the time constraints.

$$\sum_{k=1}^{K} \sum_{l=1}^{L} X_{ijkl} \geq d_{kl}, \quad \forall i = 1, \ldots, I, \quad \forall j = 1, \ldots, J \tag{1.5}$$

$$\sum_{l=1}^{L} \sum_{k=1}^{K} \sum_{i=1}^{I} t_{ijk} \cdot X_{ijkl} \leq H_j \quad \forall j = 1, \ldots, J \tag{1.6}$$

$$X_{ijkl}, d_1^+, d_1^-, d_2^+, d_2^-, d_3^+, d_3^- \geq 0, i \in I, j \in J, k \in K, l \in L \tag{1.7}$$

Constraint (1.5) ensures that the customer demands are met, whereas constraint (1.6) limits the hours available for processing on each type of milling. Constraint (1.7) ensures that all the decision variables are non-negative.

## 1.3 Transformation of Fuzzy Goals

In fuzzy goal programming (FGP), the membership function corresponding to the $k$-th fuzzy goal of type $z_k(x) \leq b_k$ $z_k(x) \leq b_k$ is defined as

$$\mu_{z_k}(x) = \begin{cases} 1 & z_k(x) \leq b_k \\ \dfrac{(b_k + t_k^u) - z_k(x)}{t_k^u} & \text{if} \quad b_k < z_k(x) \leq b_k + t_k^u \\ 0 & z_k(x) > b_k + t_k^u \end{cases}$$

where $t_k^u$ is the upper tolerance limit and $\mu_{z_k}(x) \in [0, 1], \forall k$ represents the membership grade of achieving the goal, with 0 and 1 representing the lowest and highest grades, respectively. The membership grade depends on the specified tolerance value given in the decision-making context (Sharma et al. 2007).

In the considered FGP model of the production of ethanol and various by-products problem, the minimize total cost goal [Goal 1, Equation (1.1)], reduce level of pollution goal [Goal 2, Equation (1.2)], and reduction level of waste goal [Goal 3, Equation (1.3)] are of the type $z_k(x) \leq b_k$. If the above goals are completely achieved, then no tolerances for them are needed and the grades of membership for the goals should be unity. When these goals are either perfectly or partially unachieved, tolerances for them are required. Kim and Whang (1998) used the concept of tolerance to convert an FGP model to a single-objective LP problem.

If $u_1^+$ and $u_1^-$ are the upper and lower tolerance limits, respectively, and $\lambda_1$ is the membership grade of the total cost goal, then the goal can be transformed as follows:

$$\sum_{l=1}^{L} \sum_{k=1}^{K} \sum_{j=1}^{J} \sum_{i=1}^{I} C_{ijkl} \cdot X_{ijkl} - \beta_1^+ u_1^+ + \beta_1^- u_1^- \leq TC$$

where $\beta_1^+ = 1 - \lambda_1$ and $\beta_1^- = 1 - \lambda_1$.

The reduction in the level of pollution goal can be transformed as

$$\sum_{k=1}^{K} \sum_{j=1}^{J} p_{jk} \cdot X_{ijkl} - \beta_{2,j}^+ u_{2,j}^+ + \beta_{2,j}^- u_{2,j}^- \leq PG_j, \forall j$$

where $\beta_{2,j}^+ = 1 - \lambda_{2,j}$, $\beta_{2,j}^- = 1 - \lambda_{2,j}$; $u_{2,j}^+$ and $u_{2,j}^-$ are the upper and lower tolerance limits, respectively; and $\lambda_{2,j}$ is the membership grade of the pollution reduction goals in mill $j$.

Finally, the reduction in the level of waste goal can be transformed as

$$\sum_{i=1}^{I}\sum_{j=1}^{J}\sum_{l=1}^{L} X_{ijkl} - \beta_3^+ u_3^+ + \beta_3^- u_3^- \leq TW$$

where $\beta_3^+ = 1 - \lambda_3$; $\beta_3^- = 1 - \lambda_3$; $u_3^+$ and $u_3^-$ are the upper and lower tolerance limits, respectively; and $\lambda_3$ is the membership grade of the waste reduction goal.

## 1.4 Formulation of Objective Function

The fuzzy goals for the problem are transformed to their respective linear constraint form. In this formulation, as the tolerance variables are to be minimized, the tolerances needed will be close to unity for each fuzzy goal. This causes the grade of membership to become larger. In particular, if the tolerance variables are zero, then there is no need to assign tolerances to fuzzy goals. Therefore, the objective function for the ethanol production problem is defined as (Kim and Whang 1998)

$$\text{Min:}\quad w_1^1\beta_1^+ + w_1^2\beta_1^- + \sum_{j=1}^{2} w_{2,j}^1\beta_{2,j}^+ + \sum_{j=1}^{2} w_{2,j}^2\beta_{2,j}^- + w_3^1\beta_3^+ + w_3^2\beta_3^-$$

where $w_1^1$, $w_1^2$, $w_{2,j}^1$, $w_{2,j}^2$, $w_3^1$, and $w_3^2$ are the respective weights corresponding to the fuzzy goals, and the sum of all the weights is one.

## 1.5 Final Form

The final LP form of the ethanol production problem is obtained as follows:

$$\text{Min:}\quad w_1^1\beta_1^+ + w_1^2\beta_1^- + \sum_{j=1}^{2} w_{2,j}^1\beta_{2,j}^+ + \sum_{j=1}^{2} w_{2,j}^2\beta_{2,j}^- + w_3^1\beta_3^+ + w_3^2\beta_3^- \tag{1.8}$$

$$\sum_{l=1}^{L}\sum_{k=1}^{K}\sum_{j=1}^{J}\sum_{i=1}^{I} C_{ijkl}\cdot X_{ijkl} - \beta_1^+ u_1^+ + \beta_1^- u_1^- \leq TC, \tag{1.9}$$

$$\sum_{k=1}^{K}\sum_{j=1}^{J} p_{jk}\cdot X_{ijkl} - \beta_{2,j}^+ u_{2,j}^+ + \beta_{2,j}^- u_{2,j}^- \leq PG_j, \forall j \tag{1.10}$$

$$\sum_{i=1}^{I}\sum_{j=1}^{J}\sum_{l=1}^{L} X_{ijkl} - \beta_3^+ u_3^+ \leq TW \tag{1.11}$$

$$\sum_{k=1}^{K}\sum_{l=1}^{L} X_{ijkl} \geq d_{kl} \quad \forall i = 1, \ldots, I, \quad \forall j = 1, \ldots, J \tag{1.12}$$

$$\sum_{l=1}^{L}\sum_{k=1}^{K}\sum_{i=1}^{I} t_{ijk} \cdot X_{ijkl} \leq H_j \quad \forall j = 1, \ldots, J \tag{1.13}$$

$$X_{ijkl} \geq 0, \quad i \in I, \quad j \in J, \quad k \in K, \quad l \in L \tag{1.14}$$

$$\beta_1^+, \beta_{2,j}^+, \beta_3^+ \geq 0, \forall j \tag{1.15}$$

$$w_1^1 + w_1^2 + \sum_{j=1}^{2} w_{2,j}^1 + \sum_{j=1}^{2} w_{2,j}^2 + w_3^1 + w_3^2 = 1 \tag{1.16}$$

## 1.6 Results

The model is formulated and executed using Frontline Premium Solver. The production details and goal achievement values corresponding to the preemptive GP and two different weighting structures of the fuzzy GP are presented in Table 1.1.

The results show that all three methods have comparable performances and successfully achieve the three goals. All three methods require no lower tolerance limits for each of the three goals. However, resource utilization in the case of preemptive GP is higher than that with fuzzy GP.

## 1.7 Conclusion

The objective of this study was to present an FGP model for a supply chain problem involving the production of ethanol and various by-products. The output of our research may become a useful analytical tool for ethanol producers that are using traditional LP and GP methods for recommendations to the producers on optimal land allocation for different varieties of ethanol in the planning process. In this study, we were able to demonstrate that the FGP approach is a better technique than a single-objective criterion when multiple conflicting objectives are involved. The model developed provides the best possible solution subject to the model constraints. Sensitivity analysis considering two different weighting structures of the goals has been performed to see the adaptability of the proposed model. Results may be tested and verified corresponding to other weighting structures specified by the decision maker, depending on the production planning situation.

**Table 1.1 Production Details (In 1000 Tons)**

| *Products* | *Premptive GP* | *Fuzzy GP (Equal Weights)* | *Fuzzy GP (Unequal Weights)* |
|---|---|---|---|
| X1111 | 0.00 | 0.00 | 0.00 |
| X1112 | 0.00 | 0.00 | 0.00 |
| X1121 | 200.00 | 200.00 | 200.00 |
| X1122 | 400.00 | 373.33 | 373.33 |
| X1131 | 0.00 | 0.00 | 0.00 |
| X1132 | 0.00 | 0.00 | 0.00 |
| X1141 | 150.00 | 150.00 | 150.00 |
| X1142 | 340.00 | 340.00 | 340.00 |
| X1151 | 0.00 | 400.00 | 400.00 |
| X1152 | 650.00 | 650.00 | 650.00 |
| X1161 | 153.82 | 0.00 | 0.00 |
| X1162 | 100.00 | 0.00 | 0.00 |
| X1211 | 500.00 | 0.00 | 0.00 |
| X1212 | 600.00 | 600.00 | 600.00 |
| X1221 | 0.00 | 0.00 | 0.00 |
| X1222 | 0.00 | 26.67 | 26.67 |
| X1231 | 300.00 | 300.00 | 300.00 |
| X1232 | 0.00 | 500.00 | 500.00 |
| X1241 | 0.00 | 0.00 | 0.00 |
| X1242 | 0.00 | 0.00 | 0.00 |
| X1251 | 0.00 | 0.00 | 0.00 |
| X1252 | 0.00 | 0.00 | 0.00 |
| X1261 | 0.00 | 500.00 | 500.00 |
| X1262 | 0.00 | 0.00 | 0.00 |
| X2111 | 0.00 | 0.00 | 0.00 |

**Table 1.1 Production Details (In 1000 Tons) (Continued)**

| *Products* | *Premptive GP* | *Fuzzy GP (Equal Weights)* | *Fuzzy GP (Unequal Weights)* |
|---|---|---|---|
| X2112 | 0.00 | 0.00 | 0.00 |
| X2121 | 0.00 | 0.00 | 0.00 |
| X2122 | 0.00 | 0.00 | 0.00 |
| X2131 | 0.00 | 0.00 | 0.00 |
| X2132 | 0.00 | 0.00 | 0.00 |
| X2141 | 0.00 | 0.00 | 0.00 |
| X2142 | 0.00 | 0.00 | 0.00 |
| X2151 | 400.00 | 0.00 | 0.00 |
| X2152 | 0.00 | 0.00 | 0.00 |
| X2161 | 0.00 | 0.00 | 0.00 |
| X2162 | 0.00 | 0.00 | 0.00 |
| X2211 | 0.00 | 500.00 | 500.00 |
| X2212 | 0.00 | 0.00 | 0.00 |
| X2221 | 0.00 | 0.00 | 0.00 |
| X2222 | 0.00 | 0.00 | 0.00 |
| X2231 | 0.00 | 0.00 | 0.00 |
| X2232 | 500.00 | 0.00 | 0.00 |
| X2241 | 0.00 | 0.00 | 0.00 |
| X2242 | 0.00 | 0.00 | 0.00 |
| X2251 | 0.00 | 0.00 | 0.00 |
| X2252 | 0.00 | 0.00 | 0.00 |
| X2261 | 346.18 | 0.00 | 0.00 |
| X2262 | 0.00 | 100.00 | 100.00 |

# References

Arthur, J., and Lawrence, K. (1982), Multiple goal production and logistics planning in a chemical and pharmaceutical company, *Computers & Operations Research,* 9(2), 127–137.

Bellman, R., and Zadeh, L. (1970), Decision-making in a fuzzy environment, *Management Science,* 17(4), 141–164.

EAA (2006), *Economic Impacts of Ethanol Production,* Bethesda, MD: Ethanol Across America.

Ignizio, J. (1976), *Goal Programming and Extensions,* Lanham, MD: Lexington Books.

Kim, J.S., and Whang, K. (1998), A tolerance approach to the fuzzy goal programming problems with unbalanced triangular membership function, *European Journal of Operational Research,* 107, 614–624.

Krajnc, D., Mele, M., and Glavic, P. (2007), Improving the economic and environmental performances of the beet sugar industry in Slovenia: increasing fuel efficiency and using by-products for ethanol, *Journal of Cleaner Production,* 15, 1240–1252.

Romero, C. (1991), *Handbook of Critical Issues in Goal Programming,* Oxford: Pergamon Press.

Sharma, D., Jana, R., and Gaur, A. (2007), Fuzzy goal programming for agricultural land allocation problems, *Yugoslav Journal of Operations Research,* 17(1), 31–42.

Zimmermann, H. (1978), Fuzzy programming and linear programming with several objective functions, *Fuzzy Sets and Systems,* 1, 45–55.

*Chapter 2*

# From Push to Pull: The Automation and Heuristic Optimization of a Caseless Filler Line in the Dairy Industry

Brian W. Segulin

## Contents

## 2.1 Introduction

The MicroDairy concept conceived by Superior Dairy coupled with Creative Edge Technologies located in Canton, Ohio, is the result of a desire to gain operational, production, and logistical efficiencies in the dairy industry. The coupling of a redesign of the traditional milk jug, an innovative production scheduling model, and a supervisory control system led to optimal control of the dairy manufacturing process.

The traditional gallon milk jug was redesigned. The traditional container did not possess the design or structural integrity to allow for shipping without additional support. A milk crate was used to provide the necessary support. The support was about the only beneficial thing provided by the milk crate. The milk crate contributed inefficiencies to production, shipping, and customer use.

Prior to each use, the milk crates had to be washed. The production process required an extra step of loading the filled containers into the crates. The crates added additional shipping weight, resulting in lower shipping capacities and higher shipping costs. Drivers were saddled with the task of retrieving crates and returning them to the dairy, thus increasing transportation costs. The customers (i.e., retailers) were forced to handle the product prior to presentation to consumers. The retailer also had to store the empty crates for return to the dairy.

The new container was deigned to be stackable, thus eliminating the need for the milk crate. With the elimination of the milk crate, the entire production process was open for reevaluation. Operational goals for the new production line design focused on efficiency and flexibility. Production goals specified a minimum acceptable throughput. Customer desires, such as the ability to mix flavors on a pallet, were factored into the design.

The resulting design is a licensable production facility. The production equipment was selected with an emphasis placed on local optimization. The responsibility of making the equipment work as a production line fell to the control system. Innovative production scheduling was introduced. This chapter presents the applied scheduling model, the control system architecture, and the steps taken to achieve global optimization through the developed interfaces.

## 2.2 The Literature

### 2.2.1 Production Scheduling

Batch production scheduling within other industries has been studied at length. The most immediate carryover in the literature occurs in the chemical industry. Brucker and Hurink (2000) applied a two-phase tabu search to the problem of scheduling batch production to a particular facility. The batches were scheduled in order to meet order deadlines. Production and cleaning times were examined in the tabu search approach as well as in a general job-shop scheduling approach. Wang and Guignard (2002) created an MILP (mixed-integer linear programming) formulation for continuous-time batch processing in the chemical industry. Burkard and Hatzl (2006) applied a heuristic minimizing makespan to batch scheduling problems in the chemical industry. The heuristic was an iterative construction algorithm with recommended diversification and intensification strategies to obtain good suboptimal solutions. Tang and Huang (2007) applied a neighborhood search within a two-stage heuristic to rolling batch scheduling for seamless steel tube production.

Cheng and Kovalyov (2001) solved the scheduling of multiple batches on a single machine much like the scheduling that must be performed in this dairy example. The primary objective was to minimize cost while also minimizing maximum lateness, the number of late jobs, and the weighted completion time. The authors offered a classification of computational complexities and presented efficient dynamic programming algorithms for the problem. Li and Yuan (2006) also discussed scheduling on single machines with the three hierarchical criteria of minimizing makespan, minimizing machine occupation time, and minimizing stock-out cost. Dynamic programming was also used to solve the problem. Yuan et al. (2006) applied dynamic programming to single-machine batch scheduling problems. Their objective was to minimize makespan when faced with product-family setup times and order release dates.

Continuous and discontinuous material flow scheduling in process industries was presented by Neumann et al. (2005). The interpretation of the problem as both continuous and discontinuous carries over nicely into the dairy problem. Although individual orders are being processed, a single batch may actually provide the needed product for multiple orders. Therefore, the production remains continuous but with established discontinuities. The basic scheduling problem is solved using a branch and bound technique.

Batch scheduling with identical process times was discussed by Quadt and Kuhn (2007). The production lines in use were flexible flow lines, much like those we examine in this chapter. Setup or cleaning is incurred between jobs of different product families. The authors minimize setup time and the mean flow time using two nested genetic algorithms.

Production control in the dairy industry was the subject of work by Nakhla (1995). He addressed the overall impact of efficient production scheduling within a dairy operation and examined the appropriate balance between optimization models and work rule approaches. The concept of optimization was discussed with an objective of minimizing cost and traditional constraints, including resource availability, demand, and order delivery dates. Basic rules of thumb that describe the carryover from an optimal solution to operations adjustments were presented.

Early research was published by Schuerman and Kannan (1978) on a production forecasting and planning system. After forecasting supply for 1-gallon packages of milk, the forecast acted as input to a production scheduling model. The production scheduling was performed using LP (linear programming) that minimized total cost and provided a starting solution for the scheduler to customize. Supply forecasting also figured prominently in Benseman (1986). Production planning was performed for the New Zealand Co-operative Dairy Company based on quotas determined by dairy farmers. An LP that maximized profit was generated. It used yield analysis to determine final production quantities and produced an acceptable starting production schedule. Mellalieu and Hall (1983) followed up on this work and presented a network-planning model that used the production schedule as input and made higher-level decisions for the cooperative. Killen and Keane (1978) took supply forecasting to the point of predicting the distribution of cow calving dates. They used an LP model to establish seasonal patterns in supply that improved their ability to schedule production.

The most significant overlap between our research and published literature existed in a case study performed by Classen and van Beek (1993). They developed an MILP that resolved bottlenecks in the packaging facility in order to produce an improved master schedule for a cheese packaging operation. Changeovers and order due dates were considered while examining clusters of jobs. The MILP minimized mean flow time.

## 2.3 Operation

Customer, product, and order data is transferred between the ERP (enterprise resource planning) and control system database. The supervisory control system validates the transferred data and raises action items for resolution. Customer orders are converted into pallets on transfer by the pallet logic. The scheduler utilizes the scheduling interface to generate a production schedule. The prepared schedule is made available to the supervisory PLC (programmable logic controller) for production. Figure 2.1 depicts the flow of the operation.

Milk is initially produced in two variations: skim and whole. Both 1 percent and 2 percent milk are the result of blending appropriate ratios of skim and whole milk. Therefore, a single filling line may simultaneously produce all four varieties of milk (Mans 2007). Other product families such as orange juice, carnival drinks, and buttermilk are also processed using the same filling line, although separate storage tanks are used for each product family.

Product is piped into one of two bowls in the filler. Between batches of these different product families, the equipment must be cleaned. Less extensive cleaning may be required between products within the same product family.

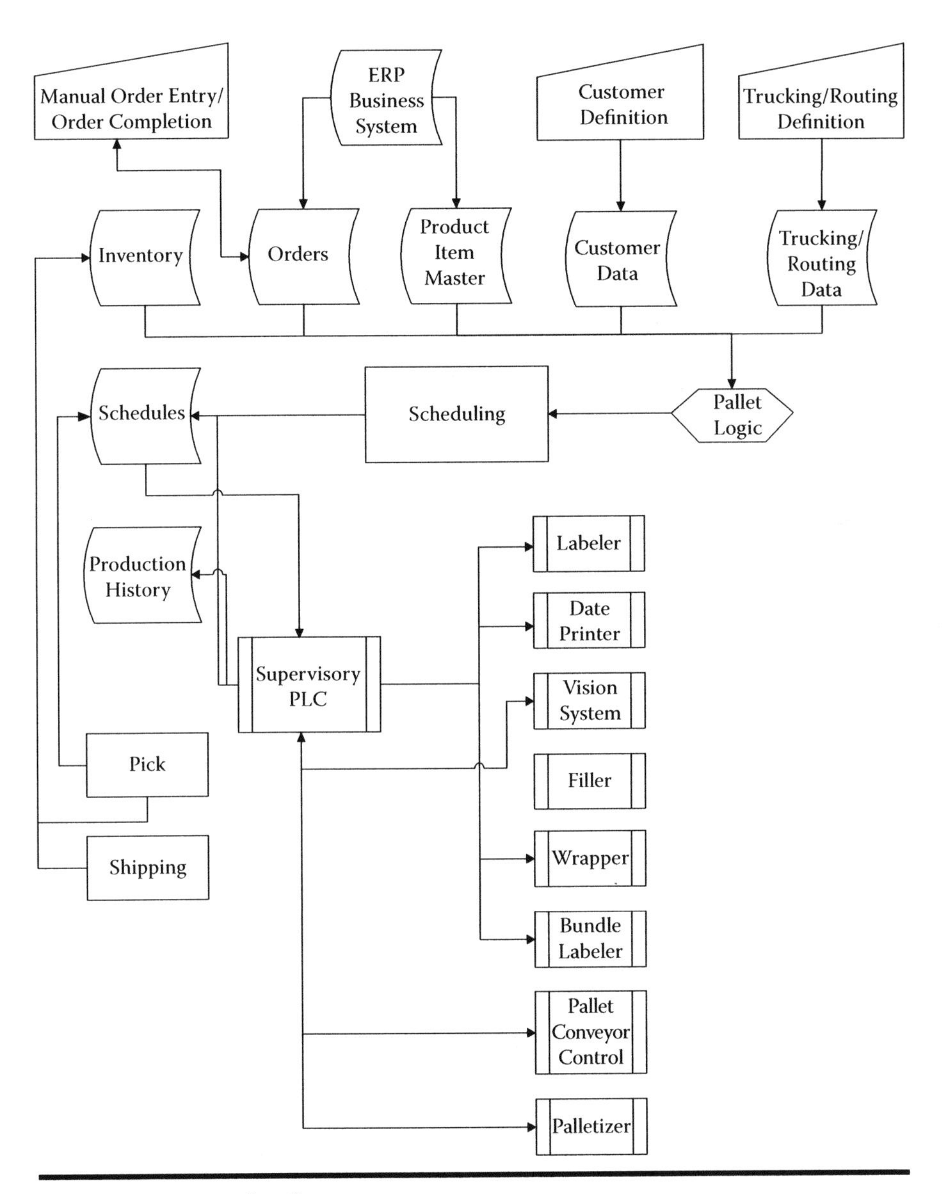


**Figure 2.1 Operation flow.**

The operator initiates production through the control system interface. The control system verifies that all equipment is ready for production and clears all equipment interfaces. Production begins with the transfer of bottles from the bottle conveyor into the four-head labeler. The control system interface is responsible for sequencing the production to meet the requirement of the order. This sequencing ensures that all bottles to be packed together are of the same flavor. The sequencing is based on the dispersion pattern used by the bundler.

When the first bottle of a pallet enters the four-head labeler, the control information for production of the pallet is downloaded to all pieces of equipment. Each piece of equipment requires specific data in a specific format. The control system interface generates each of these data items in the required format.

Production tracking occurs as the bottles move down the production line. As pallets enter production, their status is updated. Subsequently the status of the order to which the pallet is assigned is also updated. When a pallet is complete, its status is updated. If the pallet is the final pallet of an order, the order status is set to complete. This production data is transferred from the supervisory control system into the control system database.

## 2.4 Scheduling

The application of optimal methods for production scheduling in the dairy industry has been limited. The predominant need that has been addressed is the need to forecast supply. As the dairy industry implements advanced production equipment technologies and demand forecasting becomes stronger, dairies may begin to push back on suppliers and enable efficient production schedules.

### *2.4.1 Scheduling Model*

A primary goal of this scheduling approach is to facilitate the transition to a pull system. Order due dates will act as constraints in the problem formulation and ultimately guide the production order. Customers have standing due dates each week for their orders and these dates must be met. Additionally, the scheduling approach will reduce inventory held. In the past, excess inventory was carried when it was uncertain as to whether an order could be slotted into the production schedule.

All processes within Superior Dairy can be placed in one of two categories: productive time and unproductive time. Productive time includes all processes that lead to the production of a saleable product. Unproductive time includes all setup, cleaning, maintenance, and stoppages for any other reason.

Data required for scheduling includes order quantities by product, order due date, machine processing times, product (cleaning) compatibility tables, facility work hours, and transportation route planning data. An order may be an actual order or may be forecasted demand.

All schedules are created using all the information introduced above. Every productive activity is paired with some unproductive activity as a follow-on. The most straightforward example of this is the case where production shifts from milk to orange juice. A full cleaning of all equipment is necessary before changing over. This cleaning is coupled with the preceding productive activity of making and packaging milk. The type of cleaning required is determined by a compatibility table. Note that if we shift between certain types of milk, cleaning may not be necessary. In that case, we simply assign a time of zero minutes to the unproductive activity between productive activities.

Before creating the production schedule, orders are preprocessed for increased efficiency. They are aggregated by product and order completion time to create longer streams of the same product on the line. By including order completion time in the aggregation scheme, on-time delivery of every order is ensured. The productive activity in the schedule now becomes a stream, or more appropriately a "batch" of similar orders to be produced in sequence. The unproductive activity retains the same interpretation as originally stated.

An application programming interface (API) is used to pass all the information now at hand, raw data and preprocessed data, into the scheduling model. The model begins building the day's schedule by chaining together pairs (productive-unproductive) of activities in an efficient method that strives for optimal utilization of all equipment. Although the product runs are no longer order-centric, the attributes of each order are carried within the schedule to ensure the integrity of the overall schedule. If an emergency order must be placed into the schedule at the last minute, this is easily handled by the model. The model continues to ensure on-time completion of orders unless, as in the extreme case, it flags an inability to meet all orders with the inclusion of the emergency order.

The scheduling model minimizes any machine downtime that results from changeover, while simultaneously allowing orders to be completed without creating excessive inventory. Two types of inventory are under consideration: work in process (WIP) and finished product (FP). WIP inventory is floor stock required to replace missing product generated by production events.

The scheduling model can be configured with different penalty parameters to affect the algorithm. The penalty parameters provide the ability to generate a schedule that meets the most pressing needs on any given day. The implementation of production scheduling provides gains in multiple areas of the supply chain.

## 2.5 Control System Architecture

The MicroDairy is defined to be a make-to-order production system. The control system implements the logic required to go from order entry to truck loading. An order is broken down into pallets, with each pallet's configuration being based on customer specifications. Pallets can be configured to hold single, double, four, six, or eight packs. All bottles in a pack will be of like flavored milk. Only one pack size will be loaded on each pallet. A pallet can be loaded with up to four different flavors of milk. The different flavors will be loaded into separate rows. Scheduling of the caseless line is done on a pallet basis.

The MicroDairy Control System is a plant floor control system that can be sold as a stand-alone system requiring minimal integration effort to interface with an existing plant ERP system. The control system is used to schedule and provide supervisory control of the caseless filling line. One of the major design criteria is the segregation of the ERP business system from the actual control system. Figure 2.2 defines the architecture of the control system.

### 2.5.1 Supervisory PLC: Plant Equipment Interface

The *supervisory programmable logic controller* (PLC) is responsible for the control of the caseless line. Production events and scheduling information are exchanged between the supervisory PLC and the enterprise database server. A set of transactions have been defined that will:

- Exchange scheduling/pallet production data
- Update label requirements
- Record significant production events
- Synchronize the supervisory PLC clock with the database server's clock

The control system is designed to minimize production interruptions caused by communication losses between the supervisory PLC and the database server. The PLC requests ten pallets worth of production information. This buffer will be updated each time a new pallet enters production. If the PLC is unable to communicate production events back to the database, the events are buffered until communication is restored. Once communication is restored, the event data will be transmitted to the database in the order required to maintain tracking integrity.

### 2.5.2 Schedule Display

The *production schedule* for a day will be automatically downloaded when completed by the scheduler using the scheduling model. The plant floor production schedule is updated throughout the day any time there is a schedule change or occurrence of a production event. The operator also has the ability to request a copy of the schedule. The supervisory PLC schedule displays the next ten pallets that are to be produced. The operator can view the entire day's production schedule by selecting the View Daily Schedule tab. The daily production schedule will be viewed via a browser popup that will display the user interface schedule interface.

The *pallet schedule* provides the order in which pallets are to be produced to meet production requirements. Production requirements are based on inventory,

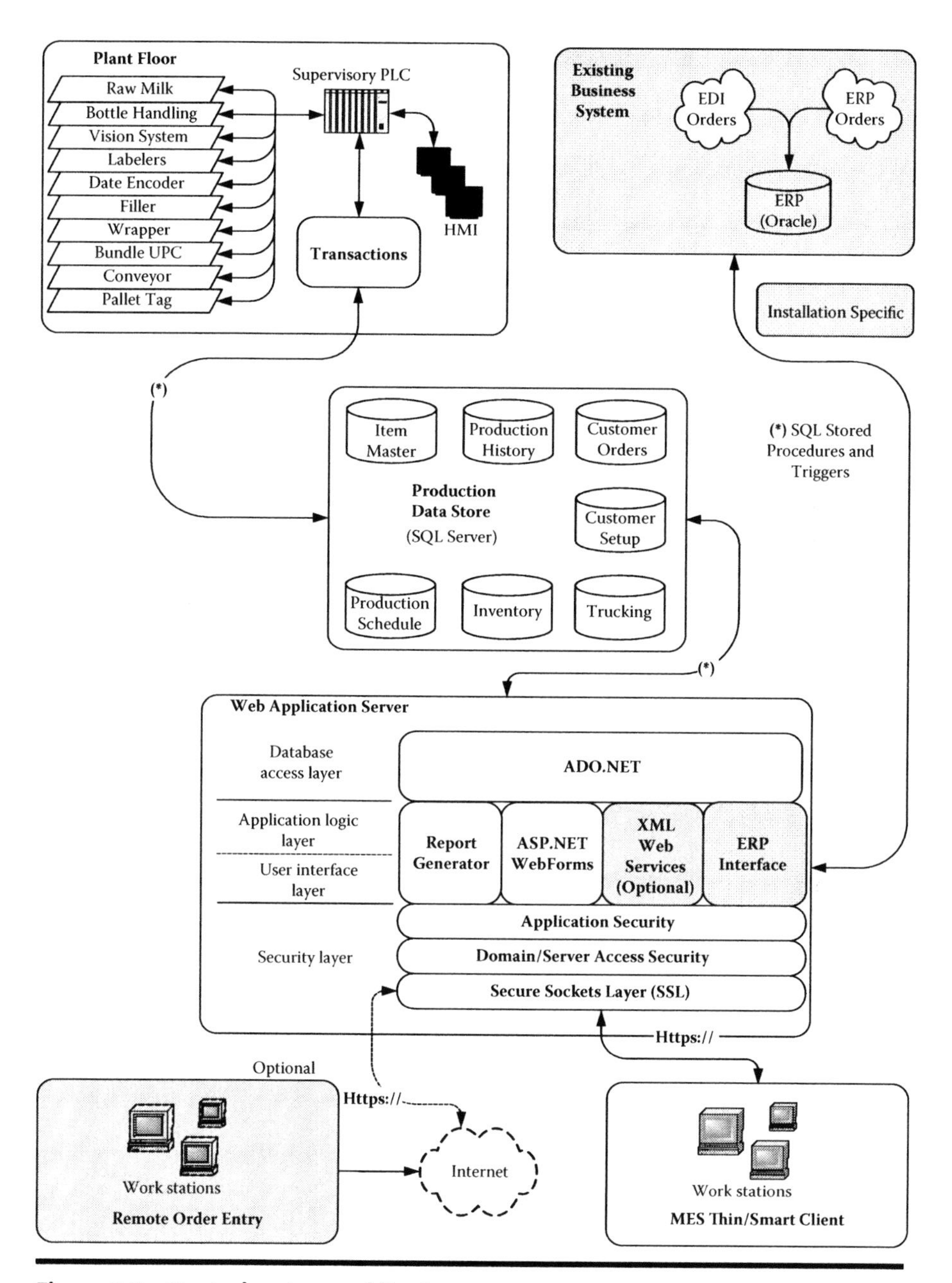


**Figure 2.2 Control system architecture.**

schedule, forecasting, and optimization. The production schedule also includes pallet characteristics specifying the pallet type, number of cases, units per case, and if a bundle UPC is required.

## 2.5.3 Line Start/Stop

After the initial download of the daily schedule, the supervisory PLC requests the production sequence of the first pallet. This detail specifies the bottle order, in terms of UPC, required to build the pallet. Following receipt of the pallet detail, the supervisory PLC requests the label requirements of the pallet. The first four required labels for the pallet will be returned. The labels will be defined by the UPC and the item description taken from the Item Master table. Following loading of the required labels, the labeler operator must enter the label-to-label head assignment. This information allows the supervisory PLC to know which label head needs to be enabled for each bottle going through the labeler. If desired, the operator may load the same label into multiple label heads. Once the labels have been assigned to the labeler heads, the operator can start production.

## 2.5.4 Bottle Conveyor System

The supervisory PLC interfaces with the *bottle conveyor* that is responsible for transferring bottles from blow molding through labeling and filling to bundling. Optimal control of the bottle stream is necessary to eliminate production delays. The conveyor system relies on a series of photo eyes and logic to stop and start the conveyor as required. As sections of conveyor fill, previous sections are stopped in a controlled manner, thus ensuring the proper operation of upstream equipment. This control provides a constant stream of bottles to the equipment, thereby improving overall equipment effectiveness.

## 2.5.5 Four-Head Labeler

The *four-head labeler* is controlled by a series of five PLCs, one for each label head and a main one to control the pacing belt and conveyor. There are a series of photo eyes located on the labeler. One photo eye is located just after the pacing belt. This eye is used to sequence through the bottles entering the labeler. There is a photo eye located at each label head. These photo eyes are used by the label head to know when a bottle is present, referred to as product detect. If a label head is enabled at product detect, a label is applied. Feedback from these photo eyes will be used by the supervisory PLC to keep track of the location of each bottle as it progresses through the labeler.

A set of registers is maintained by the supervisory PLC for each label head. This set consists of an enable count register and a disable count register. These counts are incremented and decremented based on the label head state as bottles move through the labeler. The values held in these registers determine the setting of the label head state by the supervisory PLC. A label head is enabled when the enable count is greater than zero and the disable count is equal to zero. A label head is disabled when the enable count is equal to zero.

When the supervisory PLC signals the main PLC to start the pacing belt, a bottle is fed into the labeler. As a bottle passes the pacing belt photo eye, the supervisory PLC determines which label head contains the label to be applied. The check is done starting at Head 1. The enable count for the determined label head is incremented and the disable count for all other label heads is decremented. If this is the first bottle of production, the supervisory PLC enables the determined label head. At product detect, the state of the label head is determined. If the head is enabled, the enable count is decremented; if the label head is disabled, the disable count is decremented. Following the count adjustment, the supervisory PLC determines if a label head state change is required.

An alarm is displayed when a label head encounters a low label condition. To respond to the low label alarm, the operator places the label head in manual mode. The pacing belt is stopped. The labeler conveyor continues to run, flushing all queued bottles from the labeler. Once all bottles are out of the conveyor, the label head is disabled and the operator signaled to replenish the supply. The operator then places the label head back into auto mode. The labeler conveyor is started and the pacing belt is started.

If an out-of-label condition is encountered, the supervisory PLC determines if another label head containing the required label exists down-conveyor from the out-of-label head. If so, the disable count of the found head is decreased by the enable count of the out-of-label head and the enable count is set to the enable count of the out-of-label head. If the disable count of the found head is zero, the head is enabled. If no other down-conveyor head contains the required label, the pacing belt is stopped. The labeler conveyor is stopped. If a broken label condition or some other label head malfunction is detected, the pacing belt and labeler conveyor are stopped.

Stopping of the labeler conveyor may result in at least one bottle having no label applied. The unlabeled bottles will be detected by the *label detect vision system* and subsequently removed by the operator. The operator must replace any missing bottles.

### 2.5.6 Date Printing

A *date printer* is positioned along the four-head labeler conveyor. The printer prints the product expiration date, a two-dimensional data matrix bar code, a bottle serial number, a bottle of batch count, and a batch count on each bottle. The data matrix bar code contains the production date, product expiration date, product UPC, pallet number, and bottle number.

In addition to determining the label head to be used as a bottle passes the pacing belt photo eye, the supervisory PLC also downloads the bottle data to the printer. The printer queues the data. When the printer detects a bottle, the next entry in the queue is formatted for printing. The downloaded bottle data consists of the product expiration date in MMM DD format, the production date in MMM DD YYYY format, the UPC symbol of the label to be applied, the pallet number, and the bottle serial number on the pallet. The expiration date, production date, and pallet number come from the pallet schedule. The UPC symbol comes from the pallet detail data. The bottle serial number is a counter maintained by the supervisory PLC that is reset at the start of each pallet.

### 2.5.7 Labeling Verification

Following exit of the four-head labeler, the *label detect vision application* inspects the bottle. The label detect application verifies

1. Placement of the label on the bottle
2. That the correct label has been applied
3. That the product expiration date is correct and readable

As a bottle exits the labeler, the expiration date and UPC are queued for download to the vision sensor. As the bottle moves along the bottle conveyor, a photo eye is encountered. Tripping of this photo eye serves as the trigger to the vision sensor to acquire an image. After acquiring the bottle image, the vision application retrieves the PLC data for use in verifying the label. The UPC symbol is read off the label and compared to the downloaded label. The two-dimensional bar code that was applied to the bottle is read. The UPC contained in the two-dimensional bar code is also compared to the downloaded UPC. The vision application verifies that the applied expiration date is the correct date and that it is readable. The results of the verification are formatted and a "job complete" signal is sent to the supervisory PLC.

If the verification fails, the four-head labeler pacing belt is stopped and an alarm is displayed. The four-head labeler finishes labeling the bottles currently queued. The supervisory PLC stops the bottle conveyor at the exit of the labeler. The operator is signaled to locate and replace the improperly labeled bottle.

### 2.5.8 Filling

The bottle *filler* reads the UPC code on the bottle label to determine the type of milk expected in the current bottle. The filler is capable of producing fat-free, 1 percent, 2 percent, and whole milk. The filler uses a two-bowl system and fills by weight. One of the bowls contains skim milk. The other bowl contains whole milk. The four flavors are produced by blending the proper amounts of milk from each bowl.

Efficiencies are gained for fat-free and whole milk production by filling both bowls in the filler with either fat-free or whole milk. Following the filling of the bottles, they are capped prior to exit from the filler.

### 2.5.9 Bottle Tracking

A vision system is positioned on the filler exit conveyor. The data matrix bar code is read from each bottle as it passes in front of the vision sensor. The UPC is extracted from the encoded data and compared to the expected UPC that has been sent to the vision system from the supervisory PLC. If the UPC does not match the expected UPC, the filler is signaled to stop, the exit conveyor is stopped, and an alarm is enunciated to the operator. A screen instructs the operator as to the type and count of bottles that were expected. The operator inserts the missing product and acknowledges the alarm. The supervisory PLC then signals the filler and conveyors to resume production.

### 2.5.10 Induction Sealer

The capped bottles pass under an *induction sealer*. The induction sealer detects any potential missed or unsealed bottles and signals the supervisory PLC. The supervisory PLC stops the conveyor and an alarm is generated for the operator.

### 2.5.11 Bundle Wrapper

Following sealing, the capped bottles are passed along the bottle conveyor to the *bundle wrapper*. The wrapper is capable of passing the single bottles through or combining bottles into two, four, six, or eight packs. The wrapper can produce two pack bundles two at a time. All other multipack bundles are done one at a time.

When a new pallet enters production, the supervisory control system communicates with the wrapper's PLC to download the pallet's configuration. This configuration exchange occurs via a load detail message. The load detail message defines the accumulation pattern required by the wrapper to produce the packs for the pallet. Basically, the load detail message instructs the bundle wrapper as to how many bottles to accumulate in the diverter, to which lane the accumulated bottles should be diverted, and the number of times, referred to as loads, that the pattern should be repeated. Because the pallet configurations are downloaded when a pallet enters production, the bundler PLC must sequence through the queued pallet configurations.

The load detail message consists of

- The pack size for the pallet
- The total number of packs on the pallet
- The number of load sets, accumulation patterns, required to produce the pallet, with a limit of two load sets per pallet load detail message

Each load consists of

- Number of loads
- Lane 1 accumulation count
- Lane 2 accumulation count
- Lane 3 accumulation count
- Lane 4 accumulation count

The second load set is required to handle pallet configurations that do not result in full layers.

## 2.5.12 Bundle Labeler

Bundles may or may not require the application of a bundle UPC. The application of a bundle UPC depends on the line item definition. All bundles loaded onto a pallet will have the same bundle UPC requirement. The bundle UPC and description will be downloaded as part of the pallet schedule. If a bundle requires a bundle UPC, the supervisory PLC will send the required print file to the *bundle labeler* when the pallet enters production. The print file defines the label to be printed, including the UPC, description, and number of bundles to which the label will be applied. The bundle labeler queues the received print files. The supervisory PLC tracks the bundles as they pass the bundle labeler. When the end of a pallet is encountered (i.e., the last bundle of a pallet), the supervisory PLC determines if the next schedule pallet requires bundle labels. If the next pallet does require labels, a contact on the bundle labeler is latched. The bundle labeler processes the next print files in its queue and begins the labeling process.

## 2.5.13 Palletizer Area

When a new pallet enters production, defined to be the first bottle of the pallet entering the four-head labeler, the pallet configuration is sent to the *palletizer*. The palletizer queues the received pallet configurations. This ensures that the next pallet's configuration is available when the previous pallet completes. The pallet configuration data consists of the pallet ID, a pallet rotation flag, the bundle type on the pallet, the number of bundles per row, and the number of bundles per pallet.

### 2.5.13.1 Pallet Conveyor Control

The pallet conveyor requires the control and interfacing of a pallet magazine, conveyor segments, palletizer, pallet wrapper, and a lift device. The supervisory PLC is responsible for coordinating pallet movement along the conveyor.

There are two conveyor sections between the pallet magazine and the palletizer. The supervisory PLC keeps these two locations populated. If no pallet exists on the first conveyor segment, the supervisory PLC starts the first conveyor segment and requests a pallet from the magazine. If no pallet exists on the second conveyor segment, the pallet is transferred between the first and second segments.

When the palletizer requests a pallet, the pallet on the second pallet segment is sent to the palletizer. Upon completion of a pallet, the palletizer signals the supervisory PLC that a pallet is ready for transfer. The supervisory PLC verifies that no pallet exists on the segment following the palletizer. The pallet is transferred and staged for the pallet wrapper.

When the pallet wrapper requests a pallet, the conveyor segment is started and the pallet is positioned for wrapping. Upon completion of the wrapping operation, the wrapper signals the supervisory PLC that a pallet is ready for transfer. The supervisory PLC will confirm that no pallet exists on the lift/transfer conveyor segment. The pallet lift/transfer segment is started and a signal is sent to the wrapper. The pallet will be transferred onto the lift/transfer segment and staged for loading onto a truck or to an inventory location.

#### *2.5.13.2 Pallet Tag Printer*

The supervisory PLC requests the printing of a *pallet tag* when a pallet enters the pallet wrapper. The pallet tag is printed on an ink-jet printer located in the palletizer area. This tag is attached to the pallet as it exits the pallet wrapper.

### *2.5.14 Data Exchange*

The *exchange of production and tracking data* between the control system and the enterprise database is accomplished by a set of transactions. The transactions map control system tags to enterprise data objects. The enterprise data objects implement the business rules defined by Superior Dairy. Any exchange of data between the control system and the database is initiated by the control system. Any data changes on the enterprise side that require a refreshing of information held by the PLC utilize the following protocol. A transaction request bit is set on the PLC, the PLC responds with the triggering of the requested transaction, data is exchanged, an optional transaction status is returned, and finally, a transaction completion bit is set. Any data requests or event data updates originating from the PLC utilize the following protocol: The PLC triggers the required transaction, data is exchanged, an optional transaction status is returned, and finally, a transaction completion bit is set.

## 2.6 Conclusion

The result of the integration of traditional batch production equipment into a production line controlled by a supervisory PLC achieves the ideals of operations research. The supervisory control system leveraged the local optimization of each piece of equipment to achieve optimal control of the dairy production process. The use of an innovative scheduling model extended the optimization by minimizing production downtime resulting from product transitions. The control system architecture provides a single point of interface to the ERP system, thus minimizing the typical touch points into the IT infrastructure.

## References

Benseman, B.R. (1986), Production planning in the New Zealand dairy industry, *The Journal of the Operational Research Society*, 37(8), 747–754.

Brucker, P., and Hurink, J. (2000), Solving a chemical batch scheduling problem by local search, *Annals of Operations Research*, 96, 17–38.

Burkard, R.E., and Hatzl, J. (2006), A complex time based construction heuristic for batch scheduling problems in the chemical industry, *European Journal of Operations Research*, 174, 1162–1183.

Cheng, T.C.E., and Kovalyov, M.Y. (2001), Single machine batch scheduling with sequential job processing, *IIE Transactions*, 33, 413–420.

Classen, G.D.H., and van Beek, P. (1993), Planning and scheduling packaging lines in food industry, *European Journal of Operational Research*, 70(2), 150–158.

Killen, L., and Keane, M. (1978), A linear programming model of seasonality in milk production, *The Journal of the Operational Research Society*, 29(7), 625–631.

Li, W., and Yuan, J. (2006), Single machine parallel batch scheduling problem with release dates and three hierarchical criteria to minimize makespan, machine occupation time and stocking cost, *International Journal of Production Economics*, 102, 143–148.

Mans, J. (2007), Computer-controlled filler runs all fat levels of milk, *Packaging Digest*, 44(2), 22–26.

Mellalieu, P.J., and Hall, K.R. (1983), An interactive planning model for the New Zealand dairy industry, *The Journal of the Operational Research Society*, 34(6), 521–532.

Nakhla, M. (1995), Production control in the food processing industry: the need for flexibility in operations scheduling, *International Journal of Operations and Production Management*, 15(8), 73–88.

Neumann, K., Schwindt, C., and Trautmann, N. (2005), Scheduling of continuous and discontinuous material flows with intermediate storage restrictions, *European Journal of Operations Research*, 165, 495–509.

Quadt, D., and Kuhn, H. (2007), Batch scheduling of jobs with identical process times on flexible flow lines, *International Journal of Production Economics*, 105, 385–401.

Schuerman, A.C., and Kannan, N.P. (1978), A production forecasting and planning system for dairy processing, *Computers and Industrial Engineering*, 2(3), 153–158.

Tang, L., and Huang, L. (2007), Optimal and near-optimal algorithms to rolling batch scheduling for seamless steel tube production, *International Journal of Production Economics*, 105, 357–371.

Wang, S., and Guignard, M. (2002), Redefining event variables for efficient modeling of continuous-time batch processing, *Annals of Operations Research*, 116(1), 113–126.

Yuan, J.J., Liu, Z.H., Ng, C.T., and Cheng, T.C.E. (2006), Single machine batch scheduling problem with family setup times and release dates to minimize makespan, *Journal of Scheduling*, (9), 499–513.

*Chapter 3*

# Optimization of Medical Services: The Supply Chain and Ethical Implications

Daniel J. Miori and Virginia M. Miori

## Contents

## 3.1 Introduction

Supply chain considerations in the healthcare industry have been thoroughly examined with respect to medical professionals, supplies, and medications. Varying approaches have been taken in these discussions, most of which rely on characterizing patient volume for determination of needs. The unifying theme through all of this work has been consideration of the patient as a customer entering a system.

In this chapter we discuss an alternate approach in which we characterize the patients as inventory, not customers. In essence, the patient comes to us as a subassembly awaiting the use of other resources. By adapting this view, we may continue to evaluate supplies, medical professionals, and medications as resources. In addition, we allow a different perspective that facilitates the use of alternative modeling approaches, and therefore develop the potential for increasingly effective practical solutions.

The approach is simple to state: a patient will be viewed as inventory. However, the implications of this statement are quite complex. The concept of humans as inventory has been discussed in a number of divergent areas of research. Industrial engineers refer to it in time and motion studies. Schools refer to it in terms of student tracking. Project managers refer to it in resource utilization, and human resource departments refer to it in staffing models. Let us not forget the greatest use of the concept—throughout history, human inventory has been taken in the form of censuses.

All these areas conjure in our minds the same ethical dilemma; they pose the possibility of dehumanizing individuals. While this may have found acceptance in certain fields, within the medical arena, thorny ethical implications are unavoidable. We cannot dehumanize individuals when they have placed their health and their lives in our hands. We therefore move forward through this research with care and endeavor to always maintain a respectful view of our patients. We hold their interests in the highest regard, and we hold ourselves responsible for compassionate treatment of the subject, resulting in compassionate and effective treatment of the patients.

The background for this chapter covers many areas. We first present related literature in these many areas. The current healthcare supply chain is discussed with the presentation of a simulation model and baseline results. The proposed inventory interpretation and potential inventory models are shown. The revised supply chain is presented with modified inventory modeling with a revised simulation model. Preliminary results and recommendations are presented along with future research directions.

## 3.2 Literature

### *3.2.1 Supply Chain Models*

Buddress and Raedels (2000) presented supply chain management tools that should be considered in healthcare. They were part of the early discussion of how to apply these tools and how to achieve significant benefits that result in control of healthcare costs.

Rivard-Royer et al. (2002) presented a discussion of stockless replenishment in medical supplies and the lessons learned in Canada. Stockless replenishment has evolved in a more effective and efficient hybrid model of replenishment resulting in healthcare cost reductions for both the distributors and the hospitals.

Supply chain concepts have achieved much greater acceptance in areas outside healthcare. Within healthcare, the concepts have begun to take hold. Runy (2005) has discussed the role of the supply chain in improving the quality of care while reducing costs. She highlighted the value of price negotiations, volume discounts, and standardization of products. Roark (2005) also examined the healthcare supply chain, although from the alternate perspective of the requisition process. He emphasized that the supply chain represents 25 percent to 30 percent of total operating expenses and that savings opportunities in this area have not begun to be exploited. The role of group purchasing organizations was discussed in detail, along with cost-saving measures that can lead to improved patient outcomes and enhancements to the bottom line.

McKone-Sweet et al. (2005) noted that the healthcare supply chain is inefficient and discussed ways to effect improvements. They noted the slow acceptance of supply chain practices in healthcare and explored the barriers to implementation through case studies. They identified barriers, ranging from a lack of executive support to the more pragmatic issue of the need for data collection and performance measures. They provided recommendations for hospitals and supply chain partners to implement workable supply chain management solutions.

The process of reducing technology and supply chain costs at Pennsylvania Hospital, run by the University of Pennsylvania Health System, was presented by Wilson et al. (2007). They embarked on a quality improvement program that redefined the processes and identified areas for cost savings.

Pan and Pokharel (2007) provided a case study of logistics operations of hospitals in Singapore. They pointed out that despite an understanding of the value of aggregating logistics functions, these organizations remain wary of including suppliers as strategic options in their supply chain. They tend to focus on outsourcing logistics and have been able to implement good stocking policies for medical supplies. Kumar et al. (2008) recently applied supply chain management principles to the healthcare industry in Singapore. They performed process reengineering on the supply chain in order to achieve cost reductions in medical services. The approach was novel in this setting, rather than focusing on individual aspects of operations, the focus was on the total delivered cost through elimination of non-value-added activities, centralized warehousing, and control of non-production goods.

Shah et al. (2008) explained effective supply chain practices in organizations with effective collaboration between independent organizations supporting the healthcare system. They examined supply chains in the context of lean principles and relational coordination theory to explain the unexpected efficiency of a high-performing healthcare supply chain.

Health policy in England differs in nature from that of the United States. Allen et al. (2009) focused specifically on this. It has primarily focused on the demand side of healthcare. The authors looked at links between procurement and the supply chain. Many aspects of these overlapping functions had been previously applied, although the role of healthcare purchasers in the management of the healthcare supply chain had not been considered.

### 3.2.2 Simulation

Lapierre and Ruiz (2007) provided an innovative approach for improving hospital logistics, addressing procurement, distribution, and inventory needs. Two modeling approaches were introduced to perform effective scheduling of purchasing, employee work, etc. A practical example using a hospital in Montreal, Canada, showed strong, efficient results, again using a conventional supply chain view.

Katsaliaki et al. (2009) compared approaches to the simulation of complex supply chains in healthcare. Distributed simulation was presented as an alternative to discrete event simulation due to the time requirements resulting from complex operations. They used the conventional healthcare supply chain model and found significant advantages resulting from the use of distributed simulation.

### 3.2.3 Humans as Inventory

Human resources departments have long looked at humans as inventory when considering staffing decisions. Gatewood and Rockmore (1986) used aspects of healthcare services in their human resource planning model. Job characteristics and worker characteristics were examined on the same scale in order to establish human resource needs. Although the research does not directly address staffing these positions as use of a human pool of inventory, like all HR models, they do rely on a pool of individuals to fulfill their needs. The cross-over to healthcare and the characterization of jobs and workers show application of supply chain concepts to the healthcare industry.

A national census is completed every 10 years in the United States. This is not a new concept. Peabody (2001) discussed the use of census in precolonial and early colonial India as the precursor for the census techniques employed today. He presented the use of human inventory as an effective tool for identifying and categorizing individuals. The use of the census was considered to be lacking in any ethical implications due to the necessity of the information gathered.

McManus (2007) further explored humans as inventory in the area of industrial engineering. He not only discussed the concept of humans as inventory, but also considered various types of human inventory such as work-in-process inventory. He discussed the reduction of waste in tracking human inventory but noted that the concept is pervasive, thus opening it up to the discussion of reducing waste.

### 3.2.4 Health Maintenance Organizations (HMOs)

HMO refers to a type of prepaid medical service in which members pay a fixed monthly fee for their healthcare. The fee is set without consideration of the amount or kind of services received (*Mosby's Medical Dictionary* 2009). Medical practices and hospitals contract with the HMO to provide the necessary medical services and, in fact, are the only source of medical services that a patient may access. Preventive medicine is stressed in order to keep costs lower. Proponents say that this strategy makes healthcare available to more people and results in earlier diagnosis and additional healthcare savings. However, the HMOs have encountered many issues relating to claims that they have refused necessary treatments and reduced care in order to make more profit. As a result, most states have enacted laws restricting HMO rules that are harmful to patients (*The Free Dictionary* 2009).

HMO members are required to select a primary physician who, when not faced with a medical emergency, may then refer the patient to other physicians as needed. The earliest form of an HMO traces back to 1910 as a prepaid health plan. Through time, these plans have evolved into the form of HMO we know now. In the 1970s, less than 40 of these organizations existed, and they were standardized with the enactment of the Health Maintenance Organization Act of 1973 (Wikipedia 2007, Norris 2009).

Towill (2006) compared the California-based non-profit HMO Kaiser Permanente (KP) to that of the National Health Service (NHS), which is comprised of four publicly funded healthcare systems in the United Kingdom. Healthcare in the United States is widely thought to be more costly and less efficient than in the United Kingdom. The focus was smooth patient flow from the onset of a problem to the completion of treatment. They found that best practices in the healthcare supply chain could be readily related to the conventional healthcare supply chain exemplified by KP. KP has been especially effective in this area due to cultural and organizational factors. A number of tools designed to help clinicians achieve best practices are identified. Keen et al. (2006) presented an alternative to the currently accepted model of healthcare delivery. They presented a model of services characterized by treatment and care needs used by KP. These services vary over time and require frequent reassessment to provide appropriate choices to patients. A network model is proposed that helps in understanding the healthcare process.

The complex nature of a healthcare supply chain and the many implications in medical and ethical arenas have resulted in the need to understand work performed in all of these areas of study. The current supply chain in our healthcare setting shares many similarities with existing studies. We examine these similarities while building a description of the current supply chain.

## 3.3 Current Supply Chain

Conventional wisdom on hospital supply chains casts them, unsurprisingly, as service operations. Patients (customers) enter the system to begin the service cycle. The patients are processed and evaluated prior to determining the path they will take through the hospital treatment network.

A complex set of services is typically developed to explain the treatments (processes) that each patient could receive. This set of services varies by provider. In this chapter we discuss a system that initially breaks down into home care, inpatient care, and outpatient resources. The services provided for each are presented in Figure 3.1.

Within the three categories of care, additional breakdowns in services have been made. These are in the areas of high-tech home care, outpatient evaluation, outpatient procedures, and outpatient education. These breakdowns are presented in Figure 3.2.

The flow of patients through the inpatient care system is displayed in Figure 3.3. The initial point of contact in the community is traditionally through a primary medical doctor (MD). The primary MD can be on a pay-per-service arrangement or through any type of organized healthcare benefits. In significant emergencies, the point of contact in the community may be emergency medical technicians (EMTs) or other emergency personnel.

From the emergency department a patient may either be released or admitted to the hospital for further care. Patients may return to the emergency room even after treatments have been completed due to complications or worsening of the medical condition.

Once admitted to the hospital, patient care may be medical, psychiatric, or an interaction between the two. The paths from hospital care are varied. If a patient begins physical rehabilitation in the hospital, the patient may be discharged and placed in the care of an outpatient physical or occupational therapist. If extended rehabilitation is required, the patient may be placed in a rehabilitation facility within

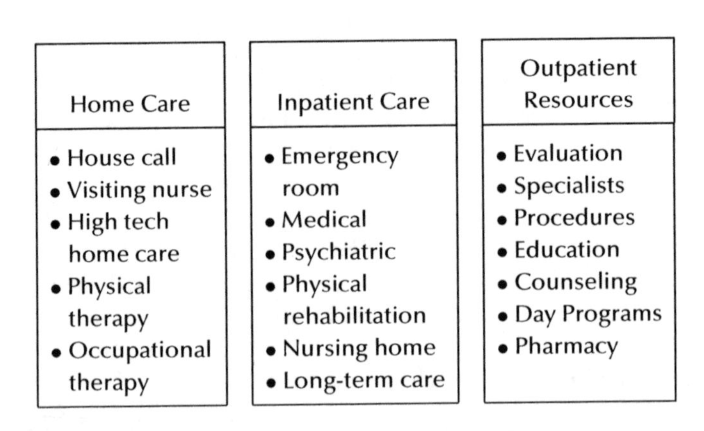


**Figure 3.1 Patient care service categories.**

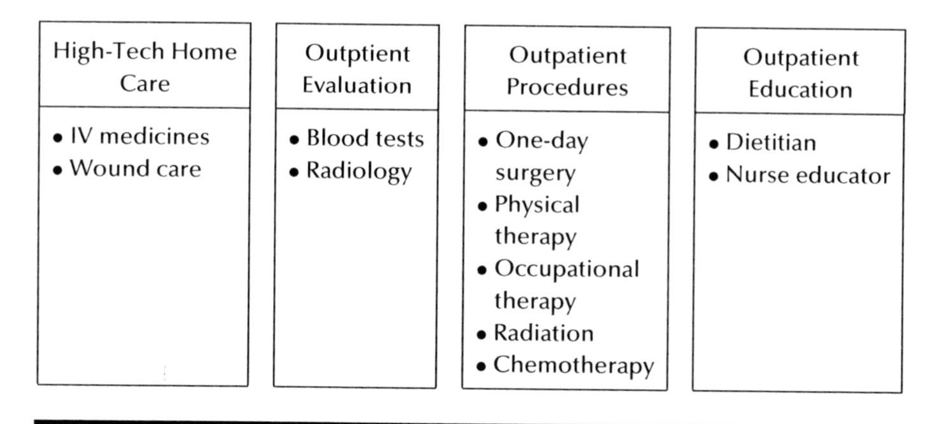


**Figure 3.2 Additional service breakdowns.**

a nursing home until ultimately released. If a patient is in a debilitated condition with limited hope of recovery, the patient may be placed in a nursing home for long-term care. Once a patient is moved to a nursing home, the patient's condition may result in a combination of rehabilitative and long-term care. Just as with the emergency room, patients may return to the hospital even after treatments have been completed due to complications or the worsening of a medical condition.

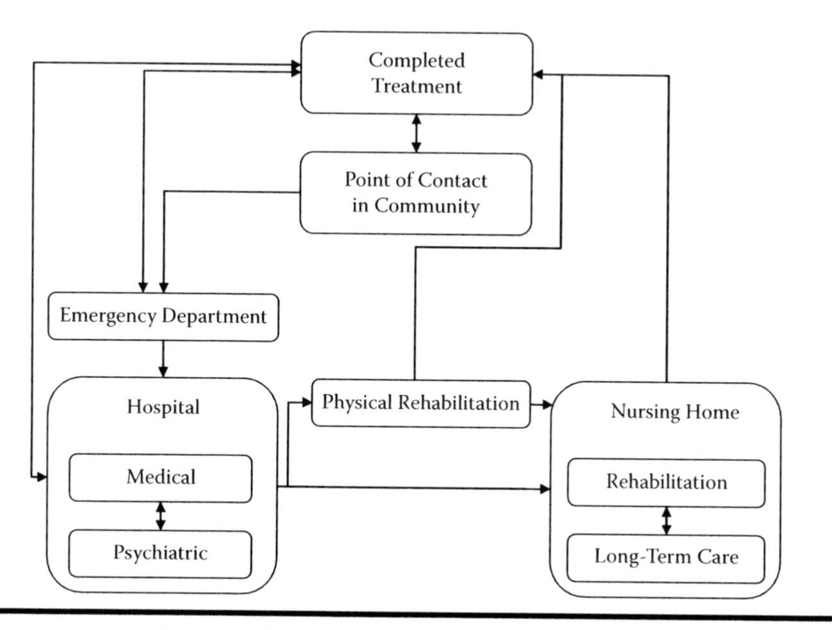


**Figure 3.3 Inpatient care flow diagram.**

Throughout this supply chain, we assume that the patient acts as a customer in a service environment. The patient therefore consumes the available resources, which include medical staff, medical supplies, pharmaceutical supplies, and hospital beds.

## 3.4 Revised Supply Chain

### 3.4.1 Basis

We propose a new supply chain structure for healthcare operations in which we view the patients as inventory. We essentially change our view of a hospital from a service operation to a production operation. We do this not to dehumanize the patients, but to emphasize them as the most important part of our supply chain. Rather than focusing on the medical staff, medical supplies, pharmaceutical supplies, and hospital beds, we focus on a method to provide improved medical care to our patients.

Years ago, the introduction of HMOs was made with the same goal in mind. Although HMOs have experienced problems through time, the tenets on which they were proposed are still sound. We present these ideas and their value in healthcare and carry them forward as a basis for the proposed approach in this chapter.

### 3.4.2 Ethical Considerations

In addition to helping identify choices in medical treatment when no clear and rational medical plan is available, the field of medical ethics is also a process by which new ideas within healthcare delivery can be evaluated. Ethics theory is not static, and in fact is constantly evolving as medical technology advances. Every day there are novel approaches to medical dilemmas and increases in understanding of existing disease. Even the well-established field of the treatment of infectious disease is under constant pressure from the ever-increasing ability of bacteria to resist treatment, a dilemma that continues to evolve at a pace greater than research can offer answers.

There are many schools of thought within medical ethics theory; one widely accepted view is that of Beauchamp and Childress (2008), who set forth their approach of principled ethics review. Rather than stipulating one single, overriding set of moral laws, they proposed that decisions can be made based on the non-absolute guidelines of what has been accepted practice, with the understanding that new ideas will come along to challenge accepted practice, and that we should be able to examine those ideas based on a given set of principles. There is a great deal of discussion as to what those principles specifically ought to be, but continuing with the Beauchamp and Childress model, they can be defined as beneficence, non-malfeasance, autonomy, and justice.

Briefly, they can be defined as working to do good things (beneficence); avoiding doing harm (non-malfeasance); keeping your patients fully informed and allowing them to make decisions regarding their care (autonomy); and striving to treat all those in your care equally (justice).

The principle of justice can be seen as speaking most directly to the equitable distribution of limited resources within a healthcare delivery system. Justice, more than any other principle, directs how limited resources should be distributed and how the stewardship of those assets should be conducted.

Having established a basis for the ethical consideration of business models of healthcare delivery, we are left with a business imperative that may be peculiar to healthcare. In a system where the patient is considered an item of inventory—albeit inventory with a vested interest in an efficient, economically viable, and humane process of manufacture—the idea that this relationship should be held to a higher standard seems intuitive. An example of this higher standard can be found in a discussion of the patient–provider relationship as a fiduciary relationship.

With the provider as trustee and the patient as beneficiary, a next step in exploring this relationship could be to propose a model of the healthcare industry as the trustee and society in general as the beneficiary. In this case, the actions to maintain this investment in the best interest of society by increasing the optimal health and function of the members of society is not simply an ideal to work toward, but could arguably become a legally binding obligation with potential consequences resulting from failure to act. The thought that adopting fiduciary responsibility as an additional principle as a constraint on the unbridled pursuit of profit above the patient's best interest may seem naive, but it may well serve as a framework for a reasoned discussion of potential dilemmas such as physician as dual agent.

### 3.4.3 Specification

Service and production supply chains are a direct reflection of the goals and strategic objectives of the organization. In a medical setting, a supply chain decision to be made is the selection of an area of specialization. It has become common, especially in larger cities, for hospitals to focus on particular areas of medical care. These areas include general medicine, heart health, trauma care, cancer treatment, neck and spinal care, pediatric care, as well as many others. In the case of this research, the focus of the hospital is general medicine.

A service supply chain follows the movement of a customer through a series of processes. The quality of service is measured on a subjective basis and can often be seen from conflicting points of view by customers and service providers. Consider the patient who is discharged from a hospital with vastly improved health, but a negative view of the hospital staff. The hospital staff conversely may view that patient as uncooperative and unkind. In a scenario like this, we are left with an inability to truly assess the quality of the care.

Production supply chains focus on the processes that must be performed to yield a finished product. The processes are measured quantitatively through measurements

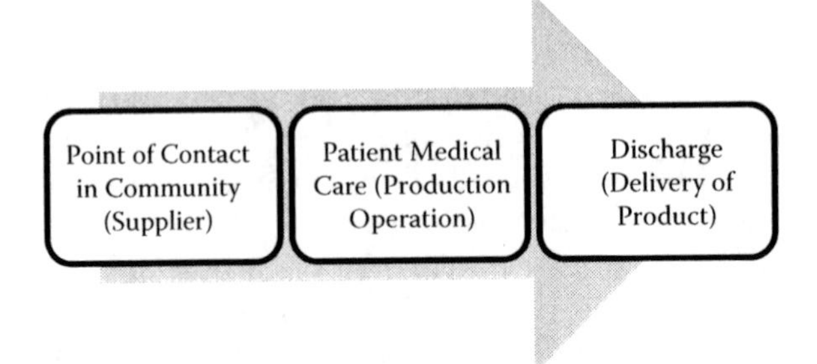


**Figure 3.4 Conversion from service to production supply chain.**

of the product and qualitatively through categorical assessments of the product. Factors associated with human interaction are measured much later, upon delivery of a finished product. The focus during the processes is on an appropriate conversion of the input (the ailing patient) to the output (the released patient).

The conceptual conversion from a service perspective to an operations perspective is shown in Figure 3.4. The point of contact in the community acts as a supplier, providing ailing patients to the system. The medical care includes what we consider to be production processes, and the final product is the discharged patient.

To more thoroughly understand and describe the healthcare supply chain (HCSC), we begin by visiting it from several perspectives. Supply chains are described as some combination of push and pull; they are also discussed with reference to their efficiency and responsiveness. Both methods are conceptually appealing in the HCSC setting.

Companies are considered efficient if their products or services are standardized. There is little, if any, variation in the end product, and therefore the product or service is highly repeatable and highly reproducible. Any number of people can be trained to effect the same result in the same way. Consumer products companies fall into this category, as do clothing staples such as underwear. In the service sector, fast-food restaurants fall into this category.

The antithesis of the efficient company is the responsive company. Responsive companies deal in high levels of specialization and employees tend to be well trained in a single area. The resulting products or services are generally unique to a particular customer. Companies such as Dell fall into this category. A base model computer is selected but it may be completely reconfigured by the customer. Healthcare requires a great deal of specialization and is a highly responsive industry. Our new supply chain, therefore, will be characterized as a responsive supply chain as shown in Figure 3.5.

The discussion of efficiency versus responsiveness provides an excellent segue into the examination of the push/pull boundary within the HCSC. An entirely push process is one that produces products to inventory. An entirely pull process is one

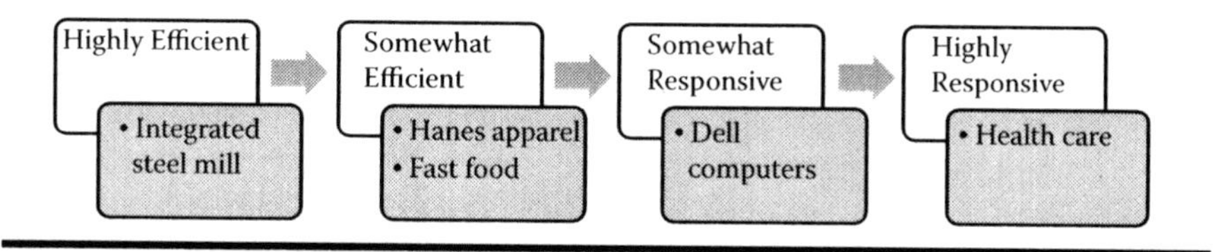


**Figure 3.5 Efficient versus responsive supply chains.**

that completely produces products to specific product design needs. The push/pull boundary provides the point at which any process converts from push to pull.

In the HCSC setting, we are tempted to say that it is an entirely pull process, but we must take note of common functions performed on all patients. Much to our dismay, the first step is completion of paperwork, no matter what condition a patient is in. Coincident with the paperwork, triage is performed on every patient to assess his or her needs. At this point, the HCSC becomes a pull process. All treatments are dictated by the condition of the individual patient. We first expand the diagram presented in Figure 3.4 and then provide a cycle view of the HCSC in Figure 3.6. In the cycle view, we begin to break down aspects of the care into logical cycles. We then use these cycles to describe the push/pull boundary in Figure 3.7.

Through the characterization of the supply chain as a responsive and heavily pull process, we have exemplified the goals of the healthcare provider. This was essential in developing our final supply chain strategy.

An interesting and anecdotal verification of the supply chain classification comes from examining the current nature of the HCSC. Medical professionals are hired with many specialties and assigned to areas where their specialties are most needed. Individual professionals within a certain area are considered interchangeable resources

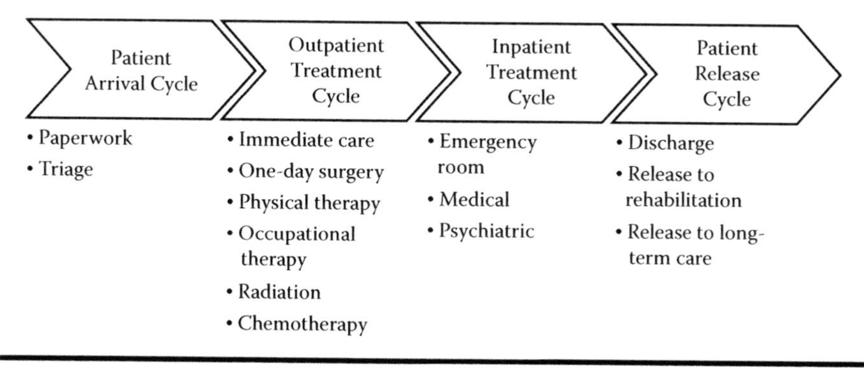


**Figure 3.6 Cycle view of healthcare supply chain.**

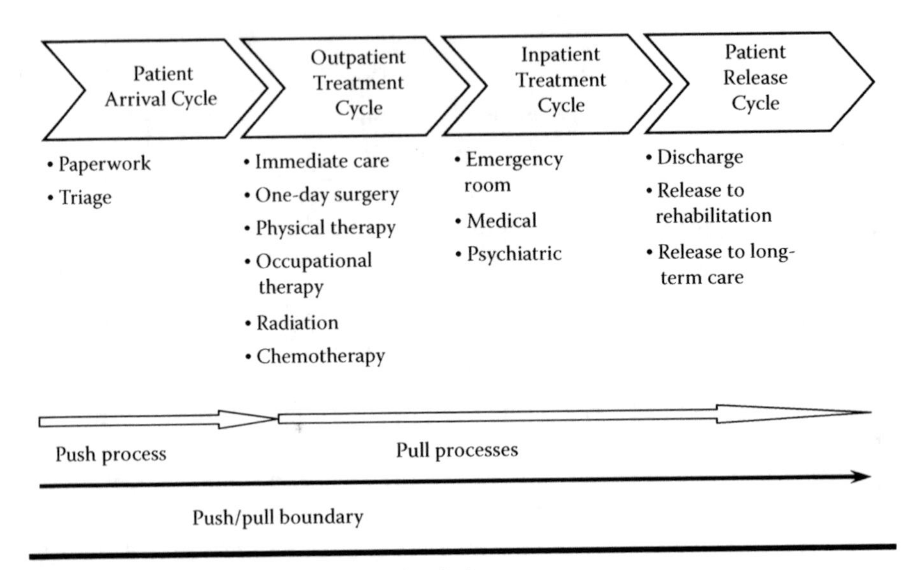


**Figure 3.7 Push/pull view of supply chain processes.**

from a manufacturing perspective. They may have different levels of seniority and therefore varying amounts of experience, but they are all trained for the same functions. Between areas of expertise, professionals may not shift unless they receive additional education and training. This is decidedly the nature of responsive supply chains.

### 3.4.4 Looking to the Future

An HCSC system, using a patient as an inventory model, should have little effect on the physical delivery of healthcare. In fact, it should not even be discernable to the individual patient in any way other than a more affordable health insurance premium. HCSC is a method to examine current systems and find the parts that work well and the parts that do not.

Using Kaiser Permanente as a marker of clinical and administrative competence does not necessitate that a fully realized system based on patient as inventory would resemble it, or in fact any HMO, in any way. The federal legislation, signed into law in the 1970s, that created the American HMO as we know it today attempted to do just that. Yet 35 years down the road, few of the for-profit HMOs remain. There was more to the formula than just imitating one successful model. An HCSC system would attempt to look at a theoretical framework for analyzing that and other models, bringing previously unavailable methods to bear. Having said this, we feel obliged to weigh in with our opinion that a cost-effective system for delivery of healthcare will not look like the decentralized patchwork of services that brought us to the mid- to late twentieth century, lacking accountability or standards. We see an ideal healthcare system being a not-for-profit entity that will fit neatly into the footprint of the existing HMO model.

An organized system for all aspects of ambulatory and inpatient treatment is, without question, a tremendous advantage when it comes to the ability to provide comprehensive medical care. Centralization of any manufacturing process allows for efficiencies of scale, standardization of process, and most importantly the ability to establish effective quality control. An ideal ambulatory model would include facilities for medical exams, x-rays, blood tests, physical therapy, pharmacy, patient education, psychiatric counseling, and medical equipment all within the same building. Walk in with your painful, swollen wrist; walk out with your fracture diagnosed, splint in place, medications, and follow-up appointment with an orthopedist for the next week without ever having left the premises.

To briefly discuss the potential shortcomings of such a centralized system of healthcare delivery, it would be appropriate to outline the differences in for-profit and not-for-profit healthcare systems. Both must make money, so the idea that not-for-profit healthcare systems should be able to sustain an ongoing operating loss is incorrect. They need to pay their bills just like the rest of us. A not-for-profit system's main advantage is that it has additional revenue streams open to it that a for-profit system would not. This would include income in the form of charitable contributions, and access to favorable tax status and public funds in exchange for taking on a socially responsible role in the community by accepting care for patients whose care would not be deemed as profitable.

A problem that would accompany any HMO model is that of the medical provider as dual agent. Doctors as an employees of the HMO would have a fiduciary responsibility to their patients to provide the best, appropriate medical care and yet would also have a fiduciary responsibility to their employers to function as responsible stewards of the employers' resources. This has been identified as an area of conflict—particularly in for-profit healthcare systems—for several years. No good remedies have been proposed; in fact, it has been suggested that a for-profit, free-market model for healthcare delivery is unsustainable for this very reason.

Americans' insistence that we have access to the best possible healthcare without regard to cost, and the commonly held belief that European or Canadian-style healthcare provides simply adequate care in its quest for affordability, are ideas that will be hotly debated for decades to come. The only clear conclusion that can be drawn from the current debate is that healthcare delivery in the United States, using a free-market model to determine the availability of services, will remain with us for the foreseeable future.

Finally, the power of a name is certainly important, and the de-humanizing effect of considering an individual human being as an object is a concept worth considering. The capacity to anticipate all potential unethical applications of any system at the outset of a venture, however, confounds the imagination.

The thing that cannot be accounted for is the unintended effect of this theoretical framework. To attempt to predict every detail of this path of research so early in the process is outright irrational. Once the model is better established, however, it is feasible to generate parameters within which the impact of HCSC theories would fall. Taking the worst possible outcome from this process and then weighing the risk for that outcome against the benefit from a streamlined, more equitable and sustainable system of healthcare delivery would be the goal of future ethical considerations.

### 3.4.5 Simulation

To move forward with the revised supply chain and the treatment of patients as inventory, we must now verify that the system is equally well represented using this variation. This was accomplished through the development of a baseline simulation model for the current supply chain and the proposed supply chain, both utilizing Extend LT simulation software.

Current supply chain simulation assumes that patients are entering service transactions. Each step in the process is considered an operation, however, because Extend LT specifies that transactions represent deterministic processing time and operations represent stochastic processing time. This includes everything from evaluation to paperwork to all treatment options. The transactions are all preceded by priority queues and the processing time is stochastic. The type of facility required (inpatient versus outpatient) and the types of treatments required are assigned based on empirical probability distributions. These distributions currently act as estimates and over time will be replaced by more realistic evaluations of a specific HCSC.

After completing the baseline simulation, which assumes that the HCSC is a service operation with customers entering and leaving according to probabilistically determined needs, a new simulation representing the revised supply chain was embarked upon. In this simulation, inventory subassemblies (patients) were to be generated at the same arrival rate as the patients in the baseline simulation. Priority stacks were used to represent subassembly waiting times, and operations were again used to represent stochastic treatment times. The choice of facility needed, as well as the type of operation (treatment) required, were all determined based on the same probabilistic assumptions applied in the baseline simulation.

Even before the simulation runs were completed, the immense similarities between the scenarios became very clear. Although transactions and operations are different in definition, in Extend LT they behave in the same fashion. Physical resources and time are required for each. Labor resources are assigned to each and released, and given that both operations and transactions refer to medical treatments, the same resources are required for both. Due to these similarities, and the characteristics of Extend LT, the baseline and revised simulation models converged into the same simulation. For closer examination, the simulation model is presented in Appendix A.

## 3.5 Conclusions

The discussion presented in this chapter provides a clear basis for accepting a healthcare supply chain model that considers the patients as inventory. The ability to shift the paradigm was shown by various means. Maintaining the integrity of the HCSC goals in the reconfiguration of the supply chain was the first important criterion. Any supply chain must meet the goals and strategies of the organization.

The second important criterion presented was an ethical criterion. Within mathematical decision-making models, we examine the ethics of our analysis but seldom have to deal with human and medical ethics. When attempting changes to a medical supply chain that might result in a modified understanding of the importance of a patient, it was critical to ensure the ethical integrity of our work.

An ability to effectively simulate the revised supply chain was the final criterion of importance. Ultimately, the simulation of the revised supply chain fully mirrored the simulation of the current supply chain. This was due to the functions defined in the Extend LT simulation language. The resulting model shows that viewing patients as inventory does not impact our ability to effectively simulate the HCSC.

## 3.6 Future Research

The validation of our new healthcare supply chain opens the door to many new avenues in the analysis of the healthcare supply chain. The first area of interest is the development of a more detailed simulation model. This model will require the further characterization of stochastic times for more specific treatment categories. Using this model, we will be able to project the value of modified structures and facilities such as immediate care facilities. This particularly allows the opportunity to determine the value of shifts toward the HMO structure.

Inventory optimization models may be used to assess the physical capacity of facilities. Extending these with Materials Requirements Planning (MRP) concepts will also help maintain control over inventory of medical resources and medical staff. Achievements in quality control and Six Sigma topics may also be achieved through our adjusted view of patients as inventory. Ethical considerations will continue to be integral to the analyses performed.

# References

Allen, B.A., Wade, E., and Dickinson, H. (2009), Bridging the divide—commercial procurement and supply chain management: are there lessons for health care commissioning in England?, *Journal of Public Procurement,* 9(1), 79–108.

Beauchamp, T., and Childress, J. (2008), *Principles of Biomedical Ethics, 6th ed.* New York: Oxford University Press.

Buddress, L., and Raedels, A. (2000), Essential tools of supply chain management, *Hospital Materiel Management Quarterly,* 22(1), 36–41.

Gatewood, R.D., and Rockmore, B.W. (1986), Combining organization manpower and career development needs: an operational human resource planning model, *Human Resource Planning,* 9(3), 81–96.

Katsaliaki, K, Mustafee, N., Taylor, S.J.E., and Brailsford, S. (2009), Comparing conventional and distributed approaches to simulation in a complex supply chain health system, *Journal of the Operational Research Society,* 60, 43–51.

Keen, J., Moore, J., and West, R. (2006), Pathways, networks and choice in health care, *International Journal of Health Care,* 19(4), 316–327.

Kumar, A., Ozdamar, L., and Zhang, C.N. (2008), Supply chain redesign in the healthcare industry of Singapore, *Supply Chain Management: An International Journal,* 13(2), 95–103.

Lapierre, S.D., and Ruiz, A.B. (2007), Scheduling logistic activities to improve hospital supply systems, *Computers & Operations Research,* 34, 624–641.

McKone-Sweet, K.E., Hamilton, P., and Willis, S.B. (2005), The ailing healthcare supply chain: a prescription for change, *Journal of Supply Chain Management,* 41(1), 4–17.

McManus, K. (2007), Taking human inventory, *Industrial Engineer Magazine,* 39(5), 20–21.

*Mosby's Medical Dictionary,* 8th edition (2009), Elsevier.

Norris, T. (2009), Helium, Inc. www.helium.com (Accessed June 14, 2009.)

Pan, Z.X. (Thomas), and Pokharel, S. (2007), Logistics in hospitals: a case study of some Singapore hospitals, *Leadership in Health Services,* 20(3), 195–207.

Peabody, N. (2001), Cants, sense, census: human inventory in late precolonial India and early colonial India, *Society for Comparative Study of Society and History,* 0010(4175/01), 819–850.

Rivard-Royer, H., Landry, S., and Beaulieu, M. (2002), Hybrid stockless: a case study: lessons for health-care supply chain integration, *International Journal of Operations & Production Management,* 22(4), 412–424.

Roark, D.C. (2005), Managing the healthcare supply chain, *Nursing Management,* 36(2), 36–40.

Runy, L.A. (2005), The supply chain and clinical quality, *Hospitals & Health Networks;* 79(3), 57.

Shah, R., Goldstein, M., Unger, B.T., and Henry, T.D. (2008), Explaining anomalous high performance in a health care supply chain, *Decision Sciences* C, 39(4), 759–789.

*The Free Dictionary* by Farlex, (2009). www.freedictionary.com

Towill, D.R. (2006), Viewing Kaiser Permanente via the logistician lens, *International Journal of Health Care Quality Assurance,* 19(4), 296–315.

Wikipedia, *The Free Encyclopedia,* (2007). www.wikipedia.com

Wilson, D., Kumor, R., and Bar, A.H. (2007), Lasso runaway technology and supply chain costs, *Nursing Management,* 38(5), 24–30.

# Appendix A: Supply Chain Simulation Model

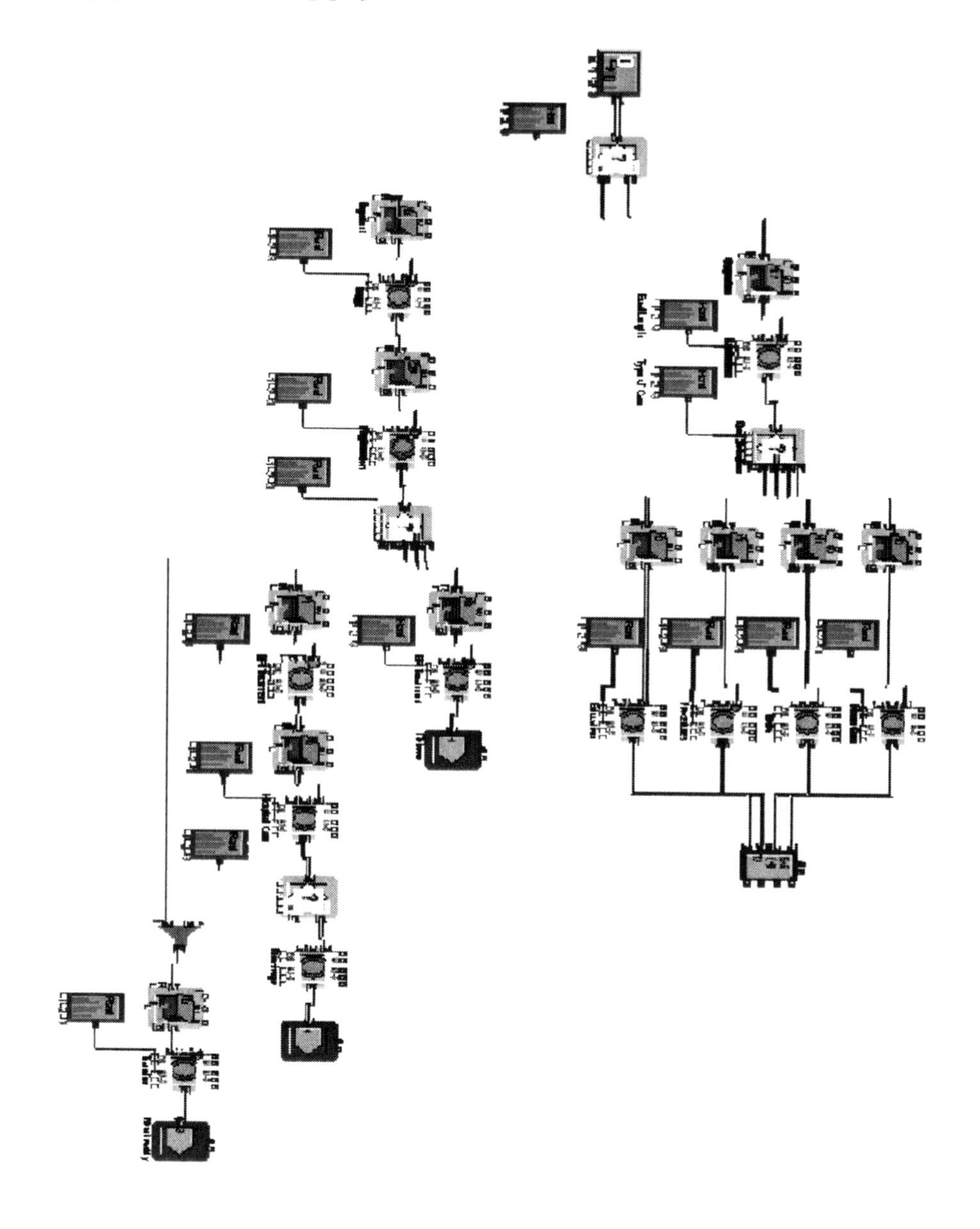

*Chapter 4*

# Using Hierarchical Planning to Exploit Supply Chain Flexibility: An Example from the Norwegian Meat Industry*

Peter Schütz, Asgeir Tomasgard, and Kristin Tolstad Uggen

## Contents



* This research is supported by the Research Council of Norway through the MeatVision project and the Center for Research-Based Innovation SFI NORMAN.

## 4.1 Introduction

Uncertainty is widely recognized as one of the most important challenges in supply chain management. For example, the supply of raw materials, demand for finished products, their prices, production costs, and capacities, are all likely to change over time. Sometimes these changes are known and predictable. More often, however, they are not. Planning and decision-making processes on all levels of the supply chain are affected by the uncertainty. Flexibility is frequently suggested as one way of coping with this problem. Upton (1994) defines it as "the ability to change or react with little penalty in time, effort, cost, or performance."

Traditionally, the research on flexibility in production networks (like supply chains) has focused on manufacturing flexibility at the plant or machine level. Gupta and Goyal (1989) reviewed different concepts of manufacturing flexibility. Gerwin (1993) discussed the strategic aspects of manufacturing flexibility. Recently, researchers have started to consider flexibility in a wider context. Zhang et al. (2002) studied value chain flexibility and how it can be used to achieve a competitive advantage, whereas Bertrand (2003) discussed the issue of flexibility in light of supply chain design. Schütz and Tomasgard (2008) examined the effect of different types of flexibility on operational supply chain planning.

The main purpose of this chapter is to discuss decision flexibility in a hierarchical setting and describe the practical implications for a case company. The models presented here are deterministic, and hence we focus on how the uncertainty can be managed based on the inherent flexibility in supply chain decisions. This reflects the optimization tools used by our case company in practice. In reality, for some of these applications the flexibility could be utilized even further by using stochastic programming models. Where such modeling approaches exist, we also discuss them briefly and point to further references.

We discuss hierarchical supply chain planning and flexibility in Section 4.2. In Section 4.3 we present the basic models for supply chain planning before we discuss their application to the supply chain planning problem in the Norwegian meat industry in Section 4.4. We conclude in Section 4.5.

## 4.2 Hierarchical Supply Chain Planning

The purpose of supply chain management is to plan, coordinate, and control a network of facilities that deals with the supply of raw materials, the transformation of these into intermediate and eventually finished products, as well as the distribution of those to the customers (see, e.g., Simchi-Levi et al. 2007). This is a very complex task even for relatively small supply chains. One approach to reducing the complexity is to split the supply chain planning problem into subproblems that resemble the hierarchical structure of the decisions: The higher-level decisions impose constraints on the lower-level decisions, whereas the lower-level decisions provide the necessary feedback to evaluate the higher-level decisions. This is one of the main ideas behind hierarchical planning (see, e.g., Hax and Candea 1984, Bitran and Tirupati 1993). Early work on hierarchical planning motivated by planning and scheduling problems can be found in Hax and Meal (1975) and Bitran and Tirupati (1993). Since then, hierarchical planning has been applied to a wide range of problems. A recent example discussing a hierarchical approach to supply chain management is given by Miller (2002).

To describe the hierarchical links between the different planning problems, we classify the planning problems according to the framework proposed by Anthony (1965). He distinguishes between decisions on strategic, tactical, and operational levels. The strategic level considers long-term decisions, the tactical level supports medium-term decisions, and the operational level deals with short-term decisions. The actual length of the time horizons on different levels can vary between different kinds of industry. While a strategic model in the oil and gas industry can have a time horizon of up to 50 years, a strategic model in the semiconductor industry, where the life span of the products is much shorter, can be only 5 to 10 years. Similar differences can be found on the tactical and operational levels.

Figure 4.1 shows the different elements of our planning hierarchy. On the strategic and tactical level we plan for the supply chain as a whole. As volume flexibility and storage flexibility are installed primarily at this level, we need the whole supply chain perspective to determine where and how flexibility is needed. On the operational planning level we include a facility planning level in addition to the supply chain. The main reason for this is that most of the operational decisions will ultimately be implemented at the facility level, so we will exploit flexibility here.

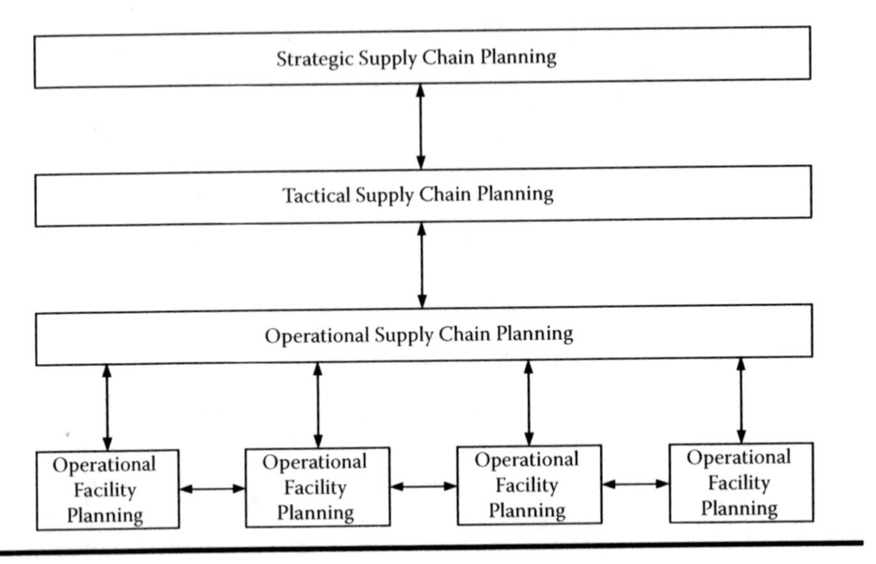


**Figure 4.1 Supply chain planning hierarchy.**

The upper-level planning problems will not only impose constraints on the lower-level problems in terms of capacity or target production volumes, but also determine the type and amount of flexibility available at the lower level. Usually, flexibility is installed at the higher level and exploited at the lower level. The results from the lower-level problems being used to evaluate the decisions made at the higher level.

### 4.2.1 Strategic Planning

The highest hierarchical level in Anthony's framework is strategic planning. In short, strategic decisions define the objectives of the organization and plan the resources used to obtain these objectives. Typical examples of decisions on the strategic level are supply chain design and facility location decisions.

Decisions at this level have a significant impact on the supply chain for a relatively long period. Once implemented, it may be impossible, or at least very expensive, to revise strategic decisions over a certain period. This implies that strategic planning has to consider a planning horizon of several years into the future. The degree of uncertainty increases with the length of the planning horizon. Models supporting strategic decisions should give solutions that are flexible when it comes to adjusting decisions according to future events in the market. Often, stochastic models are used that take into account uncertainty in, for example, market demand or market prices.

Strategic models usually aggregate future demand to reduce the number of time periods in the model. This approach captures uncertainty in the long run, e.g., changes in market size or market shares. Short-term variations, however, disappear from the dataset. Neglecting these short-term variations will underestimate the need for capacity.

Schütz et al. (2009), for example, examine the case of redesigning a supply chain given uncertain demand. They show that a stochastic model will install more capacity than the deterministic expected value problem where uncertain demand is replaced by its expected value. Also, modeling short-term demand variations instead of using aggregated demand levels results in a higher demand for flexibility. The stochastic model is able to value and install volume flexibility (i.e., excess capacity). This flexibility can be exploited in the future, as the production volumes can be increased depending on realized demand.

### 4.2.2 Tactical Planning

The next level in the planning hierarchy is tactical planning. According to Anthony (1965), the purpose of tactical planning is to ensure the efficient and effective utilization of the organization's resources. The planning horizon is usually several months long and should cover a seasonal cycle (Hax and Candea 1984, Miller 2002). The tasks include balancing supply and demand, assigning production volumes to different plants, planning capacity utilization, workforce planning, and inventory planning, among others.

Traditionally, tactical planning has focused on how to adjust the production system in order to meet fluctuating customer demand (Hax and Candea 1984, Swaminathan and Tayur 2003). These adjustments can be made, for example, by changing the workforce, or building and depleting inventory. An early example of combining inventory control and production smoothing in aggregate production planning can be found in Winters (1962). Inventory is often carried to hedge against variations in demand. These variations in demand can be known in advance (e.g., seasonal patterns) but can also be uncertain.

Aggregate production planning exploits volume flexibility in the supply chain by allocating production volumes and adjusting the workforce at the different production facilities. Using storage flexibility by building and depleting inventories allows not only for smoothing capacity utilization, but also to prepare for seasonal variations in demand. Focusing only on the production system, however, ignores the possibility of balancing supply and demand by actively managing customer demand. Combining aggregate production planning with methods from revenue management (see, e.g., Talluri and van Ryzin 2004) introduces additional flexibility: As an alternative to building inventory or reducing the rate of production, demand for finished products could be increased using a time-limited marketing campaign. This type of flexibility is very important for supply chains dealing with perishable products or other products with a limited shelf life.

### 4.2.3 Operational Planning

On the operational planning level, we focus on planning activities for the next few weeks. Capacity limits and often also supply and demand are fixed for at least the beginning of the planning horizon. The problem is to determine the best production plan for the near future. The level of detail must be very high on this level to be able to create a plan that takes into account all real-life constraints of the production processes. As a consequence, it is often impossible to solve such a problem for the whole supply chain in a reasonable period of time. We therefore suggest using moderately aggregated data for operational supply chain planning and detail these plans further by means of operational planning for each facility.

Flexibility at the operational level is provided mainly by the possibility to allocate specified production volumes to different facilities or to use different bills of material in the production process. In addition to this operational flexibility, we can exploit the flexibility designed into the production network at the higher levels of the planning hierarchy. Considering short-term variations in, for example, demand during the supply chain design phase increases the amount of volume flexibility available in the supply chain. We can therefore react to large demand variations by depleting seasonal inventories of raw materials and increasing the production volumes at different production facilities in order to satisfy observed demand.

Schütz and Tomasgard (2008) studied the impact of different types of flexibility on operational supply chain planning under uncertainty. Their results show that operational flexibility becomes more important in case little or no volume and storage flexibility are installed. They also showed that the planning process is not affected by uncertainty if enough flexibility is available in the system.

## 4.3 Modeling the Supply Chain

A supply chain is usually described as a network of production facilities, warehouses, and distribution centers. We consider it to be a network of production processes that are connected to each other by the flow of material through the network. Let us first describe the production processes before we give a mathematical representation of the problem.

### 4.3.1 Supply Chain Description

We consider two generic types of production processes: combining processes and splitting processes (see, e.g., Schütz et al. 2009, Schütz and Tomasgard 2008). In the real world, these production processes can be different production lines, technologies, or production phases. Combining processes are common in manufacturing and assembling and are often described by means of a Bill of Material (BoM). This BoM lists the type and number of input products ($p_i$) required to produce one output product ($p_o$) (see Figure 4.2a). Splitting processes are often found in the process

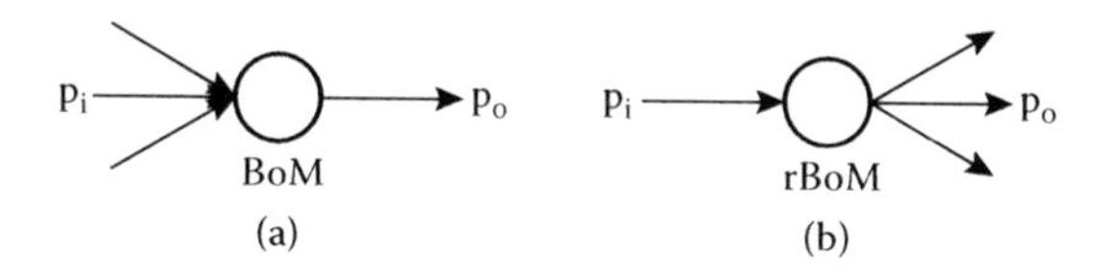


**Figure 4.2 Splitting and combining processes: (a) combining process and (b) splitting process.**

industry and are similarly described: Using a so-called reversed Bill of Material (rBoM), we define in which output products ($p_o$) one unit of input product ($p_i$) can be split (see Figure 4.2b).

If one facility houses several production processes, we have to include multiple nodes, one for each process, in the supply chain. The disadvantage of increasing the size of the network is made up for by properly modeling the material flow between the production processes. See Figure 4.3 for an example of a supply chain consisting of three nodes with two production facilities housing two splitting processes and one combining process.

Our formulation allows us to choose from different BoMs for a given production process. Each BoM is assigned to the production of a given product using a specified production process. The possibility of choosing the BoM also introduces operational flexibility into the model as we can adjust the production process, for example, based on the availability of raw materials. Correspondingly, rBoMs are assigned to the processing of one specific input product.

### *4.3.2 Optimization Models*

The optimization models are based on the supply chain's generic description as a sequence of splitting and combining processes. We first present the notation before discussing the constraints and models for the different planning problems. The notation follows Schütz et al. (2009) and Schütz and Tomasgard (2008).

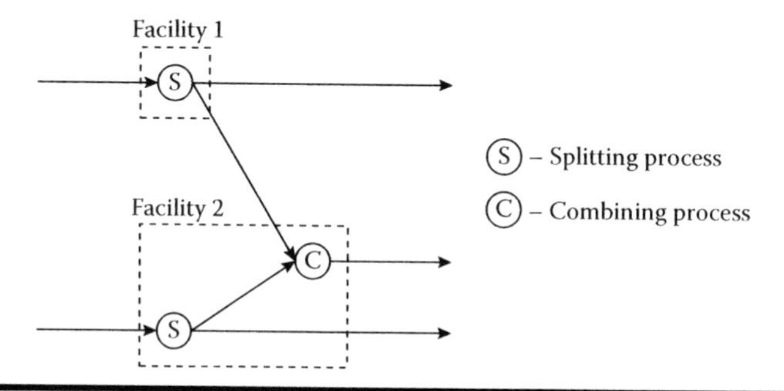


**Figure 4.3 Example of a simple supply chain with two facilities and three processes.**

#### 4.3.2.1 Notation

Let us introduce the following notation for the optimization models presented here:

- Sets:
  - $\mathcal{F}_c$ Set of possible facility locations for combining processes
  - $\mathcal{F}_s$ Set of possible facility locations for splitting processes
  - $\mathcal{W}$ Set of possible warehouse locations
  - $\mathcal{L}$ Set of all possible locations, $\mathcal{L} = \mathcal{F}_c \cup \mathcal{F}_s \cup \mathcal{W}$
  - $\mathcal{C}$ Set of customer locations
  - $\mathcal{U}(j)$ Set of upstream locations able to send products to location $j$, $j \in \mathcal{L} \cup \mathcal{C}$
  - $\mathcal{D}(j)$ Set of downstream locations able to receive products from location $j$, $j \in \mathcal{L}$
  - $\mathcal{O}_c(j)$ Set of combining processes that can be performed at location $j$, $j \in \mathcal{F}_c$
  - $\mathcal{O}_s(j)$ Set of splitting processes that can be performed at location $j$, $j \in \mathcal{F}_s$
  - $\mathcal{O}(j)$ Set of all processes that can be performed at location $j$, $\mathcal{O}(j) = \mathcal{O}_c(j) \cup \mathcal{O}_s(j)$
  - $\mathcal{O}$ Set of all processes, $\mathcal{O} = \cup_{j \in \mathcal{L}} \mathcal{O}(j)$
  - $\mathcal{B}(o)$ Set of (reversed) bills of materials that can be used for process $o$, $o \in \mathcal{O}$
  - $\mathcal{P}$ Set of products
  - $\mathcal{P}_i(o)$ Set of input products for process $o$, $o \in \mathcal{O}$
  - $\mathcal{P}_o(o)$ Set of output products of process $o$, $o \in \mathcal{O}$
  - $\mathcal{N}$ Set of event nodes in the scenario tree
  - $\mathcal{T}$ Set of time periods
- Indices and superscripts:
  - $b$ Bill of materials index, $b \in \mathcal{B}(o)$, $o \in \mathcal{O}$
  - $j, k$ Location indices, $j, k \in \mathcal{L} \cup \mathcal{C}$
  - $o$ Process index, $o \in \mathcal{O}(j)$, $j \in \mathcal{L}$
  - $p, q$ Product indices, $p \in \mathcal{P}$
  - $n$ Event node index, $n \in \mathcal{N}$
  - $t$ Time period index, $t \in \mathcal{T}$
- Parameters, constants, and coefficients:
  - $A^b_{po,pi}$ Yield of product $p_o \in \mathcal{P}_o(o)$ in case one unit $p_i \in \mathcal{P}_i(o)$ is processed with rBoMs $b \in \mathcal{B}(o)$, $o \in \mathcal{O}s$
  - $B^b_{po,pi}$ Amount of product $p_i \in \mathcal{P}_i(o)$ needed to produce one unit $p_o \in \mathcal{P}_o(o)$ using BoM $b \in \mathcal{B}(o)$, $o \in \mathcal{O}_c$
  - $C^o_{jt}$ Capacity of process $o$ at location $j$ at time $t$
  - $D^p_{jt}$ Demand for product $p$ at customer location $j$ at time $t$
  - $F^o_j$ Fixed cost of location process $o$ at location $j$
  - $H^p_{jt}$ Shortfall penalty for one unit of demand of product $p$ at customer location $j$ at time $t$
  - $I^p_{jt}$ Cost of holding one unit inventory of product $p$ at facility $j$ in period $t$

$P^{p}_{jbt}$ Cost of processing one unit of product $p$ at location $j$ using (reversed) Bill of Material $b$ at time $t$

$S^{p}_{jt}$ Supply of product $p$ at location $j$ at time $t$

$R^{p}_{jt}$ Revenue of selling one unit of product $p$ at location $j$ at time $t$

$T^{p}_{jkt}$ Cost of transporting one unit of product $p$ from location $j$ to location $k$ at time $t$

$i^{p}_{j0}$ Initial inventory of product $p$ at location $j$

$z^{p}_{j0}$ Initial backlog of product $p$ at location $j$

- Decision variables:

$i^{p}_{jt}$ Inventory of product $p$ at location $j$ at period $t$

$w^{p}_{jkt}$ Amount of product $p$ transported from location $j$ to location $k$ at time $t$

$x^{p}_{jbt}$ Amount of product $p$ processed/produced at location $j$ using Bill of Material $b$ at time $t$

$y^{o}_{j}$ 1 if process $o$ is located at facility location $j$, 0 otherwise

$z^{p}_{jt}$ Shortfall for product $p$ at customer location $j$ in period $t$

### *4.3.3 Model Formulations*

The description of the processes, inventory, and demand satisfaction constraints are the same for all models, whereas capacity constraints and objective functions depend on the problem under consideration. Note that even though the formulation of the constraint does not change when considering a different problem, the input data does. The number of time periods in the model, aggregation of products into product groups and/or families, etc., all change and become more detailed as we move from strategic to operational planning.

#### *4.3.3.1 Common Constraints*

Let us first consider the constraints that are common for all models. The splitting processes are described by the following two constraints:

$$S^{p}_{jt} + \sum_{k \in \mathcal{U}(j)} w^{p}_{kjt} = \sum_{o \in \mathcal{O}_s(j)} \sum_{b \in \mathcal{B}(o)} x^{p}_{jbt} \quad j \in \mathcal{F}_s,\ p \in \bigcup_{o \in \mathcal{O}_s(j)} \mathcal{P}_i(o),\ t \in \mathcal{T}, \tag{4.1}$$

$$\sum_{o \in \mathcal{O}_s(j)} \sum_{q \in \mathcal{P}_i(o)} \sum_{b \in \mathcal{B}(o)} A^{b}_{p,q} \cdot x^{q}_{jbt} = \sum_{k \in \mathcal{D}(j)} w^{p}_{jkt} \quad j \in \mathcal{F}_s,\ p \in \bigcup_{o \in \mathcal{O}_s(j)} \mathcal{P}_o(o),\ t \in \mathcal{T}. \tag{4.2}$$

Constraint (4.1) ensures that all products supplied by external suppliers, $S^p_{jt}$, and those sent from upstream processes, $w^p_{kjt}$, are processed, $x^p_{jbt}$. Equation (4.2) sends all produced products, $x^q_{jbt}$, to downstream nodes, $w^p_{kjt}$, that is, further processing or customers. The combining processes are described in a similar fashion:

$$S^p_{jt} + \sum_{k \in \mathcal{U}(j)} w^p_{kjt} = \sum_{o \in \mathcal{O}_c(j)} \sum_{q \in \mathcal{P}_o(o)} \sum_{b \in \mathcal{B}(o)} B^b_{q,p} \cdot x^q_{jbt}$$

$$j \in \mathcal{F}_c,\ p \in \bigcup_{o \in \mathcal{O}_c(j)} \mathcal{P}_i(o),\ t \in \mathcal{T}, \tag{4.3}$$

$$\sum_{o \in \mathcal{O}_c(j)} \sum_{b \in \mathcal{B}(o)} x^p_{jbt} = \sum_{k \in \mathcal{D}(j)} w^p_{jkt} \quad j \in \mathcal{F}_c,\ p \in \bigcup_{o \in \mathcal{O}_c(j)} \mathcal{P}_o(o),\ t \in \mathcal{T}. \tag{4.4}$$

The first restriction of these, Equation (4.3), ensures that all required input is either transported into the node or externally supplied. The second restriction, Equation (4.4), forces all produced products to be transported to downstream nodes.

The inventory mass balance is given by Equation (4.5). Inventory from the previous period, $i^p_{jt-1}$, plus products transported to the warehouse, $w^p_{kjt}$, are equal to the sum of inventory at the end of the period, $i^p_{jt}$, and the products shipped from the warehouse, $w^p_{jkt}$. Equation (4.6) makes sure that the sum of products transported into a customer node, $w^p_{kjt}$, and shortfall, $z^p_{jt}$, equals demand.

$$i^p_{jt-1} + \sum_{k \in \mathcal{U}(j)} w^p_{kjt} = i^p_{jt} + \sum_{k \in \mathcal{D}(j)} w^p_{jkt} \qquad j \in \mathcal{W},\ p \in \mathcal{P},\ t \in \mathcal{T}, \tag{4.5}$$

$$\sum_{k \in \mathcal{U}(j)} w^p_{kjt} + z^p_{jt} = D^p_{jt} \qquad j \in \mathcal{C},\ p \in \mathcal{P},\ t \in \mathcal{T}. \tag{4.6}$$

### 4.3.3.2 Strategic Models

The strategic models are for supply chain (re)design, that is, determining where to install production capacity for a certain process. We formulate the capacity constraints for splitting processes [Constraint (4.7)] and combining processes [Constraint (4.8)] as follows:

$$\sum_{p \in \mathcal{P}_i(o)} \sum_{b \in \mathcal{B}(o)} x^p_{jbt} \le C^o_{jt} \cdot y^o_j \qquad j \in \mathcal{F}_s,\ o \in \mathcal{O}_s(j),\ t \in \mathcal{T}, \tag{4.7}$$

$$\sum_{p \in \mathcal{P}_o(o)} \sum_{b \in \mathcal{B}(o)} x^p_{jbt} \le C^o_{jt} \cdot y^o_j \qquad j \in \mathcal{F}_c,\ o \in \mathcal{O}_c(j),\ t \in \mathcal{T}. \tag{4.8}$$

The objective function is to minimize the total costs, that is, the sum of facility costs and operating costs. The operating costs are defined here as the sum of production costs, transportation costs, and the shortfall penalty.

$$\min \sum_{j\in\mathcal{L}}\sum_{o\in\mathcal{O}(j)} F_j^o y_j^o + \sum_{p\in\mathcal{P}}\sum_{t\in\mathcal{T}}\left[\sum_{j\in\mathcal{L}}\left(\sum_{o\in\mathcal{O}(j)}\sum_{b\in\mathcal{B}(o,p)} P_{jbt}^p x_{jbt}^p + \sum_{k\in\mathcal{D}(j)} T_{jkt}^p w_{jkt}^p\right) + \sum_{j\in\mathcal{C}} H_{jt}^p z_{jt}^p\right] \tag{4.9}$$

The strategic models must consider the complete supply chain. The model therefore includes all constraints.

### 4.3.3.3 Tactical and Operational Models

For tactical and operational models, the production capacity is given. Constraints (4.7) and (4.8) are modified as follows to reflect this:

$$\sum_{p\in\mathcal{P}_i(o)}\sum_{b\in\mathcal{B}(o)} x_{jbt}^p \le C_{jt}^o \qquad j\in\mathcal{F}_s,\ o\in\mathcal{O}_s(j),\ t\in\mathcal{T} \tag{4.10}$$

$$\sum_{p\in\mathcal{P}_o(o)}\sum_{b\in\mathcal{B}(o)} x_{jbt}^p \le C_{jt}^o \qquad j\in\mathcal{F}_c,\ o\in\mathcal{O}_c(j),\ t\in\mathcal{T} \tag{4.11}$$

For tactical and operational planning, the objective function is given as maximizing profits, that is, the sum of sales revenues minus production costs, transportation costs, and the shortfall penalty.

$$\max \sum_{t\in\mathcal{T}}\left(\sum_{p\in\mathcal{P}}\sum_{j\in\mathcal{C}}\sum_{k\in\mathcal{U}(j)} R_{jt}^p w_{kjt}^p - \sum_{j\in\mathcal{L}}\sum_{p\in\mathcal{P}}\sum_{o\in\mathcal{O}(j)}\sum_{b\in\mathcal{B}(o)} P_{jbt}^p x_{jbt}^p - \sum_{p\in\mathcal{P}}\sum_{j\in\mathcal{L}}\sum_{k\in\mathcal{D}(j)} T_{jkt}^p w_{jkt}^p - \sum_{p\in\mathcal{P}}\sum_{j\in\mathcal{L}} I_{jt}^p i_{jt}^p - \sum_{p\in\mathcal{P}}\sum_{j\in\mathcal{C}} H_{jt}^p z_{jt}^p\right) \tag{4.12}$$

The tactical models consider the supply chain as a whole and include all constraints, whereas the operational models only consider a few facilities (if not only one) at a certain level of the supply chain. The operational models therefore include either the constraints for the splitting processes or the combining processes. Apart from the number of constraints in the problem, the main difference between the tactical and operational models is the data set used in the optimization procedure.

## 4.4 Supply Chain Planning in the Norwegian Meat Industry

We have, in close collaboration with the Norwegian meat cooperative Nortura, developed a hierarchical system of optimization models for supply chain planning. Nortura operates 41 production facilities located all over Norway (in 2008), producing more then 1500 different products. The following description of the meat supply chain is based on the one found in Tomasgard and Høeg (2002) and Tomasgard and Høeg (2005). The different optimization models are presented later.

### *4.4.1 The Meat Supply Chain*

Nortura is owned by a majority of the Norwegian farmers. It is the largest meat-producing company in Norway with the following market shares: 76 percent for slaughtering, 53 percent for cutting, 50 percent for processing, and 50 percent on sales of meat products. In addition to Nortura, there are some private enterprises of varying sizes.

Nortura's value chain consists of slaughtered animals, cut parts, and processed products. These flows again may be within a geographical region, between regions, or between Nortura and its competitors.

#### *4.4.1.1 Slaughtering*

The production process starts with the arrival of livestock at the slaughterhouses. Nortura operates 18 slaughterhouses. The size of the slaughterhouses is subject to economies of scale (Kern 1994), favoring few but large slaughterhouses. Due to legislation, however, livestock transported from a farm to the slaughterhouse can spend, at most, 8 hours on a truck before it has to be unloaded. This means that the minimum number of facilities needed to cover the geographic region of Norway would be ten or eleven (see, e.g., Van den Broek et al. 2006, Schütz et al. 2008).

At the slaughterhouse, carcasses and intestines are produced. The carcasses are either sold to competitors or sent downstream to cutting facilities. Useful intestines are sent to processing plants for further processing into finished products. There is also an external market for intestines.

#### *4.4.1.2 Cutting*

The next step in the value chain is to split the carcasses into smaller standardized pieces, and in some cases further slice the pieces into fine cuts like sirloin, T-bone steaks, and so on. This is done in accordance with some pre-decided cutting and slicing patterns. The cut pieces are either used for further processing or sales. There exists a large set of cutting patterns, all tailored to serve different situations for further processing.

In fact, this level of the supply chain is very important for exactly that reason. The more flexibility that is kept the less cutting and processing is done on the carcass. In tactical planning, this would be the level in the supply chain to consider postponing a production decision. Unfortunately, there is often a need for some of the products made from the carcass in the short-term market, but not all. So there will be a trade-off between selling some products fresh today, selling frozen products later (at a lower price), and building seasonal stock.

But the cutting decision and choice of pattern (of course being irreversible) is a strong candidate when it comes to pointing at the single point that has more effect on both value creation and flexibility in the operational decisions following thereafter. Demand for finished products and production recipes for further processing are therefore critical issues when one decides which cutting patterns should be chosen.

#### *4.4.1.3 Processing*

The processing units produce finished products for sales to the market based on the available cutting parts. Finished products are sold by the processing units to the sales units within the company. In addition, the company produces for inventory build-up because of fluctuating prices, short-term demand variations, and seasonal demand variations. Inventory build-up of finished products is limited, however, due to the limited shelf life of fresh meat products.

The most important decisions are what to produce, where to produce it, and which recipes to use. At this point, much of the production flexibility comes from using different recipes. The goal is to choose recipes that maximize overall company profit and to make sure that demand is met. This often means that the cutting parts needed for the cost-optimal recipe in one region would have better use in other products either locally or in other regions. Alternative uses exist that will give higher local costs, but higher global profit will outweigh this. To utilize the resources, one needs a global view of the available cutting parts, the production capacities, and the demand.

#### *4.4.1.4 Sales and Distribution*

One of the main tasks of the sales and distribution process is to forecast demand. Market signals are then sent backward in the supply chain to the production plants and cutting plants to make sure the correct decisions concerning recipes and cut parts are made. As the purpose of this value chain is to balance the supply side with the demand side in a combination of push and pull signals, clearly the marketing department has an important function. Through price signals and marketing, this is the only place where the demand can be influenced. The timing issue is very important: What you sell today should be considered in light of the best alternative use of the raw material in the future. The link to the cutting and processing departments is then clear. In fact, much of the strength of this particular company lies in the possibility to coordinate these functions. But again, this requires clever coordination within the chain and between the different time horizons.

### 4.4.2 The Planning Modules

The different optimization models are connected through the exchange of information that is passed from the upper-level planning problem to the lower-level planning problem. Each of the optimization models can be run as a stand-alone model, but it is the system of models that can exploit the flexibility inherent in the supply chain to deal with uncertainty. The planning hierarchy of the different models is outlined in Figure 4.4.

The number of facilities in Nortura's production network and the number of products result in significant problems. As the models for operational planning are run once a week, runtime is crucial for providing decision support. Initial tests with a subset of products indicate that it will be possible to include the majority of Nortura's products in the models described below. The size of an operational supply chain planning problem with a 12-week planning horizon and considering 550 finished products (approximately one third of the finished products) is 340,000 variables and 60,000 constraints. The strategic and tactical models are used less frequently, so runtime is not such a big issue.

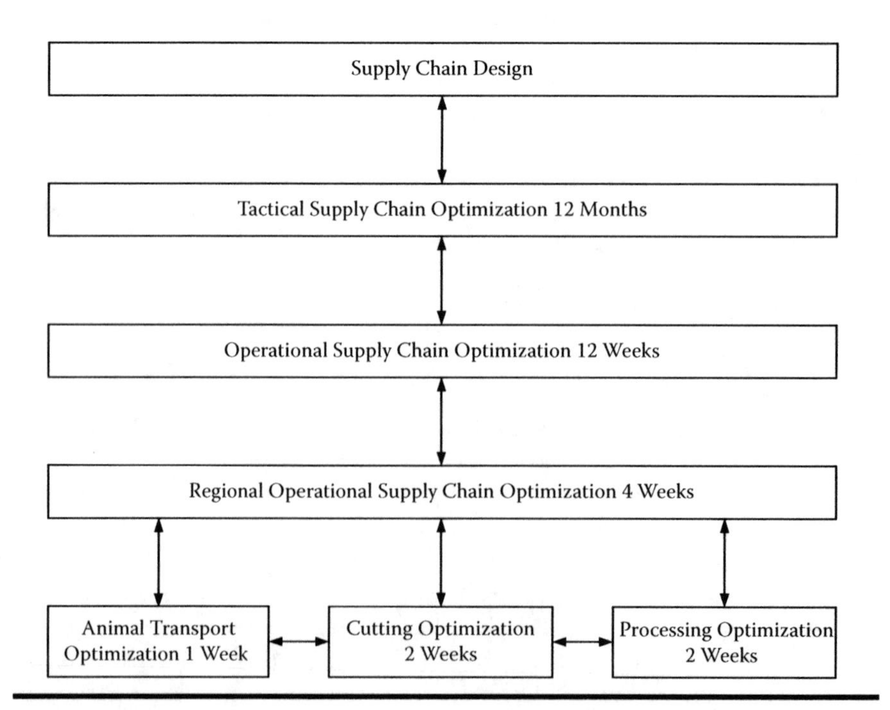


**Figure 4.4 Planning hierarchy in the Norwegian meat industry.**

#### *4.4.2.1 Strategic Supply Chain Planning*

Strategic planning models have been developed for supply chain design. The first models developed for Nortura are facility location models. They lack the supply chain perspective but focus on modeling economies of scale in total facility costs instead. Van den Broek et al. (2006) performed a deterministic analysis for locating slaughterhouses such that the total supply of livestock is slaughtered, minimizing the sum of facility and transportation costs. Studying the same problem with a stochastic supply of livestock, Schütz et al. (2008) found that the supply uncertainty required more slaughtering capacity at the slaughterhouses. The location of the slaughterhouses, however, was very similar to the one found by Van den Broek et al. (2006).

The first model considering Nortura's complete supply chain was described in detail by Schütz et al. (2009). They took a supply chain perspective for determining how to restructure the processing level given uncertain demand for finished products. The basic model is given by Constraints (4.1) through (4.8) and the Objective Function (4.9). Comparing the results of the model with uncertain demand to results from the deterministic expected value problem (where uncertain demand is replaced by its expected value), they found that the stochastic model installs more capacity, that is, more volume flexibility, than the deterministic one. They also studied the effect of different data aggregation levels and showed that the problem instances based on disaggregated data also install more production capacity.

In order to be able to exploit volume flexibility at the tactical and operational level, we have to design this flexibility into the supply chain at the strategic level (Bertrand 2003). We see from the aforementioned results that we need to consider both uncertainty and operational decisions to be able to determine the amount of volume flexibility needed in the system.

#### *4.4.2.2 Tactical Supply Chain Planning*

The model at the tactical level uses the real supply chain topology (which should be based on the results from the strategic planning models). The main task is to balance aggregated supply and demand over a 12-month planning horizon. This balance can be achieved through inventory build-up and demand management. Our current model for Nortura is based on Constraints (4.1) through (4.6), and (4.10), (4.11), and Objective Function (4.12). It focuses on inventory and production decisions but does not include the possibility of affecting customer demand, for example, as a result of marketing campaigns or pricing decisions. We have chosen to omit the modeling of demand functions, as this usually results in nonlinear models.

The decisions from the tactical model were used to assign aggregated production volumes to the different facilities. Inventory building decisions for seasonal production and production smoothing were made for each facility. Another result was the material flow within the complete supply chain from the slaughterhouses to distribution centers.

The results from the tactical planning allowed for the use of storage flexibility, that is, the possibility to transfer raw materials, components, or finished products in time (see, e.g., Schütz and Tomasgard 2008). Especially in the case of over- or undersupply of certain raw materials, storage flexibility must be exploited in order to balance supply and demand. This must be done on this planning level, as the length of seasons often require long planning horizons in order to make the right inventory decisions.

### *4.4.2.3 Operational Planning Modules*

On the operational planning level, we have two models for supply chain planning as well as models for the cutting and processing level. An example for decision support at the slaughtering level can be found in Oppen and Løkketangen (2008). They discussed the collection of livestock being subject to restrictions on production and animal welfare. The problem was formulated as a vehicle routing problem with inventory constraints and solved by a heuristic based on tabu search (Glover and Laguna 1997). At the other end of the supply is the distribution of finished products. We have not yet developed any models for supporting distribution planning for Nortura. This is an area for future research.

#### 4.4.2.3.1 Operational Supply Chain Planning

We developed two models for supply chain planning on the operational level. Both are based on Constraints (4.1) through (4.6), (4.10), and (4.11), and Objective Function (4.12). Similar to the tactical model, the main purpose of these models is to balance supply and demand in each period of the planning horizon. The main differences between the tactical model and the operational model lie in the length of the planning horizon and the level of data aggregation. The material flow in the supply chain is depicted in Figure 4.5.

The first operational model has a planning horizon of 12 weeks and considers the whole supply chain. Supply of livestock is known for the first 2 weeks and forecasted thereafter. Demand for finished products is known in the first week, but forecasted for the remaining weeks. The amounts of intermediate products demanded by industrial customers are also known in the first week. Data regarding livestock and

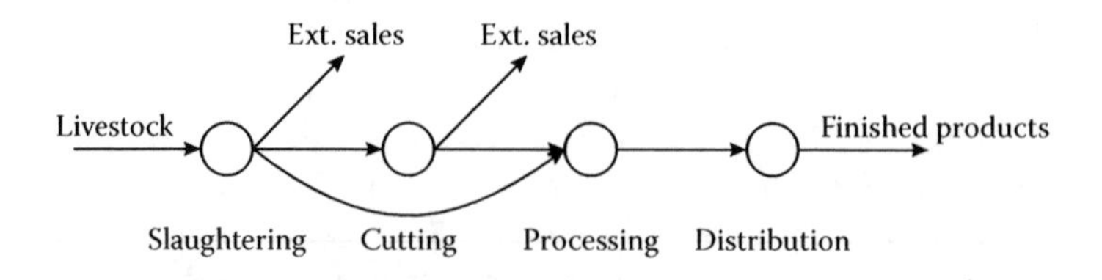


**Figure 4.5 Material flow in Nortura's supply chain.**

intermediate products is aggregated in product groups, whereas demand for finished products is described on a detailed level.

The model primarily makes decisions regarding the allocation of production to the different facilities. It uses the inventory levels and production volumes from the tactical model as input for the planning period. This data represents soft constraints, however, as newer and more detailed information of supply and demand is available. The allocation of weekly production volumes also allows for determining the volumes produced and consumed of the different intermediate products, as well as the material flow between the different facilities of the production network. The results from this model are used as input for the second model for operational supply chain planning.

The second operational supply chain model has a planning horizon of 4 weeks and considers only the facilities belonging to a given region. Nortura operates production facilities throughout Norway. Clustering these facilities into regions results in up to five regions with facilities at all levels of the supply chain. The material flow between the different regions is determined based on the solution of the first operational model. The results from the second model provide the necessary input for the models at the next hierarchy level. Both supply of raw materials and demand for finished products at the different facilities are used by the cutting optimization model and the processing optimization model.

The operational supply chain models exploit volume and storage flexibility by locating production volumes and using inventories at the different production facilities. Short-term variations in demand for finished products require the ability to increase the production volume beyond the average volumes planned for at the upper planning levels. Also, additional cut products may have to be stored in order to satisfy future demand. This flexibility is very important in order to cope with uncertain demand. Because finished products are perishable, they can only be stored for a limited amount of time.

#### 4.4.2.3.2 Operational Cutting Optimization

The part of the supply chain that consists of the transformation of whole animal carcasses into products is called cutting. In the global operational optimization model called Regional Cutting Optimization (RCO) created for this problem, we also take into account the slaughtering of animals and transportation from slaughterhouse to cutting facility. Including slaughterhouses also introduces the need to include orders for whole carcasses from external customers. What distinguishes this model from the operational supply chain model is that the supply of animals to the slaughterhouses is given, and demand for cut products is given for each cutting facility. The cut products can be sold directly to external customers as finished products or the demand can come from internal processing facilities. The model will typically be solved for all slaughterhouses and all cutting facilities in a geographical region for a 2-week time horizon. A sketch of the model can be seen in Figure 4.6 with supply, slaughtering, transportation, cutting with two splitting stages, and sales.

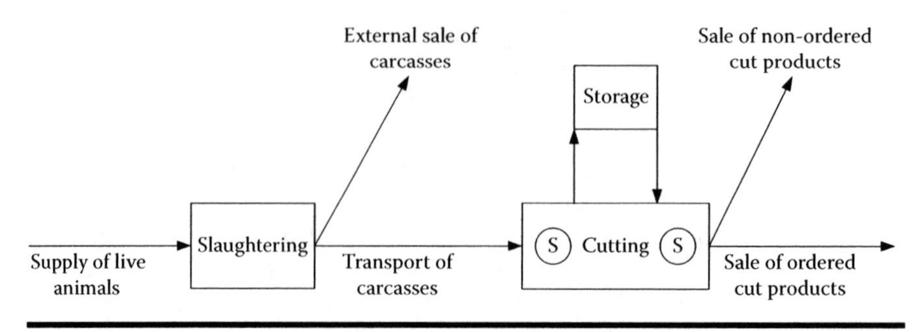


**Figure 4.6 The regional cutting optimization model.**

The main decision taken by the model is which cutting pattern to use. Cutting is a splitting process, and choosing between different cutting patterns will be the same as choosing between different splitting processes, which gives different output for the same input. Which cutting patterns that are available for each carcass depends on the type and quality of the carcass. There are two splitting stages included in the model, as can be seen in Figure 4.6. Both of these can produce finished products, but the first stage will produce larger pieces than the second. The large pieces from the first stage are input to the second stage.

The aim of this model is to maximize profit and also fulfill orders given the physical constraints and available raw material. The model will utilize flexibility in different cutting patterns and the flexibility provided by storage and in the choice of which carcasses to transport where or sell.

In addition to the constraints given by the general model for the operational level, there will be restrictions in handling travel time between slaughterhouses and cutting facilities. Also, there exists the possibility to sell products that are not ordered at a reduced price. Because the model deals with fresh food, all products and carcasses have a limited shelf life, meaning that products produced from a carcass must be sold a maximum number of days after the animal is slaughtered. This means that we must keep track of the age of the products in storage. If products exceed their shelf life, they will be wasted. These perishability constraints increase the complexity of the model, both when it comes to modeling and also computational time.

The Local Cutting Optimization (LCO) model considers only one cutting facility and the time horizon is 4 to 5 days. Here, the supply of carcasses to the cutting facility together with demand for cut products is known. Because this model is smaller than the RCO, solution time will be reduced, and the model can be run several times each day with updated input data. Flexibility in storage and the choice of cutting pattern will be utilized to obtain the best possible solution given the available raw material.

#### 4.4.2.3.3 Operational Processing Optimization

The third level in the meat supply chain is processing. In general, processing is a combining process, using different cut products for producing finished products

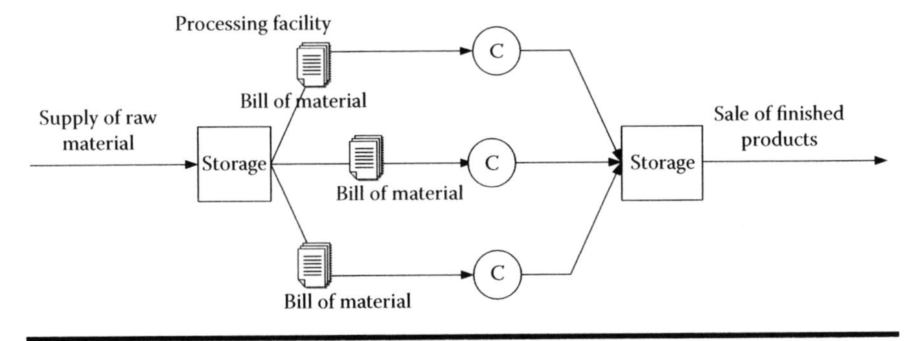


**Figure 4.7 The processing optimization model.**

such as sausages, ground meat, or cold cuts. Similar to cutting optimization, supply of cut products and demand for finished products are known for the 2-week planning horizon. We only have one model for supporting the planning of the processing level, considering a single processing facility. The facility is represented as a collection of production lines for the different finished products. Figure 4.7 illustrates the material flow in the processing optimization model.

The main goal of this model is to choose the best processing recipes for producing finished products. Basically all finished products can be produced by more than one recipe. Choosing the production recipes is very important as the different finished products compete for the same cut products. Which recipe is chosen depends on the availability of fresh cut products and the amount of stored frozen products that can be used in the production process.

The model's objective is to maximize the profit from selling finished products. The operational flexibility provided by choosing different recipes is utilized by the model to satisfy customer demand, while at the same time ensuring that the available input products are used to create the highest possible value for Nortura.

The processing optimization model builds on Constraints (4.3) through (4.6), (4.11), and Objective Function (4.12). In addition, there are constraints regarding the processing time for certain finished products (e.g., cured ham). Storing finished products at the processing facilities is not desirable as this reduces the products' shelf life in the stores. However, in order to exploit storage flexibility with our model, we add perishability constraints to the model. That way, we can keep track of the finished products' shelf life.

## 4.5 Conclusions

The hierarchical planning system presented here enables companies to install and exploit flexibility in their supply chain, helping them cope with uncertainty. It is important to connect the different types of flexibility to the decision hierarchy, because although flexibility is ultimately exploited at the operational level, each type can only be installed at a certain decision. For example, volume flexibility that is not installed by a strategic supply chain design process is not available once demand is observed and production volumes have to be assigned to the different production facilities. More research is required to better understand how to invest in flexibility. As flexibility can be designed into a supply chain at different levels, an important question is how much to spend on the correct flexibility mix.

It is, however, not an easy task to implement such a supply chain planning system for day-to-day use in a company. There are some major challenges that must be overcome in order to successfully use the different models. The biggest obstacle in our experience is often the company's IT-systems capability to collect consistent data on the aggregation level needed for each of the optimization models. High-quality input data is essential for getting high-quality results. Also, planners must be convinced to trust the results from the model. From our experience, this process is made easier by developing the different modules as if they were stand-alone models. Implementing these models one after the other allows the user to develop a better understanding of them. The success of one such model can help facilitate the development and implementation of the other models.

Throughout the process of working with optimization models in the Norwegian meat industry, we have observed that the company itself increases its understanding of the problems. Working with the abstract description of problems and processes improves the understanding of connections between and consequences of different actions and strategies.

## References

Anthony, R.N. (1965), *Planning and Control Systems: A Framework for Analysis*, Studies in Management Control, Boston, MA: Harvard University Press.

Bertrand, J.W.M. (2003), Supply chain design: flexibility considerations, in A.G. de Kok and S.C. Graves, Eds., Supply chain management: design, coordination and operation, Vol. 11 of *Handbooks in Operations Research and Management Science*, Amsterdam: Elsevier, chapter 4, pp. 133–198.

Bitran, G.R., and Tirupati, D. (1993), Hierarchical production planning, in S.C. Graves, A.H.G. Rinnooy Kan, and P.H. Zipkin, Eds., Logistics of production and inventory, Vol. 4 of *Handbooks in Operations Research and Management Science*, Amsterdam: Elsevier Science, chapter 10, pp. 523–568.

Gerwin, D. (1993), Manufacturing flexibility: a strategic perspective, *Management Science*, 39(4), 395–410.

Glover, F.W., and Laguna, M. (1997), *Tabu Search*, Norwell, MA: Kluwer Academic Publishers.

Gupta, Y.P., and Goyal, S. (1989), Flexibility of manufacturing systems: concepts and measurements, *European Journal of Operational Research*, 43(2), 119–135.

Hax, A.C., and Candea, D. (1984), *Production and Inventory Management*, Englewood Cliffs, NJ: Prentice Hall.

Hax, A.C., and Meal, H.C. (1975), Hierarchical intergration of product planning and scheduling, in M.A. Geisler, Ed., *Studies in Management Sciences*, Vol. 1: Logistics, New York, NY: Elsevier.

Kern, C. (1994), *Optimale Größe von Schlachtbetrieben unter ausschließlicher Berücksichtigung der Schlacht-und Erfassungkosten*, Frankfurt: Buchedition Agrimedia, Verlag Alfred Strothe.

Miller, T.C. (2002), *Hierarchical Operations and Supply Chain Planning*, London: Springer Verlag.

Oppen, J., and Løkketangen, A. (2008), A tabu search approach for the livestock collection problem, *Computers & Operations Research*, 35(10), 3213–3229.

Schütz, P., Stougie, L., and Tomasgard, A. (2008), Stochastic facility location with general long-run costs and convex short-run costs, *Computers & Operations Research*, 35(9), 2988–3000.

Schütz, P. and Tomasgard, A. (2008), The Impact of Flexibility on Operational Supply Chain Planning, Technical Report, Norwegian University of Science and Technology.

Schütz, P., Tomasgard, A., and Ahmed, S. (2009), Supply chain design under uncertainty using sample average approximation and dual decomposition, *European Journal of Operational Reseach*, 199(2), 409–419.

Simchi-Levi, D., Kaminsky, P., and Simchi-Levi, E. (2007), *Designing and Managing the Supply Chain: Concepts, Strategies and Case Studies*, 3rd ed., New York, NY: McGraw-Hill.

Swaminathan, J.M., and Tayur, S.R. (2003), Tactical planning model for supply chain management, in A.G. de Kok and S.C. Graves, Eds., Supply chain management: design, coordination and operation, Vol. 11 of *Handbooks in Operations Research and Management Science*, Amsterdam: Elsevier, chapter 8, pp. 423–454.

Talluri, K.T., and van Ryzin, G.J. (2004), *The Theory and Practice of Revenue Management*, Berlin: Springer Science+Business Media.

Tomasgard, A., and Høeg, E. (2002), Supply chain optimization and demand forecasting, in J. Olhager, F. Persson, E. Seldin, and J. Wikner, Eds., *Produktionslogistik 2002—Modeller for Effektiv Produktionslogistik*, Linköpings tekniske högskola.

Tomasgard, A., and Høeg, E. (2005), A supply chain optimization model for the Norwegian meat cooperative, in W.T. Ziemba and S.W. Wallace, Eds., Applications of stochastic programming, Vol. 5 of *MPS-SIAM Series on Optimization*, Society for Industrial and Applied Mathematics, Philadelphia, PA, chapter 14, pp. 253–276.

Upton, D.M. (1994), The management of manufacturing flexibility, *California Management Review*, 36(2), 72–89.

Van den Broek, J., Schütz, P., Stougie, L., and Tomasgard, A. (2006), Location of slaughterhouses under economies of scale, *European Journal of Operational Research*, 175(2), 740–750.

Winters, P.R. (1962), Constrained inventory rules for production smoothing, *Management Science*, 8(4), 470–481.

Zhang, Q., Vonderembse, M.A., and Lim, J.-S. (2002), Value chain flexibility: a dichotomy of competence and capability, *International Journal of Production Research*, 40(3), 561–583.

# *Chapter 5*

# Transforming U.S. Army Supply Chains: An Analytical Architecture for Enterprise Management

Greg H. Parlier

## Contents

An army fights with its weapons but lives off its logistics . . .

**—Military maxim**

So now let us embark on our enquiry into what is true . . . we sometimes notice that our senses deceive us, and it is wise never to put too much trust in what has let us down . . .

**—René Descartes**

Nothing is more powerful than an idea whose time has come.

**—Victor Hugo**

## 5.1 Executive Summary

Fully engaged in the Global War on Terror, the U.S. Army is also committed to a comprehensive and ambitious "Transformation" endeavor. Fundamentally, however, without an enabling transformation in logistics, there can be no Army-wide transformation. This chapter provides a practical approach for understanding the Army's extremely complex logistics system by introducing a systems framework that is guiding an ongoing project addressing major challenges confronting Logistics Transformation. The project focus is on inventory management policy prescriptions illuminated through the prism of an enterprisewide supply chain analysis. A multi-stage conceptual model of the logistics structure is used to segment and guide the effort, and to systematically *analyze* major organizational components of the supply chain, diagnose structural disorders, and prescribe remedies (Figure 5.1). A summary of these disorders and their consequences is presented at the beginning of this chapter. *Integration* challenges are addressed using cost–benefit perspectives that incorporate supply chain objectives of efficiency, resilience, and effectiveness. The *design and evaluation* section proposes an "analytical architecture" consisting of four complementary modeling approaches, collectively referred to as "dynamic strategic logistics planning," to enable a coordinated, systemic approach for Logistics Transformation. An organizational construct is presented for an "engine for innovation" to accelerate and sustain continual improvement for Army logistics and supply chain management—a "Center for Innovation in Logistics Systems." The chapter concludes with an overview of strategic management challenges to transformation implementation.

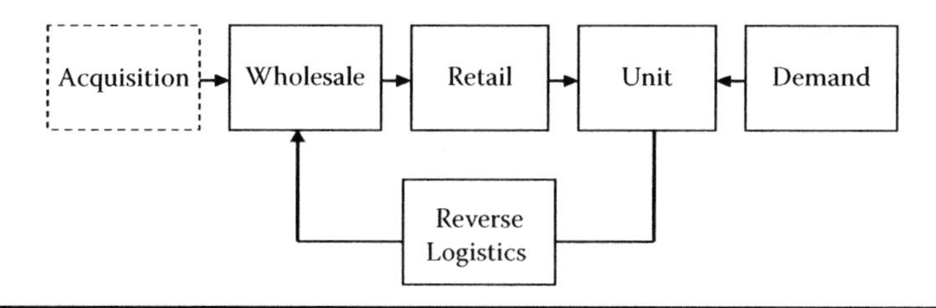


**Figure 5.1 Conceptual approach: multi-stage logistics model.**

## 5.2 Multi-Stage Analysis of Enterprise Challenges: A Summary

The U.S. Army's logistics enterprise is truly enormous in scale and scope. However, it is not merely the size and complexity of the supply chain that causes difficulty, but rather the structure and policies within the system that are the root cause of persistent problems. U.S. Army logistics have especially suffered from several disorders that are both systemic and chronic. This research project has illuminated these problems using inventory management theory, supply chain principles, and logistics systems analysis as key sources of diagnostic power. To summarize generally, these causal disorders and their respective effects include

1. Lack of an empirically measured readiness production function at the tactical unit stage, which induces both uncertainty and variability at the point of consumption in the supply chain, resulting in inappropriate planning, improper budgeting, and inadequate management to achieve readiness objectives
2. Limited understanding of mission-based operational demands, environmental effects, and associated spares consumption patterns, which contribute to poor tactical demand forecast accuracy, operational support planning and cost-ineffective retail stock policy
3. Failure to optimize retail stage stock policy to achieve cost-efficient readiness (customer) objectives, which results in inefficient procurement and reduced readiness
4. Failure to proactively synchronize and manage reverse logistics, which contributes significantly to increased requirement objectives (ROs), excess inventory, and increased delay times (order fulfillment) with reduced readiness
5. Inadequately organized wholesale stage depot repair operations, which may be creating a growing gap in essential repair capacity while simultaneously precluding the enormous potential benefits of a synchronized, closed-loop supply chain for reparable components
6. Limited visibility into and management control over disjointed and disconnected manufacturing (OEM) and key supplier procurement programs in the

acquisition stage, which are vulnerable to boom and bust cycles with extremely long lead times, high price volatility for aerospace steels and alloys, and increasing business risk to crucial, unique vendors in the industrial base, resulting in diminishing manufacturing sources of materiel supplies and growing obsolescence challenges for aging aircraft and vehicle fleets

7. Independently operating, uncoordinated, and unsynchronized stages within the supply chain, creating pernicious "bullwhip" effects including high RO, inadequate stock levels, long lead times, and declining readiness
8. Fragmented data processes and inappropriate supply chain measures focusing on interface metrics, which mask the effects of efficient and effective alternatives, and further preclude an ability to determine "readiness return on net assets" or to relate resource investment levels to readiness outcomes
9. Lack of central supply chain management and supporting analytical capacity results in multi-agency, consensus-driven, bureaucratic "solutions" hindered by lack of an Army supply chain management science and an enabling "analytical architecture" to guide Logistics Transformation
10. Lack of an "engine for innovation" to accelerate and then sustain continual improvement for a learning organization

The existing logistics structure is indeed vulnerable to the supply chain "bullwhip." While endless remedies have been adopted over the years to address visibly apparent symptoms, the fundamental underlying disease has not been adequately diagnosed or treated—much less cured. Now, to better understand these underlying causes of failure, a new approach to logistics management is required for the U.S. Army.

The analytical challenge is to conquer unpredictability: to better understand and then attack the root causes of variability and uncertainty within each stage and their collective contributions to volatility across the system of stages—the "bullwhip effect." By improving demand forecasting and reducing supply-side variability and inefficiencies within each of the stages, logistics system performance is moving toward an efficient frontier in the cost–availability trade space.

The first step in suppressing the bullwhip effect is to isolate, detect, and quantify inefficiencies within each stage and their respective contributions to systemwide aggregate inventory ROs. The next step is to use this knowledge to drive inventory policy. Since U.S. Army inventories are managed to these computed ROs, reducing the value of the RO is critical in eliminating unnecessary inventory. As prescriptions for improved performance recommended by this project are implemented in each of the stages, their respective contributions to reducing RO—while sustaining or actually improving readiness performance—can be measured, compared, and assessed within a rational cost–performance framework (Figure 5.2).

In general, these various contributions to aggregate systemwide ROs—induced by the bullwhip effect—can be isolated, quantified, and then systematically reduced by understanding and attacking root causes: reducing demand uncertainty by

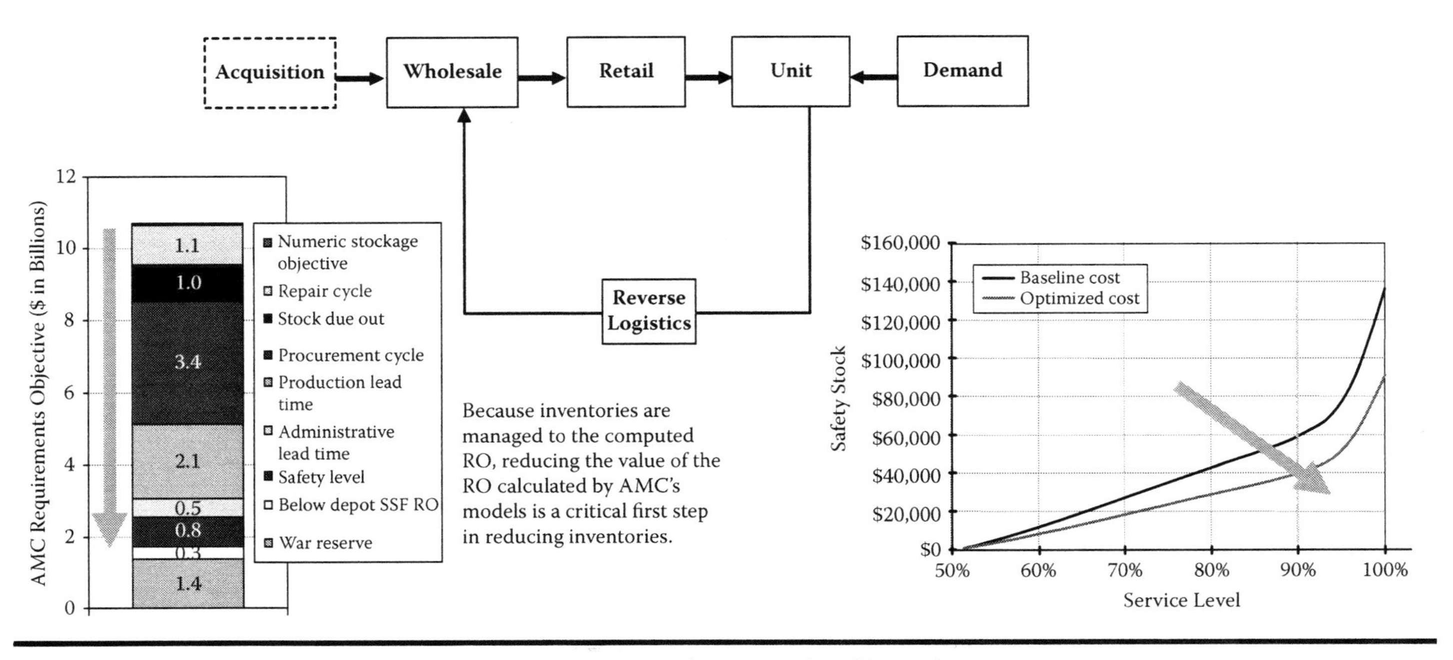


**Figure 5.2 Improving system efficiency: across the system of stages and within each stage.**

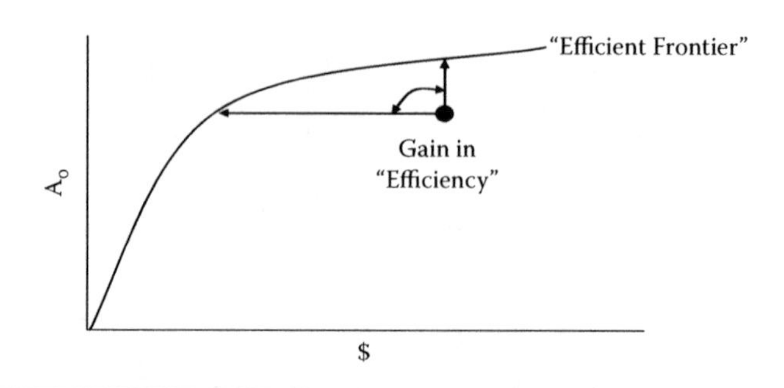


**Figure 5.3 Achieving "efficiency" in the cost–availability trade space.**

adopting empirically derived, mission-based demand forecasting; reducing supply-side lead times and their associated variability; and improving order fulfillment while reducing backorders and requisition wait times by implementing readiness-based sparing (RBS) stock policies, inventory pooling, and ultimately, tactical-level demand-driven supply networks.

An especially compelling and urgent need, and also one with lucrative potential benefits, is the reverse pipeline: as retrograde operations become more responsive and contribute to a synchronized closed-loop supply chain, it becomes possible to reduce ROs and safety stock for specific depot-level reparable components (DLRs) while simultaneously reducing backorders and *increasing readiness* ($A_o$). As these efforts are systematically pursued, the logistics system becomes more efficient: RO (safety stock, etc.) is reduced while performance (backorders and $A_o$) is increased, thereby moving toward the "efficient frontier" in the cost–performance trade space (Figure 5.3).

## 5.3 Multi-Stage Integration for Efficiency, Resilience, and Effectiveness

Simply recognizing that these "bullwhip" conditions exist does not guarantee that needed changes will actually be made. Moreover, these debilitating and persistent effects can be avoided only if long-term organizational behavior and management processes are addressed.

In addition to reducing demand uncertainty, and identifying the causes and reducing the effects of supply and demand variability within each of the logistics stages, the stages must also be "integrated"—linked together in meaningful ways—in order to enable credible cause-and-effect relationships to be identified among new initiatives, Department of the Army resource allocation investment levels, and readiness-oriented tactical outcomes.

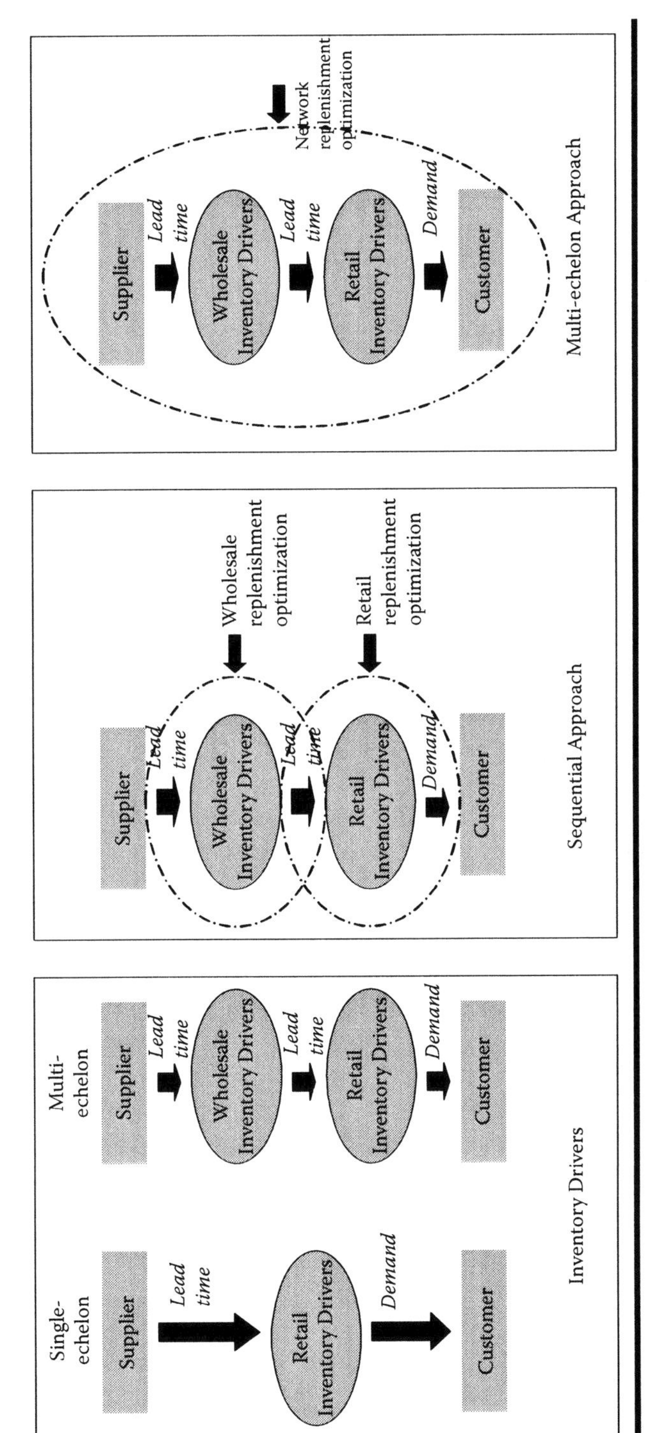


**Figure 5.4 Multi-echelon integration: a prerequisite for enterprise transformation.**

### 5.3.1 Achieving Efficiency: An Integrated Multi-Echelon Inventory Solution

A recurring management challenge in complex supply chains is determining where, and in what quantities, to hold safety stock in a network to protect against variability and to ensure that target customer service levels are met. In an effort to improve supply chain efficiency, an appreciation for the interdependencies of the various stages is required in order to fully understand how inventory management decisions in one particular stage or location impact other stages throughout the supply chain.

For military and aerospace logistics systems, optimizing these decisions requires a decision support system that captures multi-echelon, multi-item, multi-indenture interactions and also the dynamics of the reverse flows for reparable components. Such a decision support system must also be linked to the various supply information transactions, depot repair and overhaul, and long-term planning systems that affect the overall responsiveness, support adequacy, and capacity of the fleet supply chain enterprise—the "readiness" of the entire, globally dispersed logistics support system.

Consequently, an integrated, multi-echelon network offers several opportunities for supply chain efficiency:

- Multiple, independent forecasts in each of the stages are avoided.
- Variability in both demand and lead time supply can be accounted for.
- The bullwhip effect can be observed, monitored, and managed.
- Its various root causes can be identified and their effects measured, corrected, and tracked.
- Common visibility across the supply chain stages reduces uncertainty, improving demand forecasting and inventory requirements planning.
- Order cycles can be synchronized (this has special significance for DLRs in the retrograde and depot repair stages).
- Differentiated service levels (e.g., $A_o$ targets for different units) can be accommodated.
- Action can be taken to reduce unnecessary inventory and operational costs while simultaneously improving readiness-oriented performance.

Although the computations to incorporate key variables, their relationships, and associated costs are certainly not trivial, they can nonetheless be performed using advanced analytical methods, including RBS optimization methods. Improved results are then possible and the organization will have far greater confidence that it is operating closer to the efficient frontier within the cost–performance trade space.

For military aircraft, it has also been demonstrated that DLRs most directly relate to aircraft performance, and, in general, minimizing the sum of DLR backorders is equivalent to maximizing aircraft availability. Significant effort has also been placed on determining optimal stock levels and locations for reparable components in a multi-echelon system. While the subsequent extension of this theory has been widespread, the focus of practical implementation within the DoD has been on fixed-wing aircraft in the Navy and the Air Force rather than rotary wing aircraft in the Army.

Another structural constraint that previously precluded an integrated multi-echelon approach for Army supply systems was the existence of separate stock funds used by the Army financial management system for retail and wholesale operations. In recent years, however, these separate funds have been combined into one "revolving fund," the "Single Stock Fund" (SSF). In theory, this should both facilitate and encourage adoption of an integrated multi-echelon approach. For example, the wholesale stage now has both visibility into the retail stage and more control over stock policy in the wholesale *and* retail stages, which it previously did not have for Class IX (repair parts). It now becomes possible for AMC to incorporate multi-echelon optimization for wholesale stock levels, *in addition to retail RBS solutions*, to be directly related to readiness (Figure 5.4).

It is not possible to truly "optimize" performance output from large-scale, complex systems if they have not first been "integrated." The key integrating enabler for improved efficiency in all Army weapon system supply chains—and the more complex the system, the more crucial the enabler—is multi-echelon readiness-based sparing. Indeed, this is a *precondition* for Army Logistics Transformation.

### 5.3.2 Designing for Resilience: Adaptive Logistics Network Concepts

The intent is certainly not to blindly adopt the latest management "fad" inundating the corporate world but rather to consider adapting proven concepts to the unique needs and challenges that the Army faces. The opposite result could occur with "just-in-time" methods. Lean manufacturing concepts have certainly helped firms become more competitive through the application of "just-in-time" principles that exchange "industrial age" mass for "information age" velocity. And many of the original lean manufacturing concepts, especially the focus on reducing "stagnant" work-in-progress inventory, have been successfully adapted for supply chain management across the entire enterprise. However, the idea of "integration," when achieved by reducing slack or "waste" in the system, does not necessarily enable greater flexibility.

Furthermore, "just-in-time" concepts, although a powerful inventory reduction method, need stable, predictable supply chains for maximum efficiency. Even when enabled by IT, lean supply chains can be fragile, vulnerable to disruption, and unable to meet the surge requirements needed to accommodate an immediate increase in demand. In fact, recent official documents describe exactly such a condition for Army logistics in recent years. Under greater duress and the compounding stress of ongoing operations, the military logistics system has indeed resulted in a lean supply chain without the benefit of either an improved distribution system or an enhanced information system.

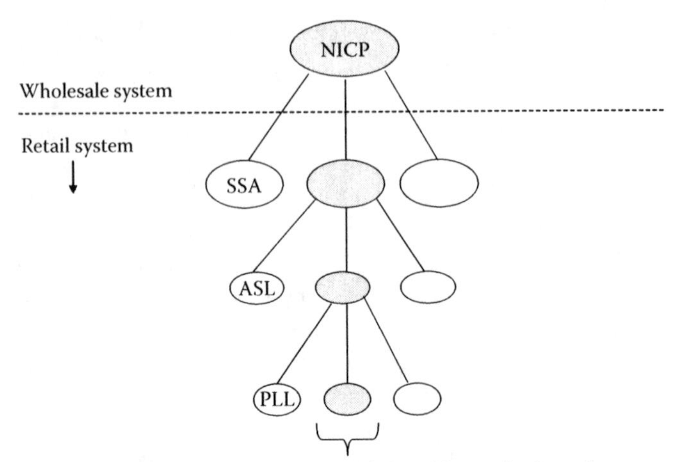


**Figure 5.5 Current structure: arborescence.**

A more appropriate analogy for Army logistics is a flexible, robust logistics "network"; not a serial "chain" or hierarchical arborescence (Figure 5.5), but rather a network "web"—as in spider web—which is then enabled by a strong analytical foundation with supporting information technology to achieve an integrated, flexible, efficient, and effective logistics capability.

The research and subsequent understanding of self-organizing systems has been rapidly advancing in recent decades, extending originally from cybernetics to incorporate growing knowledge in cognitive science, evolutionary biology, dynamical systems, stochastic processes, computational theory, and culminating now in "complex adaptive systems." For military operations, this "network-centric" future force will be linked and synchronized in time and purpose, allowing dispersed forces to communicate and maneuver independently while sharing a common operating picture. Conceptually, the traditional mandate for overwhelming physical "mass," in the form of a linear array of land combat forces converging at the decisive place and time, is replaced by attaining comparable "effects" derived from dispersed and disparate forces operating throughout a nonlinear battlespace.

Our ability to logistically support these concepts, especially the notion of an agile supply network at the theater and tactical levels for Army and joint logistics distribution, may be much closer at hand now than previously recognized. At the tactical level, for example, the demand-driven supply network (DDSN), which includes mission-based forecasting on the demand side and RBS, lateral supply, and

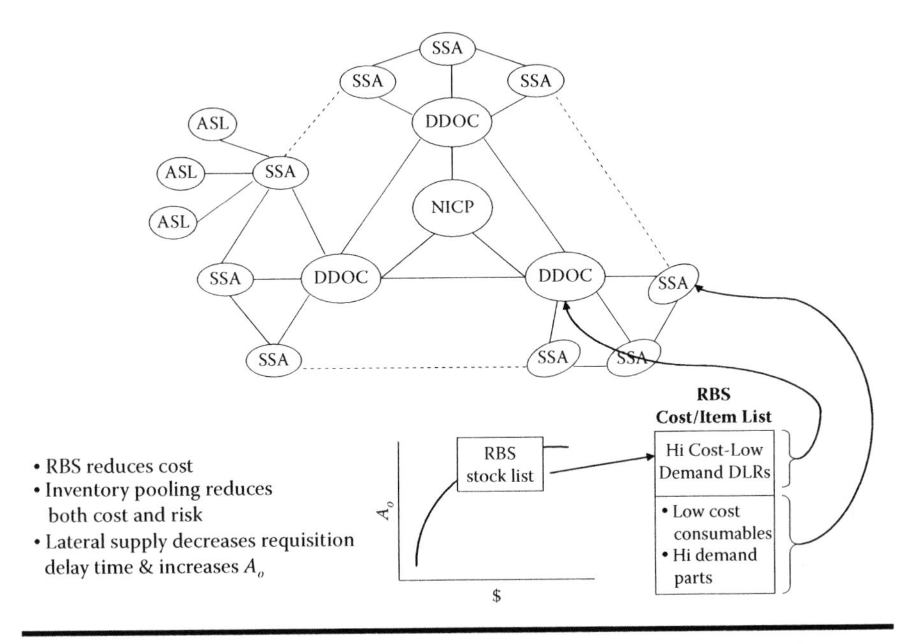


**Figure 5.6 Demand-driven supply network (DDSN).**

risk-pooling (especially for DLRs) on the supply side, provides the foundational basis for a more agile and resilient network "web" (Figure 5.6).

Through theoretical development corroborated by recent field tests, this DDSN concept has also been shown to attain *both* improved effectiveness ($A_0$) and, as total asset visibility (TAV) and in-transit visibility (ITV) IT-based technologies are incorporated, increasingly better efficiency. Such a tactical-level DDSN is not only effective and efficient then, but also both resilient and adaptive, enabling a rapid transition away from the traditional hierarchical arborescence structure, which required "mountains of iron" to buffer uncertainty, inefficiencies, and rigidity, toward an adaptive network design consistent with Sense and Respond Logistics.

By applying design principles for supply chain resilience, a supply chain operating a large-scale (global), demand-driven ("pull") system under stable and predictable demand can quickly adapt to support localized, temporary requirements that may involve considerable uncertainty, but which must be "pushed" to the customer (combat units) to achieve maximum effectiveness (mission $A_0$ in this case). Resilient design concepts include the identification of "push-pull" boundaries separating "base" from "surge" demand using decoupling points for the placement and use of strategic capacity and inventory.

These concepts suggest, first, creating pre-positioned, mission-tailored support packages designed using RBS in conjunction with mission-based forecasting. These tailored mission support packages can then accommodate replacement part needs at deployed locations where existing (e.g., host nation) sustainment is not immediately or readily available. This is an example of defining a "decoupling" point in the existing supply chain and creating additional slack inventory to accommodate a short-term surge that the existing logistics supply network infrastructure cannot immediately support.

Second, to accommodate sustained, rather than temporary, higher demand for extended operations, resilient supply chain design principles suggest creating additional capacity, or relocating existing capacity, closer to the demand source. This strategic supply chain concept shifts "decoupling" points and push-pull boundaries by dynamically changing the supply chain configuration. Hence, the logistics network responds quickly to initially accommodate a short-term need with built-in (pre-positioned) slack inventory, and then adapts, if and when necessary, by actually changing its configuration to sustain increased longer-term requirements by relocating maintenance and supply support capacity closer to the source of demand.

In summary, effort for attaining resilience must focus on strategically designing and structuring supply chains to respond to the changing dynamics of globally positioned and engaged forces, conducting different operational missions under a wide range of environmental conditions. Ultimately, this necessitates supply chain *management innovation.*

### 5.3.3 Improving Effectiveness: Pushing the Logistics Performance Envelope

Economists commonly make a distinction between efficiency and productivity: efficiency refers to the output achieved from inputs using a given technology, while productivity also incorporates the results of changes in technology. By "efficient" we refer to those methods (whether policies, techniques, procedures, or technologies) that reduce uncertainty and/or variability both within any particular stage as well as across the "system of stages" that comprise the multi-stage logistics enterprise. Using these methods would have the effect of moving toward the "efficient frontier" in the cost–availability trade space (Figure 5.7). Achieving an "efficient" solution

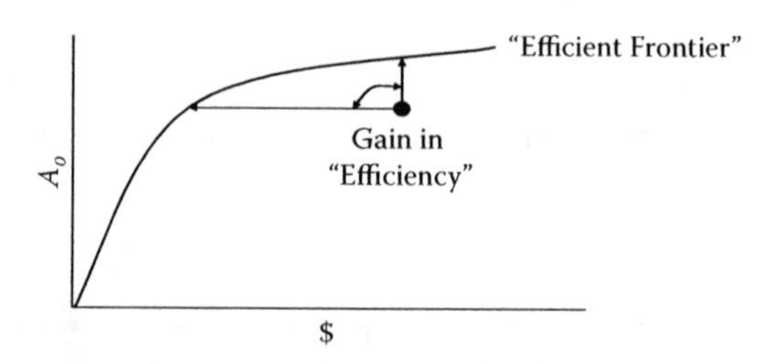


**Figure 5.7 Achieving "efficiency" in the cost–availability trade space.**

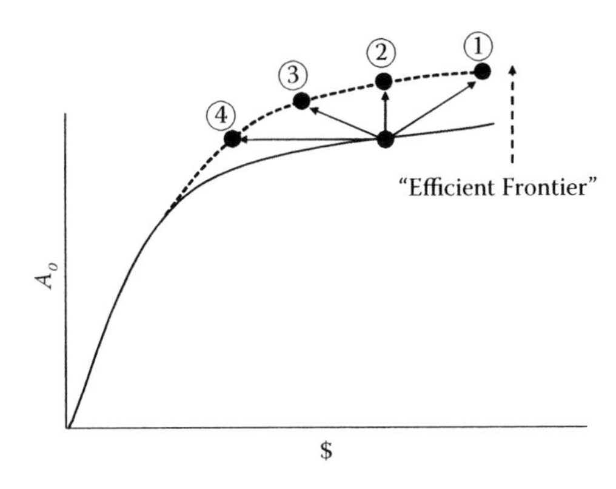


**Figure 5.8 Increasing "effectiveness" in the cost–availability trade space.**

results in operating on the existing efficient frontier and implies the best possible use of existing resources *within the constraints of the current system design and business practices* using existing technology.

In contrast, a more "effective" ("productive") method is one that actually shifts the existing efficient frontier representing an improved "operating curve." This reflects the fact that current business practices have actually changed: new or different technologies are being exploited. Cost–benefit analyses can be performed on various initiatives that yield improved, but different, results (Figure 5.8).

Finally, the ultimate goal to is sustain continual improvement and progress over time through "innovation" in all its various forms ... the notion of "pushing the envelope" (Figure 5.9). This is the essence of "productivity gain" and differentiates, in competitive markets, those commercial firms that successfully compete, survive, and flourish over extended periods from those that do not.

For a governmental activity, an "engine for innovation" is needed to compensate for the lack of competitive marketplace pressures typically driven by consumer demand and customer loyalty. The most obvious "engine" for a military organization is imminent or evident failure on the battlefield. Failure in battle, especially if sufficient to cause the loss of a major war, clearly constitutes an "unmet military challenge," which is one of several key historical prerequisites for a "revolution in military affairs" (RMA).

In the absence of imminent or evident failure resulting in battlefield losses that threaten the nation's interests and/or values, an alternative "engine for innovation" is an extensive experimentation capacity providing an ability to "see" the impact of alternative concepts, policies and procedures, doctrine, tactics, and organizational design—a "virtual" or "synthetic" environment that can realistically illuminate a better way, thereby possibly preempting future failure. This experimentation capacity must also be complemented by institutional means to incorporate positive results into

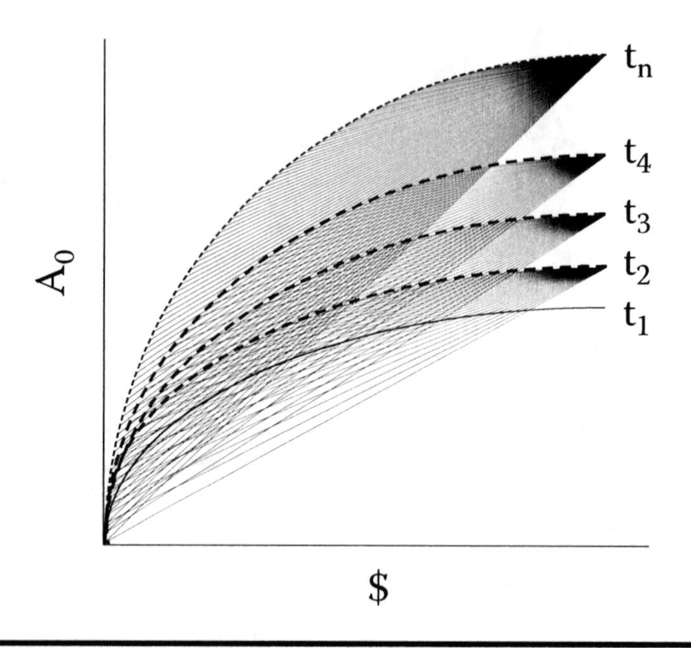


**Figure 5.9 Pushing the envelope: innovation to sustain continual improvement.**

new or existing policies, doctrine, and resource programs—in short, an organizational capacity to encourage improvement, and both stimulate and accommodate positive change.

## 5.4 Design and Evaluation: An "Analytical Architecture"to Guide Logistics Transformation

A viable strategy is now needed to transition from the existing state of affairs toward a desired outcome defined by those enterprise characteristics described previously: efficiency, resilience, and effectiveness. Inherent in developing such a strategy, are the need to (1) optimize the allocation of limited resources, and (2) understand and anticipate in advance the consequences, likely outcomes, and risks associated with an unlimited array of tasks that must be selected, sequenced, and synchronized for implementation.

These two analytical approaches—optimization modeling to efficiently allocate constrained resources toward desired objectives, and predictive modeling, including testing, experimentation, and simulation, to anticipate likely outcomes and effects within a complex system—must be used together in a complementary manner to illuminate a viable plan for implementation. They provide an analytically based strategy to link means (resources) with ways (concepts and plans) to achieve desired ends (objectives): an "analytical architecture" to guide Logistics Transformation.

The modeling and simulation methodology outlined in the next sections would provide this much-needed analytical capacity and could constitute a "dynamic strategic planning" capability for Logistics Transformation. The purpose of this engine for innovation, regardless of the form it ultimately takes, is to provide large-scale systems simulation, modeling and analytical expertise, and experimentation capacity? needed to serve as a credible test bed. This capability will generate the compelling analytical arguments needed to induce, organize, sequence, and synchronize the many changes needed to gain momentum and then accelerate transformation for Army logistics, including those identified and described previously. The purpose, function, and relationships of key components of this enabling "analytical architecture" are described next.

### 5.4.1 Multi-Stage Supply Chain Optimization

Evolutionary progress for an Army Logistics Transformation trajectory can be easily imagined along a spectrum transitioning from legacy-reactive to future-anticipatory concepts:

- Reactive, cumbersome, World War II–era mass-based, order and ship concept where "days of supply" is the primary metric
- Modern supply chain management incorporating velocity-based, sense and respond concept where "flow time" is the metric
- Adaptive and dynamic, inference-based, autonomic logistics network concept to anticipate and lead, where the metrics are "speed and quality of effects"

However, a clearly defined implementation scheme for "transformation" is certainly not self-evident. Analytical methodologies are needed to properly sequence the vast array of new initiatives, modern technologies, process changes, and innovative management policies in cost-effective ways: Which ones depend on others as "enablers" for their success? How many can be done in parallel? For those that can be, will it be possible to identify and quantify the different effects of their respective contributions? Will the synergistic consequences of interactions among complementary initiatives be measurable? Which ones can be precluded by combinations of other, more cost-effective options? And how can we be assured that these various initiatives are not inadvertently discarded because their potentially positive effects on readiness are "lost" in the existing "noise" of such a complex, massive supply chain? In short, how can cause and effect be "disentangled" as transformation proceeds?

The earlier use of a multi-stage conceptual model to analyze the Army's logistic structure naturally lends itself to the use of dynamic programming (DP) or a comparable problem-solving technique. In this multi-stage, graphical example, the challenge is to determine the optimal allocation of a defined budget across a range of initiatives associated with these several logistics stages. Consideration must also be given to various constraints that may be imposed within each of the stages. The overall goal is to maximize output from the "system of stages"—readiness (i.e., $A_o$).

From a practical perspective, this approach also reinforces the crucial importance of developing a clearly defined, empirically measured readiness production function, and adopting RBS stock policies as enabling prerequisites to realize further cost-effective improvements to the system. For example, if the link between the unit stage (where readiness is produced for specific capabilities) and the retail stage inventory levels has not been optimized to desired readiness objectives ($A_o$) by adopting RBS, then the potential positive effects of a wide range of other improvements throughout the supply chain will not be clearly visible and fully realized. Their real effects will simply be lost in the downstream "noise" of a very volatile, disconnected, and inefficient supply chain. Additionally, potential investments should not be chosen on an individual basis but rather on how they interact with each other.

### 5.4.2 Dynamic Strategic Planning

Second, use of a *multi-period* model must be incorporated into Logistics Transformation to accommodate both the extensive and extended nature of this enormous undertaking. As events occur and a transformation trajectory evolves, a mechanism is needed to routinely update the "optimal" solution that, inevitably, will change over time due to (1) the inability to perfectly forecast future conditions, (2) consequences of past decisions that do not always reveal the results expected, and (3) the opportunities provided by adaptation and innovation as they materialize and offer improved solutions requiring new decisions.

This dynamic strategic planning (DSP) approach is, in essence, a multi-period decision analysis challenge that also encourages and assists in identifying, clarifying, and quantifying risk to the transformation effort. Risk "assessment," a precursor to risk "management," is needed to reduce and mitigate the inevitably disruptive consequences of any major transformative effort with all the uncertainties surrounding significant change.

Most planning methods generate a precise, "optimized" design based on a set of very specific conditions, assumptions, and forecasts. In contrast, DSP presumes that forecasts are inherently inaccurate ("the forecast is always wrong") and therefore "builds in" flexibility as part of the design process. This engineering systems approach incorporates and extends earlier best practices, including systems optimization and decision analysis. DSP allows for the optimal solution—more precisely, optimal "policy"—which cannot be preordained at the beginning of the undertaking, to reveal itself over time while incorporating risk management: a set of "if-then-else" decision options that evolve as various conditions unfold, which, even when anticipated, cannot be predicted with certainty. This planning method yields more robust and resilient system designs that can accommodate a wider range of scenarios and future outcomes than those more narrowly optimized to a set of specific conditions. Although perhaps easier to engineer and manage, traditional "optimal" designs can quickly degenerate toward instability when such conditions no longer exist.

These observations suggest that large-scale transformational endeavors are much more than conventional "construction" engineering efforts. They represent a major human enterprise where effective managerial decision making requires a thorough understanding of the evolution and dynamics of the change undertaken. New software tools now make it possible for managers to actively participate in the development of these system dynamics models, so-called management "flight simulators," that have become the basis for learning laboratories in many organizations.

Army Logistics Transformation will benefit enormously from such an application. Because supply chain behavior often exhibits persistent and costly instability, a "stock management structure" can be used to model and explain these effects. Because this structure involves multiple chains of materiel stocks, information, and financial flows, with resulting time delays, and because decision rules often create important feedback loops among the interacting operations of the supply chain, system dynamics is well suited for modeling and policy design (Figure 5.10).

Much of the management literature in business process reengineering emphasizes finding, then relaxing, major bottlenecks in the existing manufacturing or operations process. Focusing improvement effort on the current bottleneck immediately boosts throughput, while effort on non-bottleneck activities is wasted. However, relaxing one constraint simply enables another to develop as time progresses. Obviously,

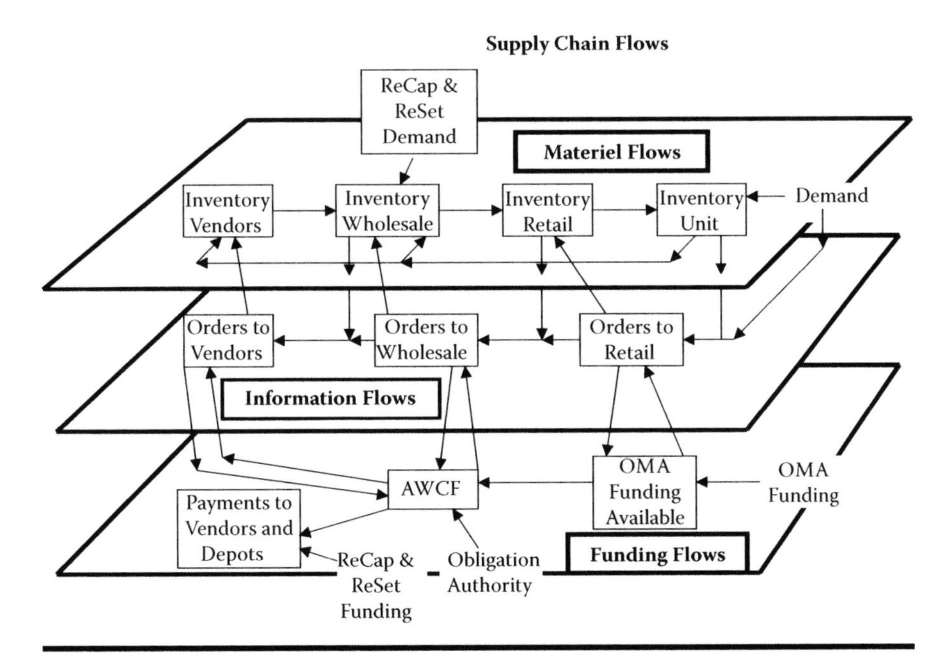


**Figure 5.10 Supply chain flows.**

waiting for each successive bottleneck to occur would prolong and retard rather than accelerate continuous improvement. The value of system dynamics modeling is accelerating this understanding by exploring the implementation of different sequences in a synthetic (simulated) environment. Using the model to anticipate and accelerate this shifting sequence of bottlenecks, a *prioritization* scheme for these many initiatives can be developed. For the Army, a system dynamics model of the supply chain has the potential to guide and help accelerate Logistics Transformation by optimally sequencing and synchronizing the vast array of initiatives that have been suggested for implementation.

Decisions Analysis, the second major analytical component in the evolution of DSP, enables structuring the combination of system dynamics-enabled design choices so they can be made in stages as a system evolves over time. Cost-effective options can be evaluated to determine the best pattern for system development, depending on how uncertainties, both within the system and external to it, are resolved over time. Thus, DSP defines an optimal *strategy* or policy rather than a fixed plan.

The most recent DSP improvements have focused on incorporating means to evaluate and build "flexibility" into designs. These include "real options" and "robust design" methods that enable calculation of the value of flexibility, which was not previously considered. Consequently, flexibility as an attribute of engineering systems design was systematically neglected. "Real options" applied to "real" physical systems is an adaptation of "options analysis," which was developed for and has been applied extensively in financial markets. Recent and ongoing applications of this newest aspect of DSP indicate the approach leads to substantial improvements in design. Also, embedding flexibility into diverse systems already "optimized" for performance under traditional deterministic concepts is leading to substantial savings in many cases.

### *5.4.3 Operational and Organizational Risk Evaluation*

Third, in conjunction with DSP, a wide variety of analytical methods should be used to understand, evaluate, and reduce "risk" during Logistics Transformation. Risk can take on different connotations, depending on the application. Accordingly, we address two concepts here: (1) the operational risk faced by the logistics system responding to various shocks, supply chain disruptions, and mission requirements that may not have been anticipated; and (2) the organizational risk to the Army logistics community, including the combination of investment or programmatic risk associated with new project undertakings and the larger impacts induced by transformation uncertainties associated with organizational change at a difficult and challenging time. Operational risk, in this decision analysis context, consists of assessing both the likelihood of a particular adverse outcome as well as the consequences of that outcome.

Practical management frameworks have recently been developed to systematically identify supply chain vulnerabilities, assess risk, and then formulate strategies to reduce those vulnerabilities and mitigate risk. Various sources and potential causes of disruption are then bundled into associated risk categories. Analytical "toolkits" can be applied to examine specific effects and larger consequences for these risk categories, and then supply chain modeling and simulation is used to analyze, evaluate, and compare alternative operational strategies and their respective costs.

Those strategies that reduce disruptive risk and enhance supply chain resilience, *while simultaneously improving both efficiency and effectiveness*, are ideal candidates for accelerated implementation. Two practical risk mitigation strategies that impact all three supply chain system performance objectives—efficiency, resilience, and effectiveness—are (1) a demand-driven supply network (DDSN) that reduces buffer inventory, improves readiness, and provides tactical agility, and (2) theater-level "decoupling points" to enhance operational agility and flexibility by providing, respectively, "slack inventory" for short, specific mission surge needs (e.g., humanitarian operations) and, when necessary, "slack capacity" for long-term increases in demand to sustain in-theater operations.

To address organizational (rather than operational) risk for Army Logistics Transformation, a variety of virtual, constructive, and live simulation methods, including analytical demonstrations, field testing, and experimentation, can identify, early on, which technologies or new methods warrant further consideration. This process enables differentiating those appropriate or sufficiently mature for implementation from others that are not. In this context, organizational risk consists of the combined effects of both uncertainty of outcomes—simply not knowing the impacts of various alleged improvements on the logistics system—and also the uncertainty of future costs incurred as a consequence of either adopting or failing to adopt particular courses of action.

An example of this accelerating "crawl-walk-run" approach is the sequence of experimentation and testing adopted by this project to first demonstrate, through rigorous analytical experimentation using the UH60 aircraft in the 101st Airborne Division, the potential value of adopting RBS as aviation retail stock policy. These insightful, positive results then provided impetus for more widespread field testing with several aircraft types in an operational training environment at Fort Rucker, the Army's rotary-wing training center and school.

Confidence and credibility in a new, different method have been gained through experience while significantly reducing the uncertainty initially surrounding the new initiative. And return-on-investment results clearly reveal reduced investment costs with improved performance, indeed exceeding aircraft training availability goals.

### 5.4.4 Logistics System Readiness and Program Development

The fourth and final enabling analytical component includes the development, refinement, and use of econometric/transfer function models. This capability is needed so that OSD- and HQDA-level budget planners and resource programmers can relate budget and program investment levels with associated performance effects, including future capability needs and desired readiness outcomes. New impetus for this long-recognized need is now provided by DoD Directive 7730.65, which requires developing and implementing a new "Defense Readiness Reporting System" (DRRS).

Under Title 10, United States Code, the Armed Services, as "force providers," generate and maintain military forces and capabilities that are then allocated to the regional joint force commanders to accomplish assigned missions. Each Title 10 "function" consists of significant institutional resources, organizations, and programs that collectively define "systems." Hence, a measure of each system's ability to achieve its respective goal can be defined as its "readiness" (e.g., logistics system readiness).

Application of this systems approach using supply chain management concepts will help identify constraints and "weak links" that are inhibiting desired readiness output (e.g., $A_o$), thus reducing the overall strength of the logistics chain. Marginal investment resources should then be spent on strengthening these weak links. OSD and the Services are pursuing many logistics initiatives, but as the supply chain structure is improved and refined, the logical next step is to understand and monitor the ability and capacity of the chain to generate output commensurate with its purpose.

New supply chain management concepts are incorporating geospatial sensors and automatic identification technologies (AIT) to enable "total asset visibility" (TAV) and the transition toward adaptive supply chains. In particular, radio frequency identification (RFID) is expected to significantly reduce transaction error rates while also providing near-real-time, high-volume data. Although these new technologies hold great potential, it is unlikely that legacy software and enterprise resource planning (ERP) systems will be able to provide improved decision support and fully extract all the potentially useful information contained in these high-volume data streams.

Recent forecasting advances for financial markets, which exhibit similar volatility, have yielded more accurate and precise results. These models, described as *generalized autoregressive conditional heteroskedasticity* (GARCH), are able to significantly reduce the error term by better quantifying interaction and lag effects among the explanatory variables and time series within the model. As the volume of data increases, the ability of GARCH techniques to better disentangle and explain cause and effect relationships *while reducing forecasting error* (unexplained model variance) improves. One project initiative involves examining the application of GARCH to RFID-generated supply and demand data for units engaged in ongoing military operations in Iraq. Early results are promising, indicating that GARCH is yielding order-of-magnitude improvements for predictive performance compared to standard methods.

In the near term, however, driven by the new DRRS mandate and enabled by supply chain concepts, econometric modeling, and dynamic forecasting to understand, measure, and monitor Army logistics as a readiness-producing system, a conceptual framework has emerged for a "Logistics Readiness and Early Warning

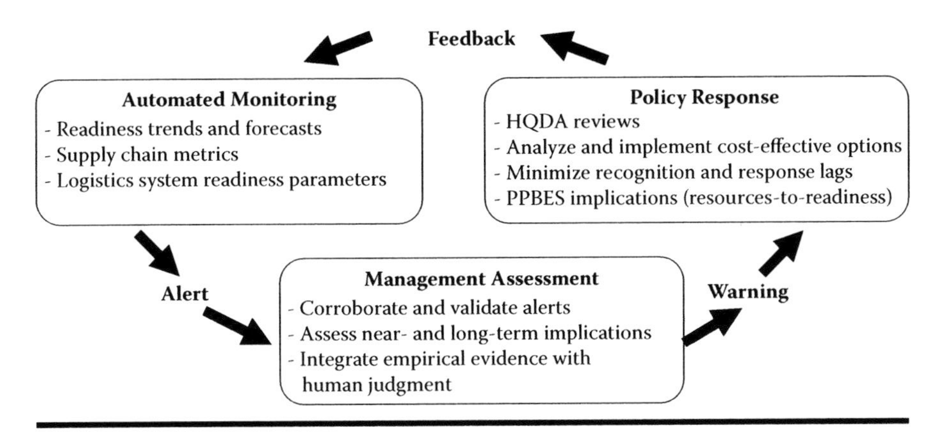


**Figure 5.11 Logistics readiness and early warning system.**

System." The purpose is not only to assess and monitor supply chain capacity to efficiently and effectively support current requirements, but also to anticipate its ability to responsively meet a range of future capabilities-based requirements. The objective is to overcome funding-induced instability manifested in periodic "boom and bust" cycles.

As Figure 5.11 portrays, three elements would interact in a "feedback-alert-warning" cycle. "Automated Monitoring" continuously tracks and forecasts both tactical readiness (e.g., $A_0$) and supply chain parameters, then signals an alert if there is a decline in projected readiness or an adverse trend in metrics. "Management Assessment" then validates an alert, quickly evaluates the potential problem, and assesses the impact of current and planned resource allocation as well as other technical initiatives that might mitigate or improve the logistics projection. After HQDA-level policy analysis and review, "Policy Response" acts to prevent a shortfall while minimizing recognition and resource response lags. This responsive link to program development is absolutely crucial to an adaptive demand network. Historically, however, this response has significantly lagged or been missing altogether, causing "boom and bust" cycles in resource programming and thus precluding viable resource-to-readiness frameworks for management decisions.

Further developed and refined over time, these forecasting models can increasingly be used for future capability forecasting, program requirements determination, and readiness prediction. These models should constitute part of a "Logistics Readiness and Early Warning System" contributing toward the DoD mandate for a larger Defense Readiness Reporting System by linking Army PPBES (resource planning system) to operational planning systems (readiness). The goal is to relate planning guidance, funding decisions, and execution performance in meaningful ways, all of which are informed by this supply chain "health monitoring and management" concept.

## 5.4.5 Accelerating Transformation: An "Engine for Innovation"

Several agencies and organizations with logistics modeling and supply chain simulation capabilities should now be integrated into a more formal research consortium to better coordinate their efforts and reinforce their respective strengths. This synergistic effort will facilitate properly sequenced field tests, experiments, and evaluation with supporting modeling, simulation, and analysis. Furthermore, these organizations should form the nucleus of an "engine for innovation" for Logistics Transformation: a "Center for Innovation in Logistics Systems" (CILS).

The CILS organizational construct consists of three components that essentially comprise the core competencies (mission essential tasks) for the center (Figure 5.12):

1. An R&D model and supporting framework to function as a generator, magnet, conduit, clearinghouse, and database for "good ideas"
2. A modeling, simulation, and analysis component that contains a rigorous analytical capacity to evaluate and assess the improved performance, contributions, and associated costs that promising "good ideas" might have on large-scale logistics systems

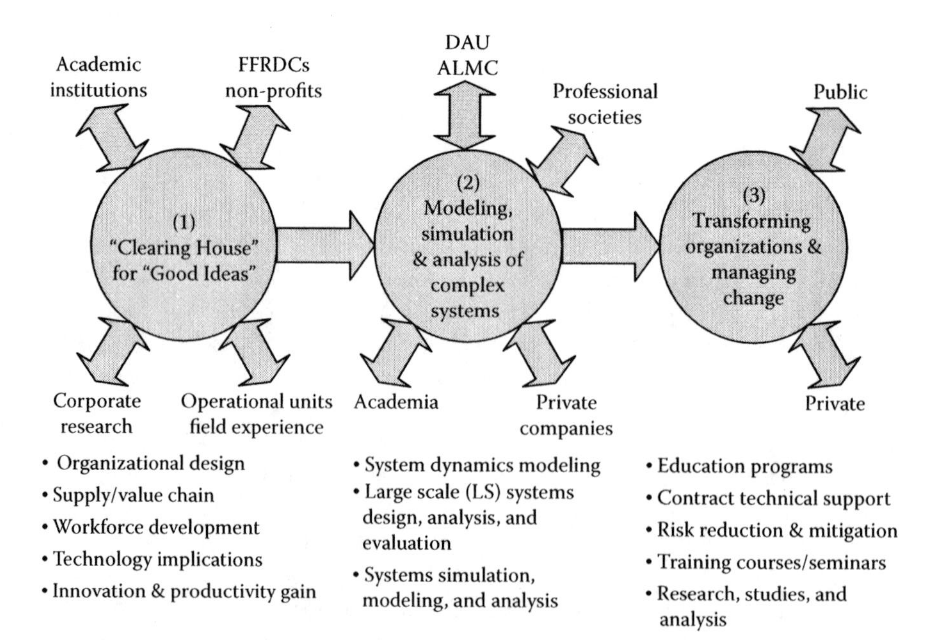


**Figure 5.12 Center for innovation in logistics systems.**

3. An organizational implementation component that then enables the transition of promising concepts into existing organizations, agencies, and companies by providing training, education, technical support, and risk reduction/mitigation methods to reduce implementation risk

These three components serve to:

1. Encourage and capture a wide variety of "inventions."
2. "Incubate" those great ideas and concepts within virtual organizations to test, evaluate, refine, and assess their potential costs, system effects, and contributions in a non-intrusive manner.
3. Transition those most promising into actual commercial and/or governmental practice.

Hence the term "innovation" is deliberately in the center's title to express the notion of an "engine for innovation" to support major transformation endeavors in the government and private sectors driven by an increasingly recognized necessity for change.

These four modeling approaches—multi-stage optimization, dynamic strategic planning, risk management, and program development—should be used in unified and complementary ways to constitute a "dynamic strategic logistics planning" (DSLP) capability. DSLP can take, as input, both the empirical evidence of ongoing operational evidence (real-world results) and also the potential contribution of new opportunities derived from an "engine for innovation" (synthetic results), and then guide—as output—Logistics Transformation toward strategic goals and objectives: an efficient, increasingly effective, yet resilient global military supply network. DSLP constitutes the "analytical architecture" needed to sustain continual improvement for Logistics Transformation (Figure 5.13).

Collectively, CILS and DSLP have the potential to accelerate the process of management innovation by building a capacity for low-risk experimentation using a credible, synthetic environment. This cycle sustains continuous improvement through a deliberative process of incremental innovation achieved through experimentation, prototyping, and field testing.

## 5.5 Final Thoughts

Tactical units in the U.S. Army are renowned for pioneering and refining the After Action Review (AAR) concept as a continuous learning method to reveal, diagnose, and correct deficiencies in order to improve performance and sustain operational excellence. Yet comparable diagnostic effort has not been prevalent at strategic levels within the institutional Army bureaucracy. Because analytically

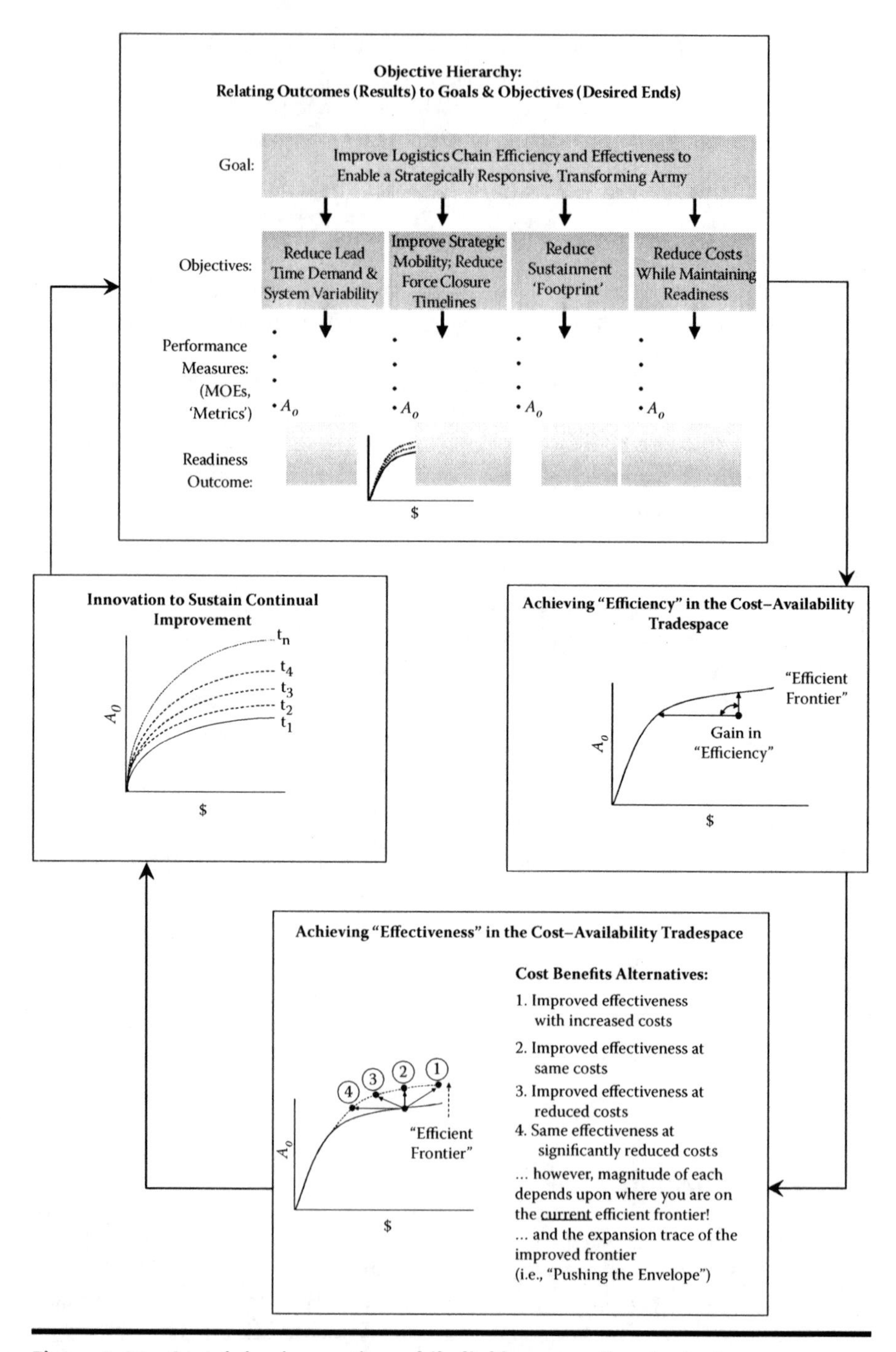


**Figure 5.13 Sustaining innovation while linking execution to strategy.**

**Figure 5.14 Logistics transformation framework: linking strategy to measurable results.**

rigorous "autopsies"—"dissection" for root cause diagnosis, understanding, and response—on management issues are not routinely performed to uncover "ground truth" and learn from mistakes, reactive "firefighting" has been the standard response to visible symptoms.

As with any complex, large-scale systems challenge, key implementing concepts will be essential to ensure a successful Army Logistics Transformation endeavor. These organizational, analytical, information systems, technology, and management concepts should all be guided by a clear understanding of the ultimate purpose for which the enterprise exists, an organizational vision for the future, and a supporting strategy to realize the vision (Figure 5.14).

This strategy must focus the effects of transformative change on capabilities-based, readiness-oriented outcomes. The development of strategic planning and management frameworks is also essential to enable learning within organizations. "Transformation" will indeed require disturbing existing cultural paradigms, causing an inevitably disruptive period of significant change. Despite the inexorable advance of technology, it will be improved management and decision support systems that ultimately enable the realization of innovation potential. Finally, this endeavor should embrace that of a Learning Organization. This will be a crucial enabler for sustaining continuous improvement.

We hope this endeavor will serve as a catalyst for an intellectual and professional resurgence in military logistics systems analysis. We are certainly encouraged by our empirical research results, which continue to reinforce and corroborate many of the intuitive concepts and ideas presented in this chapter. Consequently, we have engaged the larger military operations research and professional logistics communities and continue to encourage the participation of all those interested in collectively pursuing this enormous challenge.

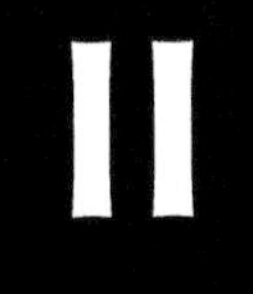


# ANALYTIC PROBABILISTIC MODELS OF SUPPLY CHAIN PROBLEMS

# Chapter 6

# A Determination of the Optimal Level of Collaboration between a Contractor and Its Suppliers under Demand Uncertainty

Seong-Hyun Nam, John Vitton, and Hisashi Kurata

## Contents

## 6.1 Introduction

Global markets are growing rapidly and competition is taking place not just between individual businesses, but along the entire value chain. Effectively managing these increasingly complex global relationships requires tight collaboration between business partners. Collaborative supply chain management is one of the most important driving forces of effective supply chain management and is recognized as a significant process that impacts profit creation, market share, competitive position, and the value of contractors and their suppliers. Collaborative supply chain management is therefore a rapidly evolving area of interest to academics and business practitioners.

Collaboration is defined by the ways in which supply chain members share information, improved practices, risks, and rewards (Lambert et al. 1998). Logistic alliances among members to improve customer service and lower operating costs was studied by Bowersox (1990). The ways in which members cooperate in order to share resources and capabilities to meet their customers' needs was the subject of research by Narus and Anderson (1996). Depending on their focus, supply chain collaborators may have different definitions. However, the general purpose of collaboration among supply chain partners is to achieve competitive advantage over other supply chains (La Londe and Masters 1994, Mentzer et al. 2000). In this chapter, a combination of the Mentzer et al. (2000) and Simatupang and Sridharan (2004) definitions of supply chain collaboration is adopted. Collaboration in this chapter refers to a contractor and its suppliers working together to achieve common goals to create competitive advantage by sharing information, making joint decisions, and sharing benefits under a long-term contract.

Close coordination enables alliances to improve demand forecasting by means of a structured process of information, knowledge, and skill sharing. This results in cost savings, stemming from a reduction in inventory-associated costs. Both the contractor and its suppliers derive the economic benefits of collaboration when it is successfully implemented. Companies such as Lucent Technologies, Sun Microsystems, Hewlett-Packard, IBM, Procter & Gamble, Wal-Mart, and Dell (Barratt and Oliveira 2001, Callioni and Billington 2001, Lee 2000) are well recognized and publicized for their supply chain efficiencies.

Lack of information sharing and a tendency for supply chain members to optimize their own operations are the primary reasons for lower overall supply chain performance. This is especially the case with short life-cycle products that make it difficult for suppliers to meet demand. To effectively match supply with demand, the contractor and its suppliers need to collaborate in the supply chain. Supply chain efficiency improves through collaborative forecasting (Aviv 2002, Zhao 2002) by reducing uncertainty and increasing the quality of business decisions. Supply chain members who implement collaborative forecasting will likely achieve the benefits of increased responsiveness, decreased supply risk, and reduced inventory costs, and therefore increased earnings and improved corporate performance. In this chapter, a collaborative forecasting model is derived that demonstrates how the level of collaboration affects the accuracy of demand forecasting and supply chain economic costs.

In practice, the information technology (IT) revolution has led to closer collaboration between buyers and suppliers (Sherman 1998). However, different numbers of suppliers will give rise to different associated costs of technology (Fliedner 2003, McCarthy and Golicic 2002, Sherman 1998, Smaros 2005). The economic and related costs associated with the level of collaboration are important because they affect the level of economic activity or "vertical coordination." Therefore, this chapter focuses on three main areas: (1) how collaborative levels in the supply chain affect the accuracy of demand forecasting using stochastic differential equations; (2) the various types of economic costs—transaction costs, inventory costs, and supply chain risk costs—that depend on both the number of suppliers and collaborative level; and (3) to develop a model that can find the optimal level of collaboration for a supply chain member, based on the number of suppliers and supply chain management costs.

The remainder of the chapter is organized as follows. A brief literature review is contained in Section 6.2. Section 6.3 discusses the collaboration forecast and its associated economic cost. Section 6.4 discusses the details of the model formulation and its associated theorems to determine the optimal level of collaboration. Section 6.5 presents numerical examples. Finally, Section 6.6 contains conclusions, together with a brief discussion of managerial guidelines and recommendations for future research.

## 6.2 Literature Review

The initial level of collaboration is achieved when supply chain members share information relating to demand, inventory, capacity positions, and suppliers' data—across the supply chain (Lee 2000). Chen (1998), Gavirneni et al. (1999), Cachon and Fisher (2000), Lee et al. (2000), and Gaur et al. (2005) provided a full characterization of conditions of demand information sharing when the demand process is based on an autoregressive moving average (ARMA) model. Deshpande and Schwarz (2005) demonstrated that supply chain forecasting and inventory replenishment can be achieved through collaboration without disclosing private information of either the supplier or retailer. McCarthy and Golicic (2002) found that forecasting collaboration can be very beneficial in increasing responsiveness and improving product availability, as well as optimizing inventory and cost. Collaboration also has an opportunistic value that can effectively improve supply chain management

(Bauknight 2000) and reduce supply chain costs (Mentzer et al. 2000, McLaren et al. 2002).

The second level of collaboration is achieved when supply chain members exchange knowledge (Lee 2000). This level requires close collaboration and deep mutual trust among the supply chain partners (e.g., Wal-Mart and Pfizer, 7-Eleven and its suppliers in Japan). Simchi-Levi et al. (2008) developed a collaborative logistics process that is a joint decision-making model for assortment planning, forecasting, inventory management, and replenishment to create competitive advantage from the customers' point of view. The Collaborative Planning, Forecasting, and Replenishment (CPFR) concept introduced in 1995 belongs to this category. Since its introduction, much scholarly literature has focused on CPFR. A case study by McCarthy and Golicic (2002) suggested that the CPFR process model may be too complicated and requires too high an investment, and that simpler collaboration models would work better. Nøekkentved (2000) concluded that CPFR is not suitable for all business-to-business relationships. Kim and Mahoney (2006) analyzed CPFR from an incomplete contracting perspective.

The third level of collaboration involves coordination and resource sharing. Supply chain members coordinate by exchanging decision rights, work, and resources. Continuous Replenishment Plan (CRP) and Vendor Managed Inventory (VMI) programs are examples of this category of collaboration. Collaboration has resulted in the reduction of inventory levels and in increased sales for both the buying companies and suppliers (Aviv 2005, Schwarz 2004). Through collaboration, buying companies and suppliers jointly develop forecast-sharing information such as inventory, promotions, and point-of-sales data (Terwiesch et al. 2005).

The fourth level of collaboration is based on the best interest of the overall supply chain. Lee et al. (1997) suggested that channel coordination be used to improve overall supply chain performance. Lee (2000) provided a model of supply chain integration through information sharing, logistics coordination, and organizational relationship linkage. Swaminathan and Shanthikumar (1999), Jagdev and Thoben (2001), and Zipkin et al. (2000) focused on collaboration as a balanced approach to the overall organization of the supply chain. Both Govindarajan and Gupta (2001) and Holweg and Pil (2001) postulated how collaboration improves supply chain performance using Dell as an example.

Pyke et al. (2000) tested Lee's (2000) approach empirically. McCarthy and Golicic (2002) claimed that collaborative forecasting yields substantial improvement in company and supply chain performance. Simatupang and Sridharan (2004) proposed three collaboration indices—information sharing, decision synchronization, and incentive alignment. Empirical tests show that there is a significant correlation between the collaboration index and operational performance. Empirical tests also show that firms in general have a positive attitude toward inter-organizational collaboration (Skjoett-Larsen et al. 2003). However, it appears that the literature does not address the optimal level of supplier collaboration, considering issues such as demand uncertainty, supply chain costs in the supply base, and number of suppliers.

Therefore, the objective of this chapter is to explore not only the demand uncertainty that results from ever-changing customer needs, but also to investigate aspects of supply uncertainty emanating from transaction costs, supply risk, supplier responsiveness, supply innovation, and the number of suppliers.

## 6.3 Demand Forecast Collaboration

The first level of integration is the sharing of demand information among supply chain members. Demand forecast accuracy is very important because it reduces the "bullwhip effect" that negatively affects the rest of the chain and induces uncertainty into the supply chain. Supply chains that apply collaborative forecasting will likely achieve the benefits of increased market responsiveness, reduced inventory and other associated costs, and increased supply chain profit. The problem of demand distortion in a supply chain can be reduced through a joint forecast that is created by incorporating past observations of demand as well as contractor and supplier signals regarding future demand (Aviv 2001, 2002). In this chapter it is assumed that all suppliers possess nearly identical technology, have sufficient production capacity, and contractors and suppliers seek to minimize the whole supply chain cost. After searching and contacting all available suppliers, the contractor deals with $n \in [1, \infty)$ suppliers with a long-term contract based on the contractor's needs. Let $\{D(t), t \geq 0\}$ denote the actual market demand at time $t$ and $\{F(r, t), t \geq 0\}$ denote the underline process of demand forecasting in the supply chain based on a collaboration level of $r(t)$. Hence, the contractor decides order quantities based on $F(r, t)$ and the number of suppliers. To measure the level of accuracy of $F(r, t)$, assume that $D(t)$ is a deterministic-bounded continuous real-value function with respect to time $t$. Contractors and suppliers collaborate through frequent information and knowledge sharing to improve the accuracy of demand forecasts. If a deviation of $D(t) - F(r, t)$ occurs, which is the difference between actual demand and forecasted demand at time $t$, then an adjustment must be made in order to rectify the current deviation from the targeted demand through collaboration of those involved. Hence, the contractor and suppliers continuously monitor a forecasting error $D(t) - F(r, t)$ and dynamically adjust the differences through their collaboration efforts. It is assumed that the deviation is to be corrected (a *measurable* control variable) through controlling actions, which involve supply chain collaboration between a contractor and its suppliers. Here, $\alpha \cdot r(t) \in [0, \infty)$, where $r(t) \in [0, 1]$ and $\alpha \in [0, \infty)$, is defined as the actual effect of collaboration between the contractor and suppliers to accurately forecast demand. Here, $\alpha$ is the coefficient of the actual collaboration effect and $r(t)$ is the collaboration level at time $t$. Supply chain members' limited forecasting capabilities are a key obstacle to collaborative forecasting (Smaros 2005). The accuracy of a forecast varies with the technological complexity of the industry and forecasting capabilities. Through collaboration, a contractor and its suppliers can measure how much the actual collaboration level influences the accuracy of forecasts and subsequent improvement of forecasts. [Here, $r(t) = 1$ implies that the maximum level of collaboration required between the contractor and each supplier is based on their full effort.] In an infinitesimal time interval, after the adjustment is applied, the resulting improvement is expected to be $\alpha \cdot r(t)\{D(t) - F(r, t)\}dt$. A higher level of collaboration guarantees a more accurate forecast. However, due to demand uncertainty and forecasting error, the disturbance is assumed to be a Wiener type in the form of $\sigma dW(t)$. The actual changes in the estimate of demand $dF(r, t)$ contain an expected correction (*drift*) superimposed upon uncertainty (*disturbance*), that is,

$$dF(r, t) = \alpha \cdot r(t)\{D(t) - F(r, t)\}dt + \sigma dW(t) \tag{6.1}$$

where $\alpha \cdot r(t) \in [0, \infty)$ is controllable and $dW(t)$ is a standard Brownian motion. Because Equation (6.1) is the continuous-time version of the first-order autoregressive process in discrete time, the parameters of Equation (6.1) can be estimated using discrete-time data by running the regression $F(r, t) - F(r, t-1) = a + bF(r, t-1) + \varepsilon(t)$ [For more details on this process, see Arnold (1974), Dixit and Pindyck (1994)]. If the value of $F(r, t)$ is currently $F_s = F(r, t = s)$, it is assumed that $F_S$ is either normally distributed or constant. If $F(r, t)$ follows Equation (6.1), then the integral (in the Ito sense) of the infinitesimal diffusion as defined in Equation (6.1) can be obtained as follows [using the method given by Theorem 8.4.2 in Arnold (1974)]:

$$F(r, t) = \exp\{-R(t)\}\left\{F_s + \int_s^t \alpha r(x) D(x) \exp\{R(x)\}dx + \sigma \int_s^t \exp\{R(x)\} dW(x)\right\} \tag{6.2}$$

where $R(t) = \alpha \int_s^t r(x)dx$ (i.e., cumulative adjustment through collaboration). Then the first two moments of $F(r, t)$—$K(r, t) = E[F(r, t)]$ and $G(r, t) = E[F^2(r, t)]$—can be calculated from the explicit form of Equation (6.2).

**Proposition 6.1**

$$K(r, t) = E[F(r, t)] = e^{-R(t)}\left\{K_s + \int_s^t \alpha r(x) D(x) e^{R(x)} dx\right\} \text{ with } K_S = E[F(r, s)]$$

$$G(r, t) = E[F^2(r, t)] = e^{-2R(t)}\left[G_S + \int_s^t \{2\alpha K(r, x) r(x) D(x) + \sigma^2\} e^{2R(x)} dx\right] \text{ with}$$

$$G_S = E[F^2(r, s)]$$

**Proof** Using a similar calculation as in Arnold (1974), the function $K(r, t)$ must satisfy the following:

$\dot{K}(r, t) = -\alpha r(t) K(r, t) + \alpha r(t) D(t)$ is associated with an initial condition of $K_S = E\{F(r, t = s)\}$. Hence,

$$K(r, t) = E[F(r, t)] = e^{-R(t)}\left\{K_s + \int_s^t \alpha r(x) D(x) e^{R(x)} dx\right\} \tag{6.3}$$

Let $G(r, t) = E\{F^2(r, t)\}$; then $\dot{G}(r, t) = -2\alpha r(t) G(r, t) + 2\alpha K(r, t) r(t) D(t) + \sigma^2$ is associated with an initial condition $G(s) = G_S$. Hence,

$$G(r, t) = E[F^2(r, t)] = e^{-2R(t)} \times \left[G_S + \int_s^t \{2\alpha K(r, x) r(x) D(x) + \sigma^2\} e^{2R(x)} dx\right] \; (Q.E.D.) \tag{6.4}$$

Therefore, the variance of $F(r, t)$ can be expressed as follows:

$$V(r, t) = E[F(r, t) - K(r, t)]^2 = G(r, t) - \{K(r, t)\}^2 \tag{6.5}$$

## 6.3.1 Associated Supply Chain Collaboration Costs

In this chapter, a model is proposed that considers the optimum level of collaboration associated with a given number of suppliers, the expected inventory cost, and the supply chain management costs under a long-term partnership. Because the level of collaboration affects these costs, a contractor can reach the lowest cost position with a certain level of collaboration. A high level of collaboration enables the contractor to respond more efficiently to demand uncertainty, but involves more complexity due to an additional exercise of collaboration and control than would result with a low level of collaboration.

### 6.3.1.1 Inventory Cost

The deviation of the actual quantity ordered from the target causes economic losses (i.e., holding and shortage costs). Unit holding costs of $C^h$ for each unsold unit in inventory and shortage costs of $C^S$ for each unit of unsatisfied demand were set. The salvage value at the end of each period is assumed to equal zero. Let $I_{\{D(t) > F(r,t)\}} = 1$ if $D(t) > F(r, t)$ and $I_{\{D(t) > F(r,t)\}} = 0$ if $D(t) < F(r, t)$; then the terms $C^S\{D(t) - F(r, t)\} I_{\{D(t) > F(r,t)\}}$ impose penalties for a stock out (a loss of goodwill) and the term $C^h\{F(r, t) - D(t)\} I_{\{D(t) < F(r,t)\}}$ represents overstock (inventory holding cost). An asymmetric loss function would be appropriate if the loss differs for the values of $F(r, t)$ that are equidistant from the target $D(t)$. The total inventory loss, $l(t)$, using an asymmetric loss function could be expressed as follows:

$$l(t) = \begin{cases} C^S\{D(t) - F(r, t)\} & \text{if } D(t) > F(r, t) \\ C^h\{F(r, t) - D(t)\} & \text{if } D(t) < F(r, t) \end{cases}$$

The expected loss (penalty cost) under the underling process can be denoted by $LS(t) = E[l(t)]$. To derive $LS(t)$, the probability of the forecasted demand being less than the actual demand can be derived as follows. For notational simplicity, let $S(r, t) = P[F(r, t) < D(t)]$; then

**Proposition 6.2**

$$S(r, t) = P[F(r, t) < D(t)] = \frac{1}{\sqrt{2\pi Q(r, t)}} \left\{ \int_{-\infty}^{Z(r,t)} e^{\frac{-x^2}{2Q(r,t)}} dx \right\} \tag{6.6}$$

where

$$Q(r, t) = \int_S^t \exp\{2R(x)\} dx, \qquad R(t) = \alpha \int_s^t \tau r(x) dx$$

and

$$Z(r, t) = (1/\sigma) \left[ D(t) \exp\{R(t)\} - F_S - \int_s^t \alpha r(x) D(x) \exp\{R(x)\} dx \right]$$

**Proof**

$$\begin{aligned} S(r, t) &= P[F(r, t) < D(t)] \\ &= P\left[ \exp\{-R(t)\} \left\{ F_S + \int_S^t \alpha r(x) D(x) \exp\{R(x)\} dx \right.\right. \\ &\quad \left.\left. + \sigma \int_S^t \exp\{R(x)\} dW(x) \right\} \le D(t) \right] = P\left[ \left[ \int_s^t \exp\{R(x)\} dW(x) \right.\right. \\ &\quad \left.\left. \le (1/\sigma) \left\{ D(t) \exp\{R(t)\} - F_S - \int_s^t \alpha \cdot r(x) \cdot D(x) \cdot \exp\{R(x)\} dx \right\} \right] \right. \\ &= \frac{1}{\sqrt{2\pi Q(r, t)}} \left\{ \int_{-\infty}^{Z(r,t)} e^{\frac{-x^2}{2Q(r,t)}} dx \right\} \end{aligned}$$

using Theorem 9.2.3 in Arnold (1974) (Q.E.D.).

For a simple expression of the following equation, let

$$Q_r = \partial Q(r, t)/\partial r = 2\alpha \int_s^t (x - s) \exp\{2R(x)\} dx$$

and

$$Z_r = (\alpha/\sigma) \left[ D(t)(t - s) \exp\{R(t)\} - \int_s^t D(x) \exp\{R(x)\}\{1 + \alpha r(x)(x - s)\} dx \right]$$

**Proposition 6.3** Given $r(t) = r \in [0, 1]$

$$S_r(r, t) = \partial S(r, t)/\partial r = -(Q_T/2Q) \left[ \int_{-\infty}^{Z(r,t)} (1/\sqrt{2\pi Q}) \exp\{-y^2/2Q\} dy \right] + (2\pi Q)^{-0.5} \left[ (Q_r/2Q^2) \int_{-\infty}^{Z(r,t)} y^2 \exp\{-y^2/2Q\} dy + (Z_r) \times \exp\{-Z^2/2Q\} \right]$$

**Proof** Leibniz's rule for differentiating an integral can facilitate the proof (Q.E.D). The expected loss function at time $t$ can be derived as follows:

**Proposition 6.4**

$$LS(t) = E[l(t)] = C^S \cdot S(r, t) \cdot \{D(t) - B(t)\} + C^h \cdot \{1 - S(r, t)\} \cdot \{A(t) - D(t)\} \tag{6.7}$$

where:

$$A(t) = E[F(r, t) | F(r, t) > D(t)] = K(r, t) + V(r, t)[\phi(u)/\{1 - \Phi(u)\}]$$

$$B(t) = E[F(r, t) | F(r, t) < D(t)] = K(r, t) - V(r, t)[\phi(u)/\Phi(u)]$$

$$\varnothing(u) = (1/\sqrt{2\pi}) \exp\{-u^2(t)/2\}, \; \Phi(u) = \Phi\{u(t)\} = \int_{-\infty}^{u(t)} \varnothing(x) dx$$

$$u(t) = \{D(t) - K(r, t)\}/\sqrt{V(r, t)}$$

**Proof**

$$LS(t) = E[l(t)] = C^S \cdot E[(D - F)|D > F] \cdot P[D > F]$$

$$+ C^h \cdot E[(F - D)|D < F] \cdot P[D < F]$$

$$= C^S \cdot E[D|D > F] \cdot S - C^S \cdot E[F|D > F] \cdot S$$

$$+ C^h \cdot E[F|D < F] \cdot (1 - S) - C^h \cdot E[D|D < F] \cdot (1 - S)$$

$D(t)$ is assumed to be a bounded continuous real value function, $E[D|D > F] = D$ (Chung 1974). Therefore,

$$LS(t) = C^S \cdot S(r, t) \cdot [D(t) - E\{F(r, t)|D(t) > F(r, t)\}]$$

$$+ C^h \cdot \{1 - S(r, t)\} \cdot [E\{F(r, t)|D(t) < F(r, t)\} - D(t)]$$

Assuming that the initial $F(t = s) = F_S$ are both constant or normally distributed, $F(r, t)$ is a Gaussian stochastic process with expectation $K(r, t)$ and variance $V(r, t)$. See Theorem 8.2.10 in Arnold (1974) and Proposition 6.1. Accordingly, the expected value can be obtained using the truncated normal distribution (Ryan 2000) as follows:

Hence,

$$E\{F(r, t)|F(r, t) > D(t)\} = K(r, t) + V(r, t)\left\{\frac{\phi(u(t))}{1 - \Phi(u(t))}\right\} = A(t)$$

and

$$E\{F(r, t)|F(r, t) < D(t)\} = K(r, t) - V(r, t)\left\{\frac{\phi(u(t))}{\Phi(u(t))}\right\} = B(t)$$

Hence,

$$LS(t) = E[l(t)] = C^S \cdot S(r, t) \cdot \{D(t) - B(t)\}$$
$$+ C^h \cdot \{1 - S(r, t)\} \cdot \{A(t) - D(t)\}$$

### 6.3.1.2 The Contractor's Acquisition Price and Suppliers' Production Cost

Assume that the unit wholesale price charged by a supplier falls between an upper bound $W_U$ and a lower bound $W_L$. Because the process and production cost reductions are the benefits of supply chain collaboration (Mentzer et al. 2000, McLaren et al. 2002), in this section, Homburg and Kuester's (2001) purchasing price was used as the contractor's acquisition price. The relationship between the number of suppliers and the level of collaboration can be portrayed as follows:

$$W(n, r, t) = \theta\{r(t)\} \cdot \{W_L + (W_U - W_L)/(b^{n-1})\},\ \theta\{r(t)\} > 0,\ b > 1,\ \text{and}\ n \geqq 1.$$

The unit production cost, $C(t)$, of the supplier can be defined as a markup based on the given wholesale price. Then the retail price $P(t)$ and $C(t)$ can be calculated as follows:

$$P(n, r, t) = \beta(t)\,W(n, r, t),\ 1 \leq \beta(t) < \infty,\ C(n, r, t) = \eta(t)\,W(n, r, t),\ 0 < \eta(t) \leqq 1.$$

### 6.3.1.3 Collaborative SCM Cost

*Collaboration costs* are composed of all the costs associated with managing a relationship with specific suppliers. These costs are associated with the exchange of information, decision making, system implementation and integration, process collaboration and integration, and data transaction and integration costs (McLaren et al. 2002). Recent studies (Bakos and Brynjolfsson 1994, Homburg and Kuester 2001, Choi and Krause 2006) indicate that collaboration costs are increasing as the numbers of suppliers increase and collaboration levels become more complex. Choi and Krause's (2006) Proposition 1.1 and 1.2 suggests that there is a positive linear relationship between the number of suppliers and collaboration costs. Hence this paper defines the collaboration costs (CL) at time $t$ as a continuous function of the collaboration level $\{r(t) \in (0, 1]\}$ and the number of suppliers $\{n^*(t) = n\}$. It follows that by letting $f^{CL}(,)$ be a continuous function representing cost, then

$$CL(n, r, t) = f^{CL}\{r(t), n\},\ CL_r = \partial f^{CL}/\partial r > 0$$

and

$$CL_n = \partial f^{CL}/\partial n > 0$$

### 6.3.1.4 Transaction Risk

Other transaction costs are related to transaction risks (Clemons et al. 1993, Choi and Krause 2006) and the number of suppliers. If a contractor increases the number of suppliers, then the probability of unreliable delivery is higher (Handfield and

Nichols 1999, Nishiguchi T. 1994, Choi and Krause 2006). Additional problems between different suppliers might be that downtime results and managerial resources must be increased to maintain the contractor's goals (Homburg and Kuester 2001). In contrast, if the contractor uses fewer suppliers, then there are several possible risks involved, such as the suppliers becoming potential competitors after acquiring core technology from the contractor. The cost of changing partners is huge when the contractor needs to change suppliers, and poor supplier performance causes even more serious harm. Both Proposition 2.1 of Choi and Krause's study (2006) and operations costs cited by Homburg and Kuester (2001) suggest that there is a positive quadratic relationship between the number of suppliers and a given contractor. Supply risk could be reduced through collaboration. Supply risk cost (SR) can be formulated as follows:

$$SR(n, r, t) = f^{SR}(n, r),\ SR_{nn} = \partial^2 f^{SR}/\partial n^2 \geq 0 \text{ for all } n$$

and

$$SR_r = \partial f^{SR}/\partial r \leq 0$$

### 6.3.1.5 Loss Related to Supply Responsiveness

Response time is an important issue in supply chain management. Some benefits of fast response time are reduced cycle time, higher quality, fewer internal problems, and better service. Supply responsiveness is viewed as the suppliers' accurate and quick responses to the contractors' requests for new requirements. Previous research indicates that single sourcing has more effective responsiveness than multiple sourcing because single sourcing has a better chance of promoting a close relationship and open communication between the contractor and the supplier (Liker and Choi 2004, Treleven and Schweikhart 1988, Handfield and Bechtel 2002, Larson and Kulchitsky 1998). Choi and Krause (2006) suggested that there should be a negative association between the number of suppliers and suppliers' responsiveness. Collaboratively sharing information with supply chain members dramatically improved customer responsiveness (Horvath 2001). Assume that supplier responsiveness decreases as the number of suppliers increases, but decreases as the $r(t)$ increases. Hence, *supply responsiveness cost* (SR) related to the number of suppliers and the level of collaboration can be represented as follows:

$$RC(n, r, t) = f^{RC}\{r(t), n\} \quad \text{with} \quad RC_n = \partial f^{RC}/\partial n > 0$$

and

$$RC_r = \partial f^{RC}/\partial r < 0$$

## 6.4 Model

Using all the costs listed in the previous section, the *expected instantaneous total cost* (TC) of the contractor at time $t$ with $n$ suppliers can then be formulated as

$$TC(n, r, t) = CD(n, r, t) + SR(n, r, t) + RC(n, r, t)$$

Let $P(n, r, t)$ be the unit retail price that the contractor charges end consumers with $n$ suppliers and $r(t)$ at time $t$, and let $\pi^B(n, r, t)$ be the contractor's expected profit at time $t$ with $n$ number of suppliers under the collaboration level $r(t)$ at time $t$; then

$$\begin{aligned}\pi^B(n, r, t) &= E[D(t)\{P(n, r, t) - W(n, r, t)\} - TC(n, r, t) - l(r, t)] \\ &= D(t)\{P(n, r, t) - W(n, r, t)\} - TC(n, r, t) - LS(r, t)\end{aligned}$$

Let $\pi^S(t)$ be suppliers' expected profit at time $t$ with $n$ number of suppliers; then

$$\pi^S(n, r, t) = E[F(r, t)\{W(n, r, t) - C(t)\}] = K(r, t)\{1 - \eta(t)\}W(n, r, t)$$

Hence, the supply chain profit, $\pi(t)$, at time $t$ is

$$\pi(n, r, t) = \pi^B(n, r, t) + \pi^S(n, r, t) = DP - W\{D - K(1 - \eta)\} - TC - LS$$

Here, the *optimal collaboration level with* n *number of suppliers* (OCL) is defined. Given $F_S = F(t = s)$ and $T < \infty$ as the planning horizon, the optimal level of collaboration with $n$ number of suppliers is sought so that the expected total profit over $[0, T]$ [is maximized under equation $n(t)$, $r(t)$, $\dot{K}(r, t)$, and $\dot{G}(r, t)$]. Using a long-term contract with a fixed number of suppliers during the planned time horizon, the OCL can be expressed using the following optimal control formulation:

$$\underset{r\in[0,1]}{\text{Max}} \int_S^T \pi(t, n, r)dt \tag{6.8}$$

$$\dot{G}(r, t) = 2\alpha r(t)\{K(r, t)D(t) - G(r, t)t\} + \sigma^2 \tag{6.9}$$

$$\dot{K}(r, t) = \alpha r(t)\{D(t) - K(r, t)\} \tag{6.10}$$

$$K(r, s) = K_S, \quad \text{and} \quad G(r, s) = G_S$$

$$0 \leq r(t) \leq 1, \quad \text{and} \quad n \in [1, \infty)$$

Subsequently, when no confusion arises, the time notation $(t)$ and collaboration level $r(t)$ will be suppressed; for instance, $K(r, t)$ will be written simply as $K$.

The OCL above is an optimal control problem. Next, the theoretical properties of the optimal control policies of the OCL are discussed. The adjoint functions $\lambda_1$ and $\lambda_2$ are associated with Equations (6.9) and (6.10) and can define the current value of the Hamiltonian function as

$$H = \pi + \lambda_1\{2\alpha r(KD - G) + \sigma^2\} + \lambda_2 \alpha r(D - K) \tag{6.11}$$

The Lagrangian with multipliers $w_1$ and $w_2$ is

$$\Gamma = H + w_1(1 - r) + w_2(r) \tag{6.12}$$

To determine the optimal solution, the following equations are needed. Since

$$LS_G = \partial LS/\partial G = C^h \cdot (1 - S) \cdot A_G - C^S \cdot S \cdot B_G,$$

$$LS_K = \partial LS/\partial K = C^h \cdot (1 - S) \cdot A_K - C^S \cdot S \cdot B_K, \quad \text{and}$$

$$LS_r = \partial LS/\partial r = S_r\{(C^S + C^h)D - C^S \cdot B - C^h \cdot A\} \quad \text{by equation (6.7)}$$

hence $\lambda_1(t)$ and $\lambda_2(t)$ satisfy the following equation:

$$\dot{\lambda}_1 = -(\partial H/\partial G) = LS_G + 2\alpha r \lambda_1, \quad \text{with} \quad \lambda_1(T) = 0$$

$$\dot{\lambda}_2 = -(\partial H/\partial K) = -W(1 - \eta) + LS_K - 2\alpha r D\lambda_1 + r\alpha\lambda_2, \text{ with } \lambda_2(T) = 0$$

Then,

$$\lambda_1(r, t) = -e^{2R(t)} \int_t^T [LS_G(x)]e^{-2R(x)} dx, \tag{6.13}$$

$$\lambda_2(r, t) = e^{R(t)} \int_t^T [2\alpha r(x) D(x)\lambda_1(x) - LS_K(x) + W(n, r)(1 - \eta)]e^{-R(x)} dx \tag{6.14}$$

$$A_K(t) = \frac{\partial A}{\partial K}(t) = 1 - 2K(t)\left\{\frac{\phi(u)}{1 - \Phi(u)}\right\} + V(t)\left[\frac{\phi(u)}{\{1 - \Phi(u)\}^2}\right][J_3(t)\{1 - \phi(u)\} + J_1(t)\phi(u)]$$

$$A_G(t) = \frac{\partial A}{\partial G}(t) = \left\{\frac{\phi(u)}{1-\Phi(u)}\right\}$$

$$+V(t)\left[\frac{\phi(u)}{\{1-\Phi(u)\}^2}\right][J_4(t)\{1-\Phi(u)\}+J_2(t)\phi(u)]$$

$$B_K(t) = \frac{\partial B}{\partial K}(t) = 1 + 2K(t)\left\{\frac{\phi(u)}{\Phi(u)}\right\}$$

$$-V(t)\left[\frac{\phi(u)}{\{\Phi(u)\}^2}\right][J_3(t)\Phi(u) - J_1(t)\phi(u)]$$

$$B_G(t) = \frac{\partial B}{\partial G}(t) = -\left\{\frac{\phi(u)}{\Phi(u)}\right\} - V(t)\left[\frac{\phi(u)}{\{\Phi(u)\}^2}\right][J_4(t)\Phi(u) - J_2(t)\phi(u)]$$

$$J_1(t) = \left[\frac{K(t)D(t) - G(t)}{\{V(t)\}^{\frac{3}{2}}}\right], \quad J_2(t) = \left[\frac{K(t) - D(t)}{2\{V(t)\}^{\frac{3}{2}}}\right]$$

$$J_3(t) = \left[\frac{\{D(t) - K(t)\}\{G(t) - K(t)D(t)\}}{\{V(t)\}^2}\right], \quad J_4(t) = \left[\frac{\{D(t) - K(t)\}^2}{2\{V(t)\}^2}\right]$$

Necessary conditions for an optimal level of collaboration and number of suppliers required are

$$\partial\Gamma/\partial r = H_r + w_2 - w_1 = 0$$

$$w_1 \geq 0,\ w_1(1-r) = 0,\ w_2 \geq 0, \quad \text{and} \quad w_2 \cdot r = 0$$

where

$$H_r = \pi_r + 2\alpha\lambda_1(KD - G) + \alpha\lambda_2(D - K) \tag{6.15}$$

and

$$\pi_r = DP_r - W_r\{D - K(1-\eta)\} - TC_r - LS_r \tag{6.16}$$

Then the optimal solution $r^*(t)$ with $n$ number of suppliers satisfies the following equations:

$$\partial\Gamma/\partial r = H_r + w_2 - w_1 = 0,\ \ w_1 \geq 0,\ \ w_1(1-r) = 0,\ \ w_2 \geq 0,\ \text{and}\ w_2 r = 0 \tag{6.17}$$

For simplicity, let the first component of the function represent the number of suppliers, the second component the collaboration level needed, and the last one denote time. Then every function can be expressed in the following way: $f(n, r, t)$.

So from Equation (6.16), the optimal solution can be characterized under $n$ suppliers as follows:

**Lemma 6.1** For given $n$ suppliers,

$$r^*(t) = \begin{cases} 1 & \text{if } H_r(n, 1, t) > 0 \\ \in [0, 1] & \text{if } H_r(n, r, t) = 0 \\ 0 & \text{if } H_r(n, 0, t) < 0 \end{cases} \tag{6.18}$$

**Proof** If $H_r(n, r, t) > 0$, then Equation (6.17) requires $w_2 = 0$; hence, from Equation (6.17), $r^* = 1$. If $H_r(n, r, t) < 0$, then Equation (6.17) requires $w_1 = 0$; hence, from Equation (6.17), $r^* = 0$. Otherwise $r^* \in [0, 1]$. □

Lemma 6.1 implies that the optimal level of collaboration with $n$ suppliers can be chosen based on the marginal return with respect to the level of collaboration. If the marginal return with respect to the level of collaboration is positive, the maximum level of collaboration is the optimal solution. If the marginal return is negative then the minimum level of collaboration can be the optimal solution. Otherwise, contractors need to choose between the maximum and minimum level of collaboration required.

**Proposition 6.5** For a given $n$ number of suppliers, suppose $\pi_r(n, 1, t) > 0$ and one of the following conditions holds:

(a) $LS_G < 0$, $LS_K < W(1 - \eta)$, $D > K$, $KD - G > 0$ or
(b) $LS_G > 0$, $LS_K > W(1 - \eta)$, $D < K$,

then $r(t) = 1$ for all $t \in [0, T]$.

**Proof** If $LS_G > 0$, then $\lambda_1 < 0$ and if $LS_G < 0$, then $\lambda_1 > 0$ by Equation (6.13). From Equation (6.14), if $\lambda_1 < 0$ and $LS_K > W(1-\eta)$, then $\lambda_2 < 0$ and if $\lambda_1 > 0$ and $LS_K < W(1-\eta)$, then $\lambda_2 > 0$. Because $V = G - K^2 > 0$ by (5), $K^2 - G < 0$. Hence if $D < K$, then $KD - G < K^2 - G < 0$. Therefore the condition $D < K$ holds, and $KD - G < 0$ always occurs.

Since $H_r = \pi_r + \lambda_1(2\alpha)(KD - G) + \lambda_2(\alpha)(D - K)$, if $\pi_r > 0$ for $r = 1$ and either (a) or (b) hold, then $H_r > 0$ for all $r \in [0, 1]$ by Equation (6.15). Hence $r(t) = 1$ for all $t \in [0, T]$. □

Proposition 6.5 shows the maximum collaboration required from the beginning to the ending time in the supply chain to maximize the supply chain profit.

**Theorem 6.1** Suppose $\pi_{rr}(n, r, t) \le 0$ for $r \in (0, 1)$; if both $H_r(n, 0, T) = \pi_r(n, 0, T) < 0$ and $H_r(n, 1, 0) = \pi_r(n, 1, 0) + 2\alpha\lambda_1(1, 0)(K_0 D_0 - G_0) + \alpha\lambda_2(1, 0)(D_0 - K_0) > 0$ hold, then

$$r^*(t) = \begin{cases} 1 & \text{for } 0 < t < t_1 \\ r^0 & \text{for } t_1 < t < t_2 \\ 0 & \text{for } t_2 \le t \le T \end{cases}$$

where $t_1 = Sup[t > 0 | H_r(n, 1, t) > 0]$, $t_2 = Inf[t \ge t_1 | H_r(n, 0, t) < 0]$ $r^0 = [r(t) \in (0, 1)$ and $t_1 < t < t_2 | H_r(n, r, t) = 0]$, $\pi_r(n, 1, 0) = D_0 P_r(n, 1, 0) - W_r(n, 1)\{D_0 - K_0(1 - \eta)\} - TC_r(n, 1, 0)$, and $\pi_{rr} = DP_{rr} - W_{rr}\{D - K(1 - \eta)\} - (TC_{rr} + LS_{rr})$.

**Proof** Since $LS_r(n, r, 0) = 0$, $\pi_r(n, 1, 0) = D_0 P_r(n, 1, 0) - W_r(n, 1, 0)\{D_0 - K_0(1 - \eta)\} - TC_r(n, 1, 0)$, and $H_r(n, 1, 0) = \pi_r(n, 1, 0) + 2\alpha\lambda_1(1, 0)(K_0 D_0 - G_0) + \alpha\lambda_2(1, 0)(D_0 - K_0)$. $H_r(n, 0, t) = \pi_r(n, 0, T)$, because $\lambda_1(T) = \lambda_2(T) = 0$.

Suppose $H_r(n, 1, 0) > 0$, $H_r(n, 0, T) < 0$, and $H_{rr} = \partial^2 H / \partial r^2 \le 0$ then $H_r(n, r, t)$ is a non-increasing function with respect to $r(t)$ and should have at least two zeros at $t = t_1$ and $t = t_2$. Hence the optimal solution is $r^*(t) = 1$ for $t \in [0, t_1]$, $r^*(t) = r^0 \in (0, 1)$ for $t \in (t_1, t_2)$, and $r^*(t) = 0$ for $t \in [t_2, T]$ by Lemma 6.1. And $t_1 = \underset{t \ge 0}{Sup}\, [t | H_r(n, 1, t) > 0]$ and $t_2 = \underset{t_1 \le t \le r}{Inf}\, [t \ge t_1 | H_r\{n, 0, t\} < 0]$. □

Manufacturers can determine the expected marginal return with respect to the level of collaboration under $n$ number of suppliers at the beginning and the ending times, and the shape of the $H_r(r, t)$ for $0 < t < T$. Five situations are examined. First, if the initial marginal return with respect to the level of collaboration with $n$ number of suppliers is positive up to time $t_1$, the marginal return is zero after time $t_1$ up to time $t_1$, and after $t_2$ the marginal return with respect to the level of collaboration is negative until terminal time $T$, then contractors need to provide the maximum level of collaboration among $n$ number of suppliers up to time $t_1$, the level of collaboration is between maximum and minimum with the $n$ number of suppliers between $t_1$ and $t_2$, and no collaboration after $t_2$.

Second, if the initial marginal return is positive but gradually decreasing and negative with a marginal return guaranteed at the terminal time, then contractors need to practice a maximum level of collaboration with $n$ number of suppliers up to time $t_1$. Note that $t_1 = t_2$ in this situation. Thereafter, contractors need to switch to the minimum level of collaboration until terminal time $T$. Third, if the marginal return is always positive over time, then the manufacturers need to practice the maximum level of collaboration with $n$ number of suppliers for all times and maintain this relationship until terminal time $T$.

Fourth, if the initial marginal return is negative but gradually increasing and positive with a marginal return guaranteed at the terminal time, then the contractors need to practice a minimum level of collaboration with $n$ number of suppliers up to time $t_1$. Thereafter, they need to switch to the maximum level of collaboration until the terminal time. Finally, if the marginal return is always negative or zero, then it is better to contract with the minimum number of suppliers and maintain the contract until terminal time.

**Corollary 6.1** For given $n$ suppliers,

(a) If $\dot{H}_r(n, 1, t) \leq 0,\ H_r(n, 1, 0) \geq \pi_r(n, 1, T) > 0$ or
(b) If $\dot{H}_r(n, 1, t) \geq \pi_r(n, 1, T) \geq H_r(n, 1, 0), > 0$

Then $r^*(t) = 1$ for all $t \in [0, T]$

**Proof** Since $H_r(n, 1, t)$ is a non-decreasing function with time $t$ and both $H_r(n, 1, 0)$ and $H_r(n, 1, T)$ are positive, hence $r^*(t) = 1$ for all $t \in [0, T]$ by Lemma 6.1. □

If marginal profits are positive for all $t$, then practicing the maximum level of collaboration with $n$ number of suppliers is the optimal policy.

**Theorem 6.2** Suppose $\pi_{rr}(n, r, t) \geq 0$ for $r \in (0, 1)$, if both $H_r(n, 1, T) = \pi_r(n, 1, T) > 0$ and $H_r(n, 0, 0) = \pi_r(n, 0, 0) + 2\alpha\lambda_1(0, 0)(K_0 D_0 - G_0) + \alpha\lambda_2(0, 0)(D_0 - K_0) \leq 0$ hold, then

$$r^*(t) = \begin{cases} 0 & \text{for } 0 \leq t \leq t_1 \\ r^0 & \text{for } t_1 < t < t_2 \\ 1 & \text{for } t_2 \leq t \leq T \end{cases}$$

where $t_1 = \underset{t \geq 0}{Sup}\ [t | H_r(n, 0, 0) < 0]$, $t_2 = \underset{t_1 \leq t \leq T}{Inf}\ [t \geq t_1 | H_r(n, 0, T) > 0]$, $r^0 = [r(t) \in (0, 1) | H_r(n, r, t) = 0$ for a given $n$ and $t \in (t_1, t_2)]$, and $\pi_r(n, 0, 0) = D_0 P_r(n, 0, 0) - W_r(n, 0, 0)\{D_0 - K_0(1 - \eta)\} - TC_r(n, 0, 0)$.

**Proof** If $H_r(n, 0, 0) \leq 0$, $H_{rr}(n, r, t) \geq 0$, $H_r(n, 1, T) > 0$ then there exists unique $t_1$ and $t_2$ such that $t_1 = \underset{t \geq 0}{Sup}[t | H_r(n, 0, 0) < 0]$ and $t_2 = \underset{t_1 \leq t \leq T}{Inf}$ $[t \geq t_1 | H_r(n, 1, T) > 0]$ because $H_r(n, r, t)$ is a continuous increasing function with respect to $r$. Hence $t_1 = \underset{t \geq 0}{Sup}[t | H_r(n, 1, T) > 0]$, and $t_2 = \underset{t_1 \leq t \leq T}{Inf}[t \geq t_1 | H_r(n, 0, 0) < 0]$ using Lemma 6.1. □

At the beginning, the marginal return with respect to the level of collaboration with $n$ number of suppliers is negative up to time $t_1$. After $t_1$, the marginal return is zero up to time $t_2$. After $t_2$, the marginal return with respect to the level of collaboration increases and is positive until terminal time $T$. When contractors need to contract with the minimum level of collaboration among $n$ number of suppliers up to time $t_1$, the level of collaboration is between maximum and minimum with the $n$ number of suppliers between $t_1$ and $t_2$, and maximum collaboration after $t_2$. Another situation occurs when the marginal return is negative for all times. Then it is better to practice the minimum level of collaboration with $n$ number of suppliers and maintain the contract until terminal time.

## 6.5 Numerical Examples

In this section the optimal policy models developed in this chapter are applied under a given sequence of events. This section also focuses on how collaboration level affects the optimal decision.

### 6.5.1 Forecast Accuracy

To check the accuracy of a forecast based on the level of collaboration, apply our model both under both high and low demand uncertainty. The following equations would apply for market demand:

$$D_1(t) = 400 \cdot Cos\{1/(t+0.01)\} - 300 \cdot Sin(\pi + 10 \cdot t) + 500 \text{ and } D_2(t) = 8000 - 600 \cdot t$$

The current time horizon is $T = 7(s = 0 \leq t \leq T = 7)$, and the other parameters assume the following: $K_0^1 = 8000$, $K_0^2 = 7900$, and $\alpha = 20$.

Under the setting above, five cases are considered. Inthe first case, the collaboration level is maximum from the beginning to the end. The second case is the same as the first one but $r(t) = 0.5$. In the third case, no collaboration is required from the beginning to the end. The fourth situation depicts a collaboration level from the beginning that is not maximum but both parties gradually increase their collaboration level as time passes. The fifth case involves the contractor and suppliers coordinating the maximum level from the beginning but gradually decreasing their collaboration level as time passes. These five cases of mean square errors (MSE) are portrayed in Tables 6.1 and 6.2.

If the conversion parameters $\alpha$ and the level of collaboration are increased enough, the accuracy of $K(t)$ improves and converges to the actual demand experienced, although the actual demand is very volatile. Note that the case of $r(t) = 1$ provides the most accurate forecast and the case of $r(t) = 0$ results in the worst forecast based

**Table 6.1 Mean Square Error (MSE) of $D_1(t)$ Based on $r(t)$**

| *r(t)* | *1* | *0.5* | *0* | *0.5 + 0.06t* | 1 − 0.1*t* |
|---|---|---|---|---|---|
| ***MSE*** | 11970 | 29433 | 69938 | 19163 | 26255 |

**Table 6.2 Mean Square Error (MSE) of $D_2(t)$ Based on $r(t)$**

| *r(t)* | 1 | 0.5 | 0 | $0.5 + 0.06t$ | $1 - 0.1t$ |
|---|---|---|---|---|---|
| *MSE* | 2717 | 5867 | 7853333 | 3776 | 5758 |

on MSE. A comparison between the fourth case and the fifth case indicates that it is better to begin with moderate collaboration and then increase collaboration as time passes than to provide the maximum collaboration initially and gradually decrease collaboration under the current parameter.

Using a high value of $r(t)$ the contractor and suppliers can reach more accurate forecasts than using a lower value of $r(t)$. (The recommended model accurately measures the accuracy of underline forecast based on the collaboration level and conversion effect.)

### 6.5.2 Optimal Collaboration Level

The parameters in the example assume the following: The current time horizon is 7 ($0 \leq t \leq T = 7$), and the minimum number of suppliers is one and the maximum number of suppliers is five ($1 \leq n \leq 5$). The other parameters used in the example assume the following three cases. First, it is assumed that the demand $D(t) = 8000 - 6000 \cdot t$; then the contractor's acquisition price, retail price, and suppliers' cost can be calculated in the following manner:

$$W_L = \$300,\ W_U = \$500$$

$$W(n, r, t) = (1.3)\{300 + (200)/(b^{n-1})\}$$

$$b = 1.05$$

[The unit retail price when $p(n, r, t) = 1.5\,W(n, r, t)$ and the average cost of the supplier is $SC(n) = 0.8\,W(n, r, t)$ per unit.] Let $C^S = \$10$, $C^h = \$5$ for the loss function. Second, the collaboration cost is $CL(n, r) = 100,000 + 20,000 \cdot n \cdot r$; and third, the supply risk cost is $SR(n, r) = \{1,908n^2 - 19,036n(1 + 0.1r) + 99,600\}(2 - r)$. Finally, the supply response cost is $RC = 10,000n(2 - r)$. All parameters are fixed except $r$, $n$, and $\alpha$, as in the next two cases.

If the contractor deals with one supplier ($n = 1$), Theorem 6.1 can be applicable in this case and its optimal policy can be determined as follows:

$$r(t) = 1 \quad \text{for } 0 \leq t \leq t_1 = 0.502235,\ r(t) \in (0, 1) \text{ for}$$

$$t_1 = 0.502235 < t < t_2 = 0.5149$$

$$r(t) = 0 \quad \text{for } t_2 = 0.5149 \leq t \leq T = 7$$

If the contractor deals with five suppliers ($n = 5$), then its optimal policy can be determined using Theorem 6.1 as follows:

$$r(t) = 1 \quad \text{for } 0 \le t \le t_1 = 0.408573,$$

$$r(t) \in (0, 1) \quad \text{for } t_1 = 0.408573 < t < t_2 = 0.5004695$$

$$r(t) = 0 \quad \text{for } t_2 = 0.5004695 \le t \le T = 7$$

These numerical examples show that the optimal policies of collaboration level between contractor and suppliers has changed depending on the number of suppliers, thus supporting the model and analytical findings.

## 6.6 Conclusion

Incorrect applications of supply chain collaboration might require greater manufacturing and operation costs and thus increase the price of the product. Moreover, one of the dilemmas for a supply chain contractor involves determining the optimal level of coordination to maximize supply chain profit considering uncertain customer demand and supply chain risk costs based on the number of suppliers.

The model developed in this chapter is the first step toward understanding the complex interrelationships of levels of collaboration, forecast accuracy, and the reduction of supply chain management costs. A continuous time supply chain system profit maximization model using stochastic optimal control theory was developed in this study. The model portrays the dynamic relationships between the contractor's forecasting capability through collaboration with suppliers, consumer product demand, and expected total cost related to supply management and risk. Several analytical results derived through the model generate appropriate managerial guidelines. Among the useful insights derived from the model are the following: A contractor considering risk costs that reflect supply chain uncertainty, as well as the forecast accuracy required, can decide simultaneously the optimal level of collaboration and at a given time.

Necessary and sufficient conditions were examined to determine and demonstrate how the supply chain members can find an optimal level of collaboration. Using an analytical and numerical analysis, supply chain members can set up different levels of collaboration, depending on the initial and terminal marginal profit sought.

To measure forecast accuracy it was assumed that demand was deterministic, which is a limitation of this study. A model that addresses random demand using discrete-time optimal control theory should be developed. Also, the model will become more practical by relaxing several assumptions, such as all suppliers are nearly identical in their capacity and technologies, order quantities are split equally into $n$ contracts, and the competitive bid prices $W(n, r, t)$ and volume discount prices are the same for every supplier. These limitations call for future study.

This chapter provided a managerial decision-making model developed to determine the optimal level of collaboration required between the contractor and its suppliers depending on the demand uncertainty and the number of suppliers in a given business environment. The accuracy of the demand forecast resulting from collaboration between the contractor and its suppliers, the use of long-term contracts, and reduction of overall costs defined as transaction, supplier risk, and supplier innovation expenditures are taken into account in the model. Applying this model, managers can determine the optimal level of collaboration based on the number of suppliers under varying demand schedules and establish optimal strategic policies to maximize supply chain profit.

## References


Arnold, L. (1974), *Stochastic Differential Equations: Theory and Applications*, New York: John Wiley & Sons.

Aviv, Y. (2001), The effect of collaborative forecasting on supply chain performance, *Management Science*, 47(10), 1326–1343.

Aviv, Y. (2002), Gaining benefits from joint forecasting and replenishment processes: the case of autocorrelated demand, *Manufacturing and Service Operations Management*, 4(1), 55–74.

Aviv, Y. (2005), On the Benefits of Collaborative Forecasting Partnerships between Retailers and Manufacturers. Working paper, Olin School of Business, Washington University, St. Louis, MO.

Bakos, J.Y., Brynjolfsson, E. (1994), Information technology, incentives, and the optimal number of suppliers, Electro/*94 International, Conference Proceedings*, 10(2), 540–557.

Barratt, M., and Oliveira, A. (2001), Exploring the experiences of collaborative planning initiatives, *International Journal of Physical Distribution & Logistics Management*, 31(4), 266–289.

Bauknight, D.N. (2000), The supply chain's future in the e-economy and why many may never see it, *Supply Chain Management Review*, (March/April), 28–35.

Bowersox, D.J. (1990), The strategic benefits of logistics alliances, *Harvard Business Review*, 68(4), 36–43.

Cachon, G., and Fisher M. (2000), Supply chain inventory management and the value of shared information, *Management Science*, 46(8), 1032–1048.

Callioni, G., and Billington, C. (2001), Effective collaboration: Hewlett-Packard takes supply chain management to another level, *OM/MS Today*, 28(5), 34–39.

Chen, F. (1998), Echelon reorder points, installation reorder points, and the value of centralized demand information, *Management Science*, 44(12), 221–234.

Choi, T. Y., and Krause, D.R. (2006), The supply base and its complexity: implications for transaction costs, risks, responsiveness, and innovation, *Journal of Operations Management*, 24(5), 637–652.

Chung, K.L. (1974), *A Course in Probability Theory*, 2nd edition, New York: Academic Press.

Clemons, E.K., Reddi, S.P., and Row, M.C. (1993), The impact of information technology on the organization of economic activity: the 'move to the middle' hypothesis, *Journal of Management Information Systems*, 10(2), 9–35.

Deshpande, V., L., and Schwarz, L. (2005). Optimal Capacity Choice and Allocation in Decentralized Supply Chains, Working paper, Purdue University.

Dixit, A.K., and Pindyck, R.S. (1994), *Investment Under Uncertainty*, Princeton, NJ: Princeton University Press.

Fliedner, G. (2003), CPFR: an emerging supply chain tool, *Industrial Management & Data Systems*, 103(1), 14–21.

Gaur, V., Giloni, A., and Seshadri, S. (2005), Information sharing in a supply chain under arma demand, *Management Science*, 51, 961–969.

Gavirneni, S., Kapuscinski, R., and Tayur, S. (1999), Value of information in capacitated supply chains, *Management Science*, 45(1), 16–24.

Govindarajan, V., and Gupta, A.K. (2001), Strategic innovation: a conceptual road-map, *Business Horizons*, 44(4), 3–12.

Handfield, R.B., and Nichols, E.L. (1999), *Introduction to Supply Chain Management*, Englewood Cliffs, NJ: Prentice Hall.

Handfield, R.B., and Bechtel, C. (2002), The role of trust and relationship structure in improving supply chain responsiveness, *Industrial Marketing Management*, 31(4), 367–382.

Holweg, M., and Pil, F.K. (2001), Successful build-to-order strategies start with the customer, *Sloan Management Review*, 43(1), 74–83.

Homburg, C., and Kuester, S. (2001), Toward an improved understanding of industrial buying behavior: determinants of the number of suppliers, *Journal of Business-to-Business Marketing*, 8(2), 5–33.

Jagdev, H.S., and Thoben, K.D. (2001), Anatomy of enterprise collaborations, *Production Planning & Control*, 12(5), 437–451.

Kim, S., and Mahoney, J.T. (2006), Collaborative planning, forecasting, and replenishment (CPFR) as a relational contract: an incomplete contracting perspective, http://www.business.uiuc.edu/Working_Papers/papers/06-0102.pdf

La Londe, B.J., and Masters, J.M. (1994), Emerging logistics strategies: blueprints for the next century, *International Journal of Physical Distribution & Logistics Management*, 24(7), 35–47.

Lambert, D.M., Stock, J.R., and Ellram, L.M. (1998), *Fundamentals of Logistics Management*, Burr Ridge, IL: Irwin/McGraw-Hill.

Larson, P.D., and Kulchitsky, J.D. (1998), Single-sourcing and supplier certification: performance and relationship implications, *Industrial Marketing Management*, 27, 73–81.

Lee, H.L., (2000), Creating value through supply chain integration, *Supply Chain Management Review*, September/October, 30–37.

Lee, H.L., Padmanabhan, V., and Whang, S. (1997), The bullwhip effect in supply chains, *Sloan Management Review*, 38(3), 93–102.

Lee, H.L., So, K.C., and Tang, C.S. (2000), The value of information sharing in a two-level supply chain, *Management Science*, 46(5), 626–643.

Liker, J.K., and Choi, T.Y. (2004), Building deep supplier relationships, *Harvard Business Review*, 82(12), 104–113.

McCarthy, T., and Golicic, S. (2002), Implementing collaborative forecasting to improve supply chain performance. *International Journal of Physical Distribution & Logistics Management*, 32(6), 431–454.

McLaren, T., Head, M., and Yuan, Y. (2002), Supply chain collaboration alternatives: understanding the expected costs and benefits, *Internet Research: Electronic Networking Applications and Policy*, 12(4), 348–364.

Mentzer, J.T., Foggin, J.H., and Golicic, S.G. (2000), Supply chain collaboration: enablers, impediments, and benefits, *Supply Chain Management Review*, 4 (September-October), 52–58.

Narus, J.A., and Anderson, J.C. (1996), Rethinking distribution: adaptive channels, *Harvard Business Review*, 74(4), 112–120.

Nishiguchi, T. (1994), *Strategic Industrial Sourcing: The Japanese Advantage*, New York: Oxford University Press.

Nøekkentved, C. (2000). Collaborative Processes in E-supply Networks, European SAP Centre of Expertise, Pricewaterhouse Coopers.

Pyke, D., Robb, D., and Farley, J. (2000), Manufacturing and supply chain management in China: a survey of state-, collective-, and privately-owned enterprises, *European Management Journal*, 8(6), 577–589.

Ryan, T. P. (2000), *Statistical Methods for Quality Improvement*, 2nd edition, New York: John Wiley & Sons.

Schwarz, L.B. (2004), The state of practice in supply-chain management: a research perspective. In Geunes, J., Akcali, E., Pardalos, P.M., Romeijn, H.E., and Shen, Z.-J. (Max) (Eds.), *Applications of Supply Chain Management and E-Commerce Research*, Dordrecht, Netherlands: Kluwer Academic Publishers, pp. 325–362.

Schwartz, N.L., and Kamien, M.I. (1991), *Dynamic Optimization: The Calculus of Variations and Optimal Control in Economics and Management*, 2nd edition, New York: North-Holland.

Sherman, R. (1998). Collaborative planning, forecasting & replenishment (CPFR): realizing the promise of efficient consumer response through collaborative technology, *Journal of Marketing Theory and Practice*, 6(4), 6–9.

Simatupang, T.M., and Sridharan, R. (2004), A benchmarking scheme for supply chain collaboration, *Benchmarking: An Internal Journal*, 11(1), 9–30.

Simchi-Levi, D., Kaminsky, P., and Simchi-Levi, E. (2008), *Designing and Managing the Supply Chain*, London: McGraw-Hill.

Skjoett-Larsen, T., Thernoe, C., and Andresen, C. (2003), Supply chain collaboration: theoretical perspeective and empirical evidence, *International Journal of Physical Distribution & Logistics Management*, 33(6), 531–549.

Smaros, J. (2005), Information Sharing and Collaborative Forecasting in Retail Supply Chain, doctoral dissertation series 2005/3, Helsinki University of Technology Laboratory of Industrial Management.

Swaminathan, J.M., and Shanthikumar, J.G. (1999), Supplier diversification: effect of discrete demand, *Operations Research Letters*, 24, 213–221.

Terwiesch, C.T., Ren Z.J., Ho, T.H., and Cohen, M.A. (2005), An empirical analysis of forecast sharing in the semiconductor equipment supply chain, *Management Science*, 5(2), 208–220.

Treleven, M., and Schweikhart, S.B. (1988), A risk/benefit analysis of sourcing strategies: single vs. multiple sourcing, *Journal of Operations Management*, 7(4), 93–114.

Zhao, Y. (2002). The value of information sharing in a two-stage supply chain with production capacity constraints: the infinite horizon case, *Manufacturing & Service Operations Management*, 4(1), 21–24.

Zipkin, P., Bramel, J., and Goyal, S. (2000), Coordination of production/distribution networks with unbalanced leadtimes, *Operations Research*, 48(4), 570–577.

# *Chapter 7*

# Online Auction Models and Their Impact on Sourcing and Supply Management

John F. Kros and Christopher M. Keller

## Contents



## 7.1 Introduction

The role that supply management plays in organizations can have a direct impact on firm performance, especially in the area of cost reduction. Advances in information technology, namely the Internet, have made it possible to transform the supply management function in many companies. This transformation has been referred to as e-procurement. E-procurement is defined as the use of Internet technologies for supply management and includes many activities, such as enterprise resource planning applications, online sourcing, and e-auctions (Hartley et al. 2004, Mithas and Jones 2007). This chapter investigates the e-procurement area of online auctions—more specifically, double auctions and their potential benefit to supply managers.

In today's ever-changing business environment, few things remain the same. One of the few exceptions to this rule is the use of auctions to facilitate the transfer of market-sensitive goods in a cost-effective manner (Gallien and Wein 2005, Sanders and Manfredo 2002, Segev et al. 2001). Auctions are commonly used to assist in the sale of commodities such as livestock, petroleum-based products, minerals, and grain products (Feldman and Mehra 1993, Ma and Leung 2007). Governments use auctions to allocate controlled assets such as exploration rights for oil and mining, airwave bandwidth, securities, and for the sale of land and timber in a transparent and equitable manner (Emiliani and Stec 2004, Pekec and Rothkopf 2003, Rothkopf and Park 2001). From a business perspective, many companies use auctions to dispose of excess inventory and equipment (Bapna et al. 2001), for revenue management (Baker and Murthy 2002, Queenan et al. 2007, Caldenteg and Vulcano, in press), and to help reduce supply risk (Zsidisin 2003). At the retail level, automobile dealers, some fresh seafood restaurants, and a variety of small retailers rely on auctions as a means of acquiring or disposing of inventory. More recently a significant portion of the U.S. consumer market has become familiar with online auctions thanks to consumer auction sites such as eBay (Pekec and Rothkopf 2003, Stern and Stafford 2006). As a result, auctions—and more importantly, online auctions—have rapidly gained in popularity and importance (Wilson 2002).

While the theoretical and empirical research bases are well established for face-to-face and reverse auctions, the same cannot be said for online auctions or face-to-face auctions with an online or televised component. Past research has explored both the use and importance of linking auction theory to real transactions (Parente et al. 2004). More importantly, studies such as this have found that forward face-to-face auctions have several limitations. Of primary concern is the finding that the effectiveness of face-to-face auctions may be limited due to time and place constraints that limit participation by both buyers and sellers. Internet-based auctions overcome these limitations by providing a remote access business platform that enables the completion of transactions in a cost-effective manner that is open to more participants (Millet et al. 2004, Segev et al. 2001). Increasing the number of participants included in the bidding process increases competition. This, in turn, should lead to the increased possibility that suppliers, who hold special capabilities, overstock or carry excess inventory, or economies of scale will join the bidding process. Supply managers can, in turn, spend less time looking for suitable suppliers and concentrate on strategic sourcing.

However, to be effective, supply managers of the future are going to need to develop a comprehensive understanding of auctions and their supporting infrastructures in order to help their organizations achieve their strategic and profit directives (Kraljic 1983, Johnson and Leenders 2003). Consequently, the purpose of this chapter is to help readers, and especially supply managers, develop a more comprehensive understanding of online auctions, their theoretical underpinnings, and how online auctions can be effectively used.

## 7.2 Literature Review

Auctions are allocative market-making mechanisms that play a valuable role in the price discovery process and are most useful when the item's price is variable or when the seller is unsure of the item's true market value. Auctions can be for single units or for lots of homogeneous commodities (Feldman and Mehra 1993, Sanders and Manfredo 2002). Two of possibly the most important advances in the business world related to auctions during the past two centuries have been the development of Boards of Trade and the Internet. Boards of Trade such as the Chicago and Kansas City Boards of Trade are market-making mechanisms that facilitate the transfer of commodities by linking multiple buyers and sellers simultaneously (Kroll 1972). By simultaneously linking buyers and sellers from around the world, Boards of Trade are able to establish the true value of a given grade of a commodity in a given market. Price variances, both above and below this spot price, may arise due to differences in quality, the distance from the spot, market, and localized supply and demand factors (Kroll 1972, Sanders and Manfredo 2002).

The second most important development that has helped advance the use and importance of auctions has been the Internet. As the Internet has gained in importance, there has been a proliferation of websites dedicated to B2B auctions, thanks to their ability to potentially reduce transaction costs for both buyers and sellers, something that can enhance overall supply chain profitability (Millet et al. 2004, Parente et al. 2004).

Some past research has found that large companies throughout the United States and Europe are currently using electronic auctions to source between 5 and 25 percent of their total purchase volumes (Kaufmann and Carter 2004). In addition to favorable cost structures and information availability, auctions, in general, benefit from transparency and serve to level the playing field for large and small buyers and sellers. Auctions can also help create foundations of trust because they are transparent and all participants theoretically have similar access to information and are treated equitably (Griffiths 2003, Pekec and Rothkopf 2003).

With the advent of the Internet, automated or "online" auctions have come to the forefront of theoretical and empirical auction research (Lucking-Reiley 2000, Millet et al. 2004). Consequently, there is a long and rich history of research on auction theory (see Emiliani 2000). While this body of knowledge appears to be extensive, few researchers have investigated double auction markets (i.e., Boards of Trade) and their impact on sourcing and supply management. A few of the studies that have sought to address this weakness in the literature include those by Feldman and Mehra (1993) and Friedman and Rust (1993), who discussed the institutions and theory underlying double auctions. Smith et al. (2003) and Luckock (2003) also provided discussions on double auction models.

## 7.3 Types of Auctions

This chapter focuses on three of the most commonly used auction formats: English, Dutch, and Vickrey auctions. Of these, the most commonly known and used is the open-outcry, price-ascending English auction. Under the English auction format, an auctioneer or auction holder solicits bids of increasing value until no higher bid is received. The individual or organization that offers the highest bid wins the auction provided that the maximum bid exceeds the seller's minimum bid threshold (Feldman and Mehra 1993).

Dutch auctions are a reversed version of an English auction. These auctions are also referred to as buyers auctions and at times reverse auctions. They are open outcry, descending-price auctions. In this case, the auctioneer or auction holder solicits bids of decreasing value until no lower bid or offer to sell is received. In this case the seller who offers to sell their goods at the lowest price wins the auction and transfers title to the purchaser (Feldman and Mehra 1993). With reverse auctions, purchasing agents issue an interest in buying a certain quantity of an item or commodity on an Internet site and sellers compete by submitting successfully lower bids until there is only one seller left. Over the past decade, reverse auctions have become a high-priority area for research, and many have investigated the role that reverse auctions play in the supply management function (see Richards 2000, Tully 2000, Judge 2001, Stein et al. 2003). Some researchers have suggested that reverse auctions are effective and reduce purchase prices (Judge 2001, Reason 2001, Grant 2003, Emiliani and Stec 2005), while others have argued that reverse auctions damage supplier relationships and create distrust among suppliers (Jap 2002, Kobe 2001, Tulder and Mol 2002, Smart and Harrison 2003, Smeltzer and Carr 2003).

Vickrey auctions are sometimes referred to as second-price auctions. Under the Vickery auction format, the highest (lowest) bidder is assigned the right to buy (sell), but the price of the transaction is set at the second-highest (-lowest) price depending on whether the English or Dutch auction is used.

English, Dutch, and Vickrey auctions are structured so that either buyers or sellers have market power and are able to set the terms of the auction. An important element of any of these auction formats is the "winner's curse," which comes in one of two forms and depends on the auction format. Under the English auction format, the seller can appropriate a maximum trade surplus (a "curse" on the buyer) by having buyers alone compete (Dyer and Kagel 1996). Conversely, sellers in a Dutch auction may be similarly "cursed" because the buyer can appropriate the maximum trade surplus by having sellers alone compete. Although the topics of English, Dutch (reverse), and Vickrey auctions are hotly debated areas, little research has emerged in the area of online double auctions.

### 7.3.1 Double Auctions

In cases where there are many potential sellers and buyers that are separated by potentially great distances, none of the aforementioned auction formats may be appropriate. The advent of the Internet has nullified the issue of distance regarding auction format, and various types of online auctions have emerged. Boards of Trade (e.g., New York Stock Exchange or Chicago Board of Trade) are probably the most commonly known and understood type of double auction. Under the double auction format, both buyers and sellers submit respective offers to buy and to sell. Bids are matched either manually, in the case of face-to-face auctions (e.g., floor trading), or on a first-come, first-served basis in the case of online auctions. The process of matching buyers and sellers generates a spot or bid price, thereby establishing a market value for a given commodity (Sanders and Manfredo 2002). Double auctions are especially effective at establishing true market values for commodities but were found to be of limited value for unique or specialty goods that have limited markets in terms of the total number of buyers and sellers (Feldman and Mehra 1993).

A powerful application of this type of auction is in the e-procurement arena. Griggs and Wild (2003) discussed the utility of double auctions, specifically the double auctions' ability to (1) establish a true market price, (2) fairly represent anonymous buyers and sellers, (3) provide continuous real-time information to all participants, (4) create markets that might not have occurred otherwise, (5) provide market oversight and process management, and (6) ensure orderly transaction flow. All of the areas that Griggs and Wild proposed would be beneficial to supply managers and would tend to lower transaction costs. In the case of commodities or near commodities, there are few buyer-supplier-specific attributes aside from price that influence the transaction. Therefore, the first area, establishing a true market price, would dramatically aid supply managers. The online auction would literally aggregate supply and demand, determine a market clearing price or statistical price point that clears the market, and determine the "true" value of the product.

Current examples of online auctions include the IMX Exchange (brings mortgage brokers and wholesale lenders together in a neutral, online, real-time trading network); Exostar (matches supply and demand across the aerospace and defense industry); ChemConnect (a global over-the-counter marketplace for midstream energy, chemicals, and plastics); or the NTE (matches submitted demand for origin-to-destination route with supply and determines a price for each transaction).

In response to the need to bring multiple buyers and sellers together within the procurement and supply area, there has been a response by companies such as Freemarkets, Manhattan Associates, and CombineNet to offer support in the design and implementation of double auctions. With the emergence of more auction-based solutions, supply managers need to understand the pricing mechanism underlying the double auction market.

### 7.3.2 A Statistical Formulation of Price Theory

Auction theory can be understood using probability distributions to represent buyer and seller preferences (Keller 2006). Each individual seller, $i = 1, \ldots, S$, has an ex-ante expectation of the price of a sale, $f_i(p) \geq 0$. An aggregate sellers' price distribution can be obtained by averaging the individual seller's distributions:

$$f(p) = \frac{\sum_{i=1}^{S} f_i(p)}{S}$$

Each individual buyer, $j = 1, \ldots, B$, has an ex-ante expectation of the price of a sale ($g_i(p) \geq 0$). An aggregate buyers' price distribution can be obtained by averaging the individual buyer's distributions:

$$f(p) = \frac{\sum_{j=1}^{B} g_i(p)}{B}$$

The sellers' proportion of the total market of buyers and sellers combined is

$$\alpha = \frac{S}{S+B} \geq 0$$

The buyers' proportion of the total market is $1 - \alpha > 0$.

#### 7.3.2.1 The Market Price Statistic

The ex-ante expectations of trading price are a mixture of the expectations of sellers and buyers. The observed prices of trade are, however, not observations of either of these expectations. Rather, the observed prices of trade are observations of the simultaneity solution of the mixture model. The simultaneity solution is the equilibrium market price and a statistic of the mixture of the buyers' and sellers' distributions.

The supply function is then the ex-ante probability of trade from a seller's perspective and is represented by a proportion of the sellers' cumulative probability distribution, $\alpha F(p) \geq 0$. The demand function is the ex-ante probability of trade from a buyer's perspective and is represented by a proportion of the unitary complement of a cumulative probability distribution ()$(1 - \alpha)(1 - G(p)) \geq 0$. The supply function has a non-negative slope (i.e., $\alpha F'(p) = \alpha f(p) \geq 0$) while the demand function has a non-positive slope (i.e., $(1 - \alpha)(1 - G(p))' = -(1 - \alpha)G'(p) = -(1 - \alpha)g(p) \leq 0$). Therefore, the simultaneous solution of the demand and supply functions is

$$p^* \quad s.t. \quad \alpha F(p^*) = (1 - \alpha)(1 - G(p^*))3.7.3$$

In the individual trade model, the simultaneity solution was suggested as one possible balance between the probabilities of a single buyer and a single seller while other trade solutions were possible. In this aggregate trade model, the simultaneity solution is not so ambiguous. The simultaneity solution is the equilibrium market price since the balance is not now probabilities of single individuals but rather the balance is the expected number of individual sellers and individual buyers. That is, the simultaneity solution is the only solution in which the number of successful buyers equals the number of successful sellers, a condition that is obviously necessary in equilibrium. This model may be transformed into standard economic models of quantity supplied and quantity demanded by using the convolution of the distributions rather than the average of the distributions.

### 7.3.3 Market Structure Affects the Market Price Statistic

It is important to consider the application of the theoretical structure developed above. Using a supply function as the cumulative normal distribution with mean of 50 and standard deviation of 10, and the demand function as the unitary complement of the cumulative normal distribution with mean of 40 and standard deviation of 20,

$$
\begin{aligned}
\Pr(\mathit{buyers} \cup \mathit{sellers}) &= \Pr(\mathit{buyers}) + \Pr(\mathit{sellers}) - \Pr(\mathit{sellers} \cap \mathit{buyers}) \\
\Pr(\mathit{buyers} \cup \mathit{sellers}) &= \Pr(\mathit{buyers}) + \Pr(\mathit{sellers}) - \Pr(\mathit{buyers}) * \Pr(\mathit{sellers}) \\
\Pr(\mathit{buyers} \cup \mathit{sellers}) &= \frac{1}{2} * 36.9\% + \frac{1}{2} * 36.9\% - \frac{1}{2} * 36.9\% * \frac{1}{2} * 36.9\% \\
&= 33.5\%
\end{aligned}
$$

This result may also be obtained by simulation. Sample the ask price from the implied supply density function, $A \sim n(50, 10)$, and sample the bid price from the implied demand density function, $B \sim n(40,20)$. Match the pairs of sampled asks and bids to first determine whether or not a trade occurs (i.e., if $A \leq B$, then a trade occurs, otherwise a trade does not occur). The results for a simulation with 10,000 trials each are as follows:.

| | *Trade %* | *Average Trade Price* | *Trade Price St. Dev. (% of Price)* |
|---|---|---|---|
| English | 33.3 | \$59.65 | \$12.91 (22) |
| Double | 32.7 | \$52.43 | \$9.69 (19) |
| Dutch | 32.5 | \$45.08 | \$9.16 (20) |

In considering market structure, the above results of a series of idiosyncratic trades must be compared with a well-functioning market that operates as a clearinghouse of multiple buyers and sellers. For multiple simultaneous buyers and sellers, let m and n represent the number of offers from buyers and from sellers, respectively. These bids and asks are independently ordered, asks in ascending order and bids in descending order. The highest bid is matched with the lowest ask and a trade occurs. The price of the trade is a value in the interval $[A_l, B_l]$, where the index l is such that $A_l \leq B_l$ and $A_{l+1} \geq B_{l+1}$. For a simple introduction to this reverse sorting method, see Dixit and Skeath (1999, Chapter 17); Friedman and Rust (1993) also provide many technical variations.

Using batch sizes of 1,000 buyers and 1,000 sellers simultaneously yields the following results, in ten treatments with a total of 10,000 buyers and sellers:

| | *Trade Percentage (%)* | *Average Trade Price* | *Trade Price St. Dev. (% of Price)* |
|---|---|---|---|
| English | | \$46.48 | \$0.56 (1.2) |
| Double | 36.7 | \$46.46 | \$0.56 (1.2) |
| Dutch | | \$46.43 | \$0.56 (1.2) |

Three things are clear from comparing the simulations. The trading percentage approaches the theoretical prior probability of 36.9 percent of successful trades. This improved trading percentage suggests that clearinghouse trades (as opposed to idiosyncratic trading) increases trade. Second, the nearly identical values for the various values of k indicate that the double-auction clearinghouse mechanism mitigates any potential market distortions caused by either the buyers or the sellers being able to unilaterally control the price of trade. Third, the standard deviation of the trading price is greatly reduced. This variability reduction in prices improves the predictability and planning of prices and is a superior system.

### 7.3.4 Theoretical and Managerial Implications

From a theoretical perspective, the findings presented in this study are important for three reasons:

1. The findings empirically demonstrate the market clearing mechanism.
2. The findings explain why individual buyers and sellers generally lack the ability to manipulate the market.

3. The findings demonstrate why commodities generally trade within narrow bands and that managers can use auction prices as a guide when determining how much to ask or offer for a given commodity.

For commodities, a repeated or continuous double auction may be used (Ma and Leung 2007, Ugtelingum et al. 2008).

In general, supply managers report that auctions provide valuable market foresight, allowing for improved management of corporate spending and time (Vigoroso 1999). One of the principal insights in repeated or continuous auctions that may be used for revenue management is that buyers with fixed time requirements pay a premium over the generally available prices (Baker and Murthy 2002). For supply managers, this means that any and all purchase time requirements should be carefully scrutinized.

As illustrated by the simulation, many parameters of the structure of the auction do not affect buyer surplus. This point was confirmed by Mithas and Jones' (2007) research of more than 700 procurement auctions. In particular, strategies regarding bidding based on estimating other buyer preferences may have little effect despite its research popularity (Peters and Severinov 2006, Shen and Su 2007, Jap and Naik 2008). In addition, automated trading strategies have grown out of this lack of buyer preference effect (Beam and Segev 2008, Ma and Leung 2007).

## 7.4 Research Implications and Conclusion

In this chapter the authors have attempted to provide readers with a clearer understanding of online auctions, auction theory, the statistical formulations that enable auctions to function effectively, and how supply managers can use auctions to help their organizations achieve earnings targets more effectively. In addition to the significant managerial implications, this chapter has provided important avenues for future research. There is a need to develop training and continuing education programs to teach supply managers about how to use online auctions to minimize risk within their supply chains. In turn, this research illustrates the need to develop a better understanding of how businesses can better use real-time auction prices to facilitate strategic planning.

With the proliferation of the Internet and electronic marketplaces, the presence and use of online auctions are expected to increase. However, the decision support buyers and sellers need to use these supply channels effectively and efficiently are still being developed. Therefore, for a supply chain manager, understanding and modeling the buyer's and seller's bidding behavior are key to understanding how auctions work and if and how they may be of benefit. This research presents a general method for modeling a buyer's and seller's bidding behavior using probability of trade functions.

## References

Baker, T., and Murthy, N. (2002), A framework for estimating benefits of using auctions in revenue management, *Decision Sciences*, 33(3), 385–413.

Bapna, R., Goes, P., and Gupta, A. (2001), Insights and analysis of online auctions, *Communications of the ACM*, 44(11), 43–50.

Beam, C., and Segev, A. (2008), Automated Negotiation: A Survey of the State of the Art, working paper 08-WP-1022, available at URL: http://www.cse.iitb.ac.in/dbms/Data/Papers-Other/Ecommerce/wp-1022.pdf

Caldenteg, R., and Vulcano, G. (2007), Online auction and list price revenue management, *Management Science* 53, 795–813.

Dixit, A., and Skeath, S. (1999), *Games of Strategy*, New York: W.W. Norton & Company.

Dyer, D., and Kagel, J. H. (1996), Bidding in common value auctions: how the commercial construction industry corrects for the winner's curse, *Management Science*, 42(10), 1463–1475.

Emiliani, M.L. (2000), Business-to-business online auctions: key issues for purchasing process improvement, *Supply Chain Management: An International Journal*, 5(4), 176–186.

Emiliani, M.L., and Stec, D.F. (2004), Aerospace parts suppliers' reaction to online reverse auctions, *Supply Chain Management: An International Journal*, 9(2), 139–153.

Emiliani, M.L., and Stec, D.F. (2005), Wood pallet suppliers' reaction to online reverse auctions, *Supply Chain Management: An International Journal*, 10(4), 278–285.

Feldman, R.A., and Mehra, R. (1993), Auctions: theory and applications, *International Monetary Fund Staff Papers*, 40(3), 485–511.

Friedman, D. and Rust, J. (1993), Preface. In Friedman, D. and Rust, J. (Eds.), *The Double Auction Market: Institutions, Theories, and Evidence*, A proceedings volume in the Santa Fe Institute Studies in the Sciences of Complexity, Sante Fe, NM.

Gallien, J., and Wein, L.M. (2005), A smart market for industrial procurement with capacity constraints, *Management Science*, 51(1), 76–91.

Grant, J. (2003), The heat is on for DaimlerChrysler costs, *Financial Times*, September 3.

Griffiths, A. (2003), Trusting an auction, *Supply Chain Management: An International Journal*, 8(3), 190–194.

Griggs, K., and Wild, R. (2003), Intelligent support for sophisticated e-commerce services, *e-Service Journal*, 2(2), 87–104.

Hartley, J.L., Lane, M.D., and Hong, Y. (2004), An exploration of the adoption of E-auctions in supply management, *IEEE Transactions on Engineering Management*, 15(2) 153–161.

Jap, S. (2002), Online reverse auctions: issues, themes, and prospects for the future, *Journal of the Academy of Marketing Science*, 30(4), 506–525.

Jap, S., and Naik, P.A. (2008), Bid analysis: a method for estimation and selection of dynamic bidding models, *Marketing Science*, 27(6), 949–960.

Johnson, P.F., and Leenders, M.R. (2003), Gaining and losing pieces of the supply chain, *Journal of Supply Chain Management*, 39(1), 27–39.

Judge, P. (2001), How I saved $1000m on the web, *Fast Company*, 43, 174–181.

Kaufmann, L., and Carter, C.R. (2004), Deciding on the mode of negotiation: to auction or not to auction electronically, *The Journal of Supply Chain Management: A Global Review of Purchasing and Supply*, 40(2), 15–26.

Keller, C. (2006), Modeling dynamic sealed-offer k-double auctions using a mixture of distributions, *Mathematical and Computer Modeling*, 44, 43–48.

Kobe, G. (2001), Supplier squeeze, *Automotive Industries*, March.

Kraljic, P. (1983), Purchasing must become supply management, *Harvard Business Review,* (September–October), pp. 109–117.

Kroll, S. (1972), Commodity hedging for insurance and profit, *The CPA Journal,* 42, 303–307.

Lucking-Reiley, D. (2000), Auctions on the Internet: what's being auctioned, and how?, *Journal of Industrial Economics,* 48(3), 227–252.

Luckock, H. (2003), A steady-state model of the continuous double auction, *Quantitative Finance,* 3, 385–404.

Ma, H., and Leung, H. (2007), An adaptive attitude bidding strategy for agents in continuous double auctions, *Electronic Commerce Research and Applications,* 6(4), 383–398.

Millet, I., Parente, D.H., Fizel, J.L., and Venkataraman, R.R. (2004), Metrics for managing online procurement auctions, *Interfaces,* 34(3), 171–179.

Mithas, S., and Jones, J.L. (2007), Do auction parameters affect buyer surplus in E-auctions for procurement?, *Production and Operations Management,* 16(4), 455–470.

Parente, D.H., Venkataraman, R., Fizel, J., and Millet, I. (2004), A conceptual research framework for analyzing online auctions in a B2B environment, *Supply Chain Management: An International Journal,* 9(4), 287–294.

Pekec, A., and Rothkopf, M.H. (2003), Combinatorial auction design, *Management Science,* 49(11), 1485–1503.

Peters, M., and Severinov, S. (2006), Internet auctions with many traders, *Journal of Economic Theory,* 130, 220–245.

Queenan, C.C., Ferguson, M., Higbie, J., and Kapoor, R. (2007), A comparison of unconstraining methods to improve revenue management systems, *Production and Operations Management,* 16(6), 729–746.

Reason, T. (2001), Looking for raw deals, *CFO Magazine,* available at www.cfo.com/printarticle/0,5317,1895%7CA,00.html?f=options (accessed June 18, 2004).

Richards, B. (2000), Dear supplier: this is going to hurt you more that it hurts me . . . , *Ecompany Now,* 1(1), 136–142.

Rothkopf, M.H., and Park, S. (2001), An elementary introduction to auctions, *Interfaces,* 31(6), 83–97.

Sanders, D.R., and Manfredo, M.R. (2002), The role of value-at-risk in purchasing: an application to the foodservice industry, *Journal of Supply Chain Management: A Global Review of Purchasing and Supply,* 38(2), 38–45.

Segev, A., Beam, C., and Shanthikumar, J.G. (2001), Optimal design of Internet based auctions, *Information Technology and Management,* 2(2), 121–163.

Shen, Z., and Su, X. (2007), Customer behavior modeling in revenue management and auctions: a review of new research opportunities, *Production and Operations Management,* 16(6), 713–728.

Smart, A., and Harrison, A. (2003), Online reverse auctions and their role in buyer-supplier relationships, *Journal of Purchasing and Supply Management,* 9(5-6), 257–268.

Smeltzer, L., and Carr, A. (2003), Electronic reverse auctions: promises, risks, and conditions for success, *Industrial Marketing Management,* 32(6), 481–488.

Smith, E., Farmer, J.D., Gillemot, L., and Krishnamurthy, S. (2003), Statistical theory of the continuous double auction, *Quantitative Finance,* 3, 481–514.

Stein, A., Hawking, P., and Wyld, D. (2003), The 20% solution? A case study of the efficacy of reverse auctions, *Management Research News,* 26(5), 1–20.

Stern, B.B., and Stafford, M.R. (2006), Individual and social determinants of winning bids in online auctions, *Journal of Consumer Behavior,* 5(1), 43–55.

Tulder, R., and Mol, M. (2002), Reverse auctions or auctions reversed? First experiments by Philips, *European Management Journal,* 20(5), 447–456.

Tully, S. (2000), The B2B tool that really is changing the world, *Fortune,* 141(6), 132–145.

Ugtelingum, P., Cliff, D., and Jennings, N.R., (2008), Strategic bidding in continuous double auctions, *Artificial Intelligence,* 172, 1700–1729.

Vigoroso, M. (1999), Are Internet auctions ready to gear up?, *Purchasing,* 126, 85–86.

Wilson, T. (2002), GE expands private hub to woo users, *Internetweek,* 890, 1, 38.

Zsidisin, G.A. (2003), Managerial perceptions of supply risk, *The Journal of Supply Chain Management: A Global Review of Purchasing and Supply,* 93(1), 14–25.

# Chapter 8

# Analytical Models for Integrating Supplier Selection and Inventory Decisions

Burcu B. Keskin

## Contents



## 8.1 Introduction

With increasingly competitive world markets, companies are under intense pressure to improve their core operations. To achieve improved performance, many companies have introduced and implemented a wide array of strategies, including supplier-managed inventories, vertical integration, reduction of suppliers, tight partnerships with their suppliers with long-term purchasing contracts, integration of purchasing decisions within the production planning process, etc. Additionally, as the cost of raw materials, component parts, and services purchased from external suppliers has been increasing, the use of outsourcing by both manufacturing and service industries have garnered interest. In this context, the evaluation and selection of competent suppliers have also garnered interest for most manufacturing firms and service providers.

Evaluating potential suppliers includes several criteria on financial condition, quality, capacity, service capability, and information technology capability. Among those criteria, net price offered by the supplier after the discounts and freight considerations, quality specifications, and delivery schedules and performance history receive the highest attention (Aissaoui et al. 2007). In general, evaluation and selection of suppliers are considered the main duties of the purchasing department (Dobler et al. 1990). After one or more suppliers have passed the evaluation process, the supplier selection process begins. This process may involve bidding or supplier proposals. When a firm considers which components, systems, or services it should make and which it should buy, it should analyze the issue at two levels: strategic and tactical (or operational). However, most of the time, in the supplier selection process, next to strategic supplier selection decisions, tactical inventory and transportation decisions are overlooked in practice and in the literature. This is primarily due to *not involving* operations/logistics managers in the supplier selection process. Nevertheless, as the "make/buy" decision continues to be one of the key strategic issues, the research that investigates the interaction of supplier selection and inventory/transportation decisions will have lasting importance, especially in terms of total cost and delivery performance. In this chapter we present analytical models that highlight this interaction and discuss potential solution approaches.

The importance of this problem lies in the fact that most manufacturers have come to realize that their ability to be competitive on cost, quality, timeliness, and service depends on their suppliers' ability to contribute. Not all suppliers need development assistance, but in most major corporations such as General Electric, John Deere, Chrysler, and Honda of America, managers help their suppliers "improve quality, enhance delivery performance, and reduce costs." Since selecting and managing suppliers is a fixed cost per supplier for the companies, this actually motivates supply base reduction by reducing the variety of items procured and consolidating items previously procured from different suppliers into one supplier. Xerox reduced its supply base by 92 percent by the early 1980s, from 5,000 suppliers to 400 suppliers. Chrysler reduced its supplier base from 2,500 to 300. Applied Materials reduced its supply base from 1,200 suppliers in the early 1990s to 400 by 2001. IBM uses 50 suppliers for 85 percent of its production requirements, and Sun Microsystems uses 40 suppliers for 90 percent of its production material needs. For companies with a reduced supply base, the integration of inventory/transportation considerations with supplier selection decisions appears as an additional cost-saving opportunity.

Another important example that highlights the importance of integrated supplier selection models is from the automotive industry. The top five automotive manufacturers, on average, currently process approximately $250 billion worth of raw materials and $59 billion worth of work-in-process inventory. Currently, 30 percent of the cost of an automobile is derived from inventory staged in a warehouse, waiting to be processed (Benton 2007). In essence, the vast majority of the cost for an automotive are non-value-added inefficiencies in the manufacturing process, which is one of the primary reasons why these automakers are experiencing declining cost efficiency levels. This is another context where an integrated supplier selection inventory modeling and transportation can be useful.

Before we discuss the analytical models, we will highlight the important components of the total cost concept for integrated supplier selection and inventory/transportation models. The total cost consists of three components:

1. *Acquisition costs:* purchase price (including discounts), quality costs, planning costs, cost of managing suppliers
2. *Ownership costs:* downtime (stock-out) costs, inventory replenishment (order fulfillment costs), inventory holding (cycle stock) costs, distribution costs, risk (safety stock) costs
3. *Post-ownership costs:* environmental costs, warranty costs, customer dissatisfaction costs

In this chapter we explicitly consider the acquisition and ownership costs as well as the integrated decisions related to supplier selection, inventory control, and transportation.

## 8.2 Analytical Models

The analytical models covered in this chapter fall into two main categories: (1) multiple suppliers, single buyer models; (2) multiple suppliers, multiple buyers models. We discuss the technical and managerial aspects of each category in detail.

### *8.2.1 Multi-Supplier, Single-Buyer Models*

Multi-supplier, single-buyer models are the most common problem setting addressed in the supplier selection literature. This is mainly due to having decentralized purchasing operations within several companies of the same firm. Our integrated supplier selection and inventory model focuses on a two-stage supply chain consisting of $N$ suppliers ($j = 1, \ldots, N$) and one buying firm. All $N$ suppliers are assumed to have been prescreened by the firm and are thus included in the supplier base. Within this setting, one of the concerns while determining which supplier(s) to work with is the type of sourcing, that is, single sourcing, dual sourcing, or multi-sourcing.

Single sourcing may be justified when (1) lower total cost results from a much higher volume (economies of scale); (2) the buying firm has more influence with the supplier; (3) lower costs are incurred to source, process, expedite, and inspect; (4) significantly lower freight costs may be obtained; and (5) more reliable, shorter lead times are required (Benton 2007). When single sourcing is enforced, the underlying integrated supplier selection model boils down to the single-supplier, single-buyer model, which is the most stylistic model for the supply chain coordination models. In Section 8.2.1.1, we summarize some of those integrated inventory/transportation models.

On the other hand, dual or multiple sourcing would be more appropriate when (1) the buyer requires some protection from shortages, strikes, and other disruptions of the supplier; (2) the buyer wants to maintain competition and has a backup source; (3) the buyer is constrained by a supplier's capacity issues; (4) local content is required for, especially for more global companies, international manufacturing locations. Dual and multi-sourcing bring up the issues of order splitting, order assignments, inventory, and transportation considerations. We highlight some analytical models that address these problems in Section 8.2.1.2.

### *8.2.1.1 Single-Sourcing Models*

Single-sourcing models, despite increasing attention given to JIT systems, are still rare in the literature. The methods utilized for selecting suppliers under single-sourcing restrictions include (Ghodsypour and O'Brien, 1998)

- Linear weight point methods that depend on human judgment
- The cost ratio, where cost of each criterion for each supplier is calculated with respect to total cost
- The analytical hierarchy process that is a pairwise comparison of each criteria

Unfortunately, there are not many mathematical models dedicated to supplier selection with single sourcing except those of Benton (1991) and Akinc (1993). Benton (1991) developed a nonlinear programming formulation using the economic ordering quantity (EOQ) model to represent the inventory decisions. In that article, the buyer's objective was to minimize the sum of purchasing cost, inventory cost, and ordering costs subject to an aggregate inventory investment constraint and an aggregate storage limitation constraint by selecting a single supplier for multiple items from a set of multiple suppliers offering all-unit quantity discounts. A heuristic procedure using Lagrangian relaxation for supplier selection, lot sizing resource limitations, and all-unit quantity discounts was developed. Akinc (1993), on the other hand, developed a decision support approach for selecting suppliers under the conflicting criteria of minimizing the annual procurement costs, reducing the number of suppliers, and maximizing suppliers' delivery and quality performances. Because it is challenging to quantify the delivery and quality performances, rather than trying to estimate a utility function for these criteria, the author focused on building three simple models and assessed the different trade-offs within these models.

Generalizing the results of these two aforementioned papers, we now describe a simple mathematical model that can be utilized for an integrated supplier selection and an inventory model under single sourcing. For this purpose, we first define the following notation:

$D$ Annual demand of the buyer
$K$ Inventory ordering cost of the buyer
$h$ Annual inventory holding cost rate of the buyer
$c_j$ Unit procurement cost offered by supplier $j$, $j = 1, \ldots, N$
$f_j$ Annual contractual cost required by supplier $j$, $j = 1, \ldots, N$
$t_j$ Per-mile transportation cost from supplier $j$ to the buyer
$d_j$ Distance between supplier $j$ and the buyer
$u_j$ Per-unit transportation cost from supplier $j$ to the buyer
$W_j$ Annual throughput capacity of supplier $j$

The decision variables are

$Q_j$ Order quantity of supplier $j$, $j = 1, \ldots, N$, $Q_j \geq 0$
$X_j$ 1 if supplier $j$ is used at all, 0 otherwise

Then, the mathematical model is formulated as

$$\min \sum_{j=1}^{N} (c_j + u_j) D X_j + \sum_{j=1}^{N} f_j X_j + \sum_{j=1}^{N} \left\{ \frac{KD}{Q_j} + \frac{1}{2} h Q_j \right\} X_j$$

$$+ \sum_{j=1}^{N} \frac{t_j d_j D}{Q_j} X_j \quad \text{(MS-SB-SS)}$$

subject to

$$\sum_{j \in \mathcal{J}} X_j = 1 \tag{8.1}$$

$$D \leq W_j X_j, \qquad \forall j = 1, \ldots, N \tag{8.2}$$

$$X_j = \{0, 1\} \text{ and } Q_j \geq 0, \qquad \forall j = 1, \ldots, N \tag{8.3}$$

The objective function minimizes the total cost of selecting a supplier including

- Annual procurement and unit-based transportation costs (first term)
- Annual fixed contractual costs (second term)

- Annual inventory ordering and holding (third term)
- Annual distance-based (and trip-based) transportation costs (final term)

Constraints (8.1) and (8.3) establish the single sourcing requirements, whereas Constraints (8.2) establish that the selected supplier should have enough capacity to serve the demand of the buyer.

Note that given $X_j = 1$ for a potential supplier $j$, $j = 1, \dots, N$, the remainder of the problem becomes a generalized EOQ formulation with generalized transportation costs. The order quantity for supplier $j$ is given as

$$Q_j = \sqrt{\frac{2[K + t_j d_j] D}{hc_j}} \tag{8.4}$$

Then, the total cost of selecting supplier $j$ would be

$$C_j = (c_j + u_j) D + f_j + \sqrt{2[K + t_j d_j] Dhc_j} \tag{8.5}$$

Finally, the supplier that has enough capacity should be selected if $j^* = \arg\min_{j=1}^{N}\{C_j : D \leq W_j\}$ (i.e., $X_{j*} = 1$), granted that the capacity requirements are satisfied.

With this transformation all of the generalization models developed for EOQ, including quantity discounts, storage and space constraints, and transportation and cargo restrictions, are applicable and usable for the integrated supplier selection and inventory problem with single-sourcing restrictions.

### 8.2.1.2 Dual- and Multi-Sourcing Models

The buying firm, before implementing a multi-sourcing policy, must specify the number of suppliers to employ. This is equivalent to selecting a subset of suppliers from a predetermined set that conforms the buyer's quality, delivery, and financial requirements. The buyer then needs to determine the distribution of order quantities among selected suppliers. Some authorities have suggested that two suppliers should be used for a material category that has an annual dollar volume higher than a given predetermined level, which is also known as dual sourcing. Some others require multiple suppliers even though managing more than one source is obviously more cumbersome than dealing with a single source. Web-based SCM applications enable closer management of diverse suppliers, streamline supply chain processes, and drive down procurement costs. Furthermore, multiple suppliers provide more flexibility under dire disruption conditions. For this purpose, multi-sourcing models have received more attention than their single-sourcing counterparts.

Due to its ability to optimize the explicitly stated objective subject to a multitude of constraints, mathematical programming is the most appropriate technique that allows the decision maker to formulate such multi-sourcing integrated supplier selection and inventory problems (Aissaoui et al. 2007). It considers internal policy constraints and externally imposed system constraints placed on the buying process in order to determine an optimal ordering and inventory policy simultaneously while selecting the best combination of suppliers. Due to these advantages, substantial work has been done in this area and a great number of studies is conducted that considered different aspects and instances of the problem since the 1990s. Among all the research in this area, four articles deserve special credit, including the articles by Pan (1989), Chaudhry et al. (1993), Ghodsypour and O'Brien (2001), and Tempelmeier (2002), due to their integrated supplier selection and inventory consideration. Furthermore, the two analytical models we present in this section are based on these articles. We discuss the contributions of these works in detail next.

Pan (1989) presented a linear programming model that can be used in the simultaneous determination of the number of suppliers to utilize and the purchase quantity allocations among suppliers in a multiple sourcing system a single objective linear programming model to choose the best suppliers, in which three criteria are considered—price, quality, and service. The total cost is taken into account as an objective function, and quality and service are considered as constraints. Chaudhry et al. (1993) considered the price, delivery, and quality objectives of the buyer, as well as the production or rationing constraints of the suppliers. In that work, suppliers were assumed to offer price breaks that depend on the sizes of the order quantities. They presented linear and mixed binary integer programming models that provide unifying frameworks for models of supplier performance measures. In a similar context, Ghodsypour and O'Brien (2001) developed a single-objective supplier selection model to minimize the total cost of logistics, including aggregate price, ordering, and inventory costs, subject to suppliers' capacity constraints and the buyers' limitations on budget. Our first analytical model is based on these articles, considering the EOQ-based inventory model together with multiple-supplier selection.

The final article by Tempelmeier (2002) considered the problem of supplier selection and purchase order sizing for a single item under dynamic demand conditions. Suppliers offer all-units and/or incremental quantity discounts that may vary over time. Time-varying deterministic prices were dictated by time-varying (all-units or incremental) quantity discount structures with several discount levels. This includes rising or falling prices, starting with a specific period, as well as special prices for limited time intervals. The model formulation is based on the well-known analogy between the plant location problem and the dynamic lot sizing problem. Our second analytical model is based on Tempelmeier (2002).

#### 8.2.1.2.1 Multi-Supplier, Single-Buyer with Multi-Sourcing Model 1

In addition to the notation introduced in the previous section, we define the following decision variables:

$Q$ Total order quantity ordered from all the selected suppliers
$Q_j$ Order quantity ordered from supplier $j$, $j = 1, \ldots, N$
$Y_j$ Percentage of $Q$

The revised mathematical model is then given as

$$\min \sum_{j=1}^{N} (c_j + u_j) DY_j + \sum_{j=1}^{N} f_j X_j + \sum_{j=1}^{N} \frac{KDX_j}{Q} + \sum_{j=1}^{N} \frac{1}{2} hc_j QY_j$$

$$+ \sum_{j=1}^{N} \frac{t_j d_j D}{Q_j} X_j \quad \text{(MS-SB-MS-1)}$$

subject to

$$\sum_{j \in J} X_j = 1 \tag{8.6}$$

$$D \leq W_j X_j, \qquad \forall j = 1, \ldots, N \tag{8.7}$$

$$X_j = \{0, 1\} \text{ and } Q_j \geq 0, \qquad \forall j = 1, \ldots, N \tag{8.8}$$

Note that the total order quantity $Q$ only appears in the objective function. Furthermore, given $X_j$ and $Y_j$, the objective function is convex in $Q$. Hence, the optimal total order quantity is given as

$$Q = \sqrt{\frac{2D\left(\sum_{j=1}^{N} KX_j + \sum_{j=1}^{N} t_j d_j X_j\right)}{\sum_{j=1}^{N} hc_j Y_j}} \tag{8.9}$$

Replacing Equation (8.9) in the objective function, we obtain

$$\sum_{j=1}^{N} (c_j + u_j) DY_j + \sum_{j=1}^{N} f_j X_j$$

$$+ \sqrt{2D\left(\sum_{j=1}^{N} KX_j + \sum_{j=1}^{N} t_j d_j X_j\right) \sum_{j=1}^{N} hc_j Y_j} \tag{8.10}$$

Then, the overall formulation is a mixed-integer nonlinear programming formulation. It can be solved with any general-purpose nonlinear programming package (Ghodsypour and O'Brien 2001).

### 8.2.1.2.2 Multi-Supplier, Single-Buyer with Multi-Sourcing Model 2

In this model, instead of an average annual cost model, we consider the integrated supplier selection and inventory model over a finite horizon. For this model, we define the following parameters:

$T$ Number of periods in the planning horizon
$D_t$ Demand requirements of the buyer in period $t$, $t = 1, \ldots, T$
$K_{jt}$ Fixed ordering cost for supplier $j$ in period $t$
$h$ Holding cost percentage per period
$R_{jt}$ Number of discount levels for supplier $j$ in period $t$
$p_{jrt}$ Unit price in discount level $r$ for supplier $j$ in period $t$, $r = 1, 2, \ldots, R_{jt}$
$h_{jrtl}$ Inventory holding costs for the complete demand of period $t$ if it is delivered by supplier $j$ in period $l$ with discount level $r$, i.e., $h_{jrtl} = h \cdot p_{jrt} \cdot d_t \cdot (t - l)$
$g_{jrt}$ Upper limit of the discount level $r$ for supplier $j$ in period $t$, $g_{jot} = 0$
$a_{jt}$ 1, if supplier $j$ is available in period $t$
$u_{jt}$ Unit transportation cost for supplier $j$ in period $t$

The decision variables in this model are as follows:

$Y_{jrtl}$ Proportion of demand in period $t$ that is delivered by supplier $j$ in period $l$ with discount level $r$
$X_{jrt}$ Binary variable for selecting discount level $r$ in period $t$ for supplier $j$
$Q_{jrt}$ Quantity purchased from supplier $j$ in period $t$ with discount level $r$

For the case of all-units quantity discounts, the following mixed-integer linear programming model is formulated:

$$\min \sum_{t=1}^{T} \sum_{l=t}^{T} \sum_{j=1}^{N} \sum_{r=1}^{R_{jt}} h_{jrtl} Y_{jrtl} + \sum_{t=1}^{T} \sum_{j=1}^{N} \sum_{r=1}^{R_{jt}} k_{jt} X_{jrt} +$$

$$+ \sum_{t=1}^{T} \sum_{j=1}^{N} \sum_{r=1}^{R_{jt}} (p_{jrt} + u_{jt}) Q_{jrt} \quad \text{(MS-SB-MS-2)}$$

subject to

$$\sum_{j=1}^{N} \sum_{l=t}^{T} \sum_{r=1}^{R_{jt}} Y_{jrtl} = 1, \ \forall t = 1, \ldots, T \tag{8.11}$$

$$\sum_{l=t}^{T} Y_{jrtl} D_t = Q_{jrt}, \quad \forall j = 1, \ldots, N; \forall t = 1, \ldots, T; \forall r = 1, \ldots, R_{jt} \tag{8.12}$$

$$Q_{jrt} \leq g_{jrt} X_{jrt}, \quad \forall j = 1, \ldots, N; \ \forall t = 1, \ldots, T; \ \forall r = 1, \ldots, Rjt \tag{8.13}$$

$$Q_{jrt} \geq (g_{j,r-1,t} + 1)X_{jrt}, \ \forall j = 1, \dots, N; \forall t = 1, \dots, T; \forall r = 1, \dots, Rjt \tag{8.14}$$

$$\sum_{r=1}^{R_{jt}} X_{jrt} \leq a_{jt}, \quad \forall j = 1, \dots, N; \quad \forall t = 1, \dots, T \tag{8.15}$$

$$Y_{jrtl} \leq X_{jrt}, \quad \forall j = 1, \dots, N; \quad \forall t = 1, \dots, T; \quad \forall l = t, \dots, T; \quad \forall r = 1, \dots, Rjt \tag{8.16}$$

$$Q_{jrt} \geq 0, \quad \forall j = 1, \dots, N; \quad \forall t = 1, \dots, T; \quad \forall r = 1, \dots, Rjt \tag{8.17}$$

$$X_{jrt} \geq 0, \quad \forall j = 1, \dots, N; \quad \forall t = 1, \dots, T; \quad \forall r = 1, \dots, Rjt \tag{8.18}$$

$$Y_{jrtl} \in \{0, 1\}, \quad \forall j = 1, \dots, N; \quad \forall t = 1, \dots, T; \quad \forall l = t, \dots, T; \quad \forall r = 1, \dots, Rjt \tag{8.19}$$

The objective function minimizes the cost of inventory holding, inventory ordering, procurement, and transportation. Constraints (8.11) satisfy demand fulfillment. Constraints (8.12) define the sizes of orders. The upper and lower limits of a discount level is determined with Constraints (8.13) and (8.14), respectively. Constraints (8.15) ensure that there will be at most one active discount level per delivery period. Constraints (8.16) establish the relation between the binary ordering variables and proportion of demand satisfied. Finally, Constraints (8.17), (8.18), and (8.19) establish binary and non-negative continuous variables.

This formulation is based on the well-known analogy between plant location and the dynamic lot sizing problem. It can also be easily revised to represent incremental quantity discounts (Tempelmeier 2002). This formulation can be solved with any mixed-integer programming solver such as CPLEX.* Tempelmeier (2002) also suggested a fast and effective heuristic to solve both all-units and incremental quantity discount models.

* CPLEX is a trademark of ILOG.

### 8.2.2 Multi-Supplier, Multi-Buyer Models

In integrated supplier selection and inventory models, there are not many models that specifically consider multiple buyers—except those of Jayaraman et al. (1999), Crama et al. (2004), Keskin et al. (2010a), and Keskin et al. (2010b). However, multi-product models can be formulated and solved as multi-buyer problems. Hence, we prefer to group the research focusing on either multi-product or multi-buyer into this section. Furthermore, we again distinguish between single and multi-sourcing and discuss these models separately in Sections 8.2.2.1 and 8.2.2.2, respectively.

#### 8.2.2.1 Single-Sourcing Models

Similar to the settings in Keskin et al. (2010a) and Keskin et al. (2010b), we consider a manufacturing firm with multiple geographically dispersed buyers. More specifically, we consider a total of $M \geq 2$ buyers procuring (or producing) a single product to meet buyer-specific stochastic demand. Unit demand arrives at the buyer according to a Poisson distribution with a buyer-specific mean $\lambda_i$, $i = 1, \ldots, M$. To satisfy the demand at each buyer $i$, the firm seeks to select a subset of suppliers from a set of potential suppliers that meet initial financial, quality, and delivery criteria. Each supplier $j$ offers the firm a fixed contractual cost, $f_j$, and a per-unit cost, $c_j$, and has an associated level of quality, $q_j$, and a disruption rate, $\theta_j$. In addition, each supplier $j$ has a specific annual throughput capacity, $W_j$. Annual throughput capacity is calculated assuming the supplier is never disrupted.

The decision variables in this setting are

$X_j$ 1, if supplier $j$ is selected, 0 otherwise, $j = 1, \ldots, N$
$Y_{ij}$ 1, if supplier $j$ is assigned to buyer $i$, $i = 1, \ldots, M$ and $j = 1, \ldots, N$
$Q_i$ Order quantity of buyer $i$, $i = 1, \ldots, M$
$R_i$ Reorder point of buyer $i$, $i = 1, \ldots, M$

Under a $(Q, R)$ policy, the inventory position (inventory on hand + outstanding backorders) is continuously reviewed, and an order of fixed quantity $Q$ is placed as soon as the inventory position drops to or below a reorder point $R$ (see Federgruen and Zheng 1992). After a supplier–buyer specific lead time, $L_{ij}$, the order $Q_i$ arrives at buyer $i$ provided that supplier $j$ is not disrupted and has sufficient capacity to do so. In this problem, because the supplier can be disrupted, the lead time, and hence the lead time demand, are affected by the suppliers' availability. Furthermore, as $q_j$ represents the percentage of products provided by supplier $j$ without any defects, only an order of $q_j Q_i$ arrives at buyer $i$.

In terms of inventory system costs at buyers, we assume that each replenishment order incurs a fixed cost of $K_i$ and an inventory carrying cost of $h_i$ per unit, per unit of time. Additionally, stockouts are backlogged at a cost of $s_i$, $i = 1, \ldots, M$, per unit, per unit of time. We also explicitly account for the transportation costs between the suppliers and plants. We express the transportation cost associated with each order delivery between supplier $j$ and buyer $i$ with two specific cost parameters: (1) a fixed cost, denoted by $p_{ij}$, which represents the sum of fixed costs of ordering (setup) and transportation; and (2) a per-mile transportation cost, denoted by $r_{ij}$, which leads to a distance-based transportation cost of $r_{ij}d_{ij}$ per order delivery, where $d_{ij}$ represents the distance between supplier $j$ and buyer $i$.

Under these assumptions, the multi-supplier, multi-buyer, single-sourcing model, ignoring the impact of disruptions, can be stated as follows:

$$\min \sum_{i=1}^{M}\sum_{j=1}^{N} c_j E[D_i] Y_{ij} + \sum_{j=1}^{N} f_i X_j + \sum_{i}\sum_{j}\left\{\frac{(p_{ij} + r_{ij}d_{ij})E[D_i]Y_{ij}}{Q_i}\right\}$$

$$+\sum_{i=1}^{M}\frac{K_i E[D_i]}{Q_i} + \sum_{i=1}^{M}\sum_{j=1}^{N} h_i\left[\frac{Q_i}{2} + R_i - E[LTD_{ij}]\right] Y_{ij}$$

$$+\sum_{i=1}^{M}\sum_{j=1}^{N}\frac{s_i n_i(R_i)E[D_i]Y_{ij}}{Q_i} \qquad \text{(MS-MB-SS)}$$

subject to

$$\sum_{j=1}^{N} Y_{ij} = 1, \qquad \forall i = 1, \dots, M. \tag{8.20}$$

$$Y_{ij} \le X_j, \qquad \forall i = 1, \dots, M \text{ and } \forall j = 1, \dots, N. \tag{8.21}$$

$$\sum_{i=1}^{M} E[D_i]Y_{ij} \le W_j X_j, \quad \forall j = 1, \dots, N. \tag{8.22}$$

$$q_j X_j \ge q_i^{min} Y_{ij}, \qquad \forall i = 1, \dots, M \text{ and } \forall j = 1, \dots, N. \tag{8.23}$$

$$Q_i \ge Q_j^{min} Y_{ij}, \qquad \forall i = 1, \dots, M \text{ and } \forall j = 1, \dots, N. \tag{8.24}$$

$$X_j \in \{0, 1\} \text{ and } Y_{ij} \in \{0, 1\}, \forall i = 1, \dots, M \text{ and } \forall j = 1, \dots, N. \tag{8.25}$$

$$Q_i \ge 0, \qquad \forall i = 1, \dots, M.$$

The objective of this formulation minimizes the expected procurement costs, the fixed contractual costs, transportation costs, inventory ordering and holding costs, and shortage costs. Constraints (8.20) ensure that each buyer should be assigned to a selected supplier. Constraints (8.21) state that buyers can only be assigned to selected suppliers. Constraints (8.22) are the capacity restriction of the suppliers; that is, the total average demand assigned to a supplier cannot exceed its annual throughput capacity. Constraints (8.23) state that the selected supplier should meet the assigned buyer's minimum quality requirement. Constraints (8.24) ensure that each buyer should meet its assigned supplier's minimum order quantity requirement. Finally, Constraints (8.25) establish the single-sourcing requirements and non-negativity of continuous variables.

This model, even without disruptions, is a stochastic, nonlinear, mixed-integer programming formulation. Unfortunately, it is not possible to solve this problem with traditional optimization techniques. For the deterministic version of this formulation, Keskin et al. (2010b) developed a generalized Bender's decomposition-based special algorithm and discussed the effectiveness. For this formulation, Keskin et al. (2010a) utilized a scatter-search powered simulation-optimization approach to solve this problem. This approach was not only useful in determining the best suppliers and optimum inventory levels at the buyers, but also in evaluating the impact of quality, disruptions, and delivery performance of the suppliers. Combining the benefits of optimization and simulation, the overall approach provides an increased ability to understand the impact of dynamic events, total system behavior, and (increasingly important today) risk mitigation in the event of expected/unexpected disruptions.

### *8.2.2.2 Dual- and Multi-Sourcing Models*

Before presenting the final analytical model, which focuses on multi-supplier, multi-buyer settings with multi-sourcing, we review the integrated supplier selection and inventory research in this area. Research in this area dates back to the article by Bender et al. (1985) where a computerized mixed-integer programming technique was employed at IBM to improve purchasing contracts. Since then, a number of other researchers contributed to the analytical models in integrated supplier selection and inventory, including Rosenthal et al. (1995), Jayaraman et al. (1999), Dahel (2003), Crama et al. (2004), and Basnet and Leung (2005).

Rosenthal et al. (1995) considered a buyer that should obtain the necessary number of stocking items from various suppliers that charge different prices and have limited capacities and different levels of qualities. A mixed-integer linear programming technique is developed to evaluate different bundles offered by the suppliers. Jayaraman et al. (1999) considered a potential set of suppliers constrained by the quality level of produced and supplied by the suppliers, the lead times to store the products, and storage capacity restrictions imposed by the suppliers. Their model is quite comprehensive except for the fact that they do not only consider the impact of inventory costs. Dahel (2003), on the other hand, presented a multi-objective, mixed-integer programming technique to simultaneously determine the number of suppliers to employ and order quantities to these suppliers in a multi-product, multi-supplier competitive environment. The model takes minimizing procurement cost, maximizing product quality, and maximizing delivery reliability. Crama et al. (2004)

also described the purchasing decisions faced by a multi-plant company. The suppliers of this company offer complex discount schedules based on the total quantity (rather than cost) of ingredients purchased. Linking local and group discounts for purchases, they model the problem as a mixed-integer nonlinear programming formulation and solve it using off-the-shelf software. Finally, Basnet and Leung (2005) presented an extension of Tempelmeier's (2002) work by considering a multi-period inventory lot-sizing problem, where there are multiple products and multiple suppliers. The decision maker needs to decide what products to order, in what quantities, with which suppliers, in which periods. An enumerative search algorithm and a heuristic are presented to address the problem.

In this section we present a formulation that generalizes the models of Jayaraman et al. (1999) and Dahel (2003) that considers integrated supplier selection and inventory under multi-supplier, multi-buyer, and multi-product under multi-sourcing restrictions. For this purpose, we define the following parameters:

$i$ Index of buyers, $i = 1, \ldots, M$
$j$ Index of suppliers, $j = 1, \ldots, N$
$k$ Index of products, $k = 1, \ldots, K$
$\vartheta_k$ Set of suppliers offering product $k$
$\mathrm{K}_j$ Set of products offered by supplier $j$
$\mathrm{I}_j$ Set of buyers that can be supplied by supplier $j$
$\mathrm{I}_k$ Set of buyers demanding product $k$
$\Delta_{ik}$ Units of product $k$ demanded by buyer $i$
$c_{ijk}$ Unit price of product $k$ quoted by supplier $j$ to buyer $i$
$q_{ijk}$ Percentage of rejected product $k$ from supplier $j$ to buyer $i$
$t_{ijk}$ Percentage of product $k$ from supplier $j$ to buyer $i$
$S_{kj}$ Maximum quantity of product $k$ that may be purchased from supplier $j$ due to capacity restrictions
$u_{jr}$ Upper cutoff point of discount bracket $r$ for supplier $j$
$d_{jr}$ Discount coefficient associated with bracket $r$ of supplier $j$'s cost function

The decision variables for this problem are as follows:

$x_{ijk}$ Units of product $k$ purchased from supplier $j$ for buyer $i$
$v_{jr}$ Volume of business awarded to supplier $j$ in bracket $r$
$y_{jr}$ 1 if the volume of business awarded to supplier $j$ falls on segment $r$, 0 otherwise

Then, the mathematical formulation is presented as

$$\min \sum_{j \in \mathcal{J}} \sum_{r \in \mathcal{R}_j} (1 - d_{jr}) v_{jr} + \sum_{i \in \mathcal{I}} \sum_{j \in \mathcal{J}_i} \sum_{k \in \mathcal{K}_j} q_{ijk} x_{ijk} + \sum_{i \in \mathcal{I}} \sum_{j \in \mathcal{J}} \sum_{k \in \mathcal{K}_j} t_{ijk} x_{ijk}$$

(MS-MB-MS)

subject to

$$\sum_{j \in \mathcal{J}_i} x_{ijk} = D_{ik}, \qquad \forall i \in \mathcal{I},\ \forall k \in \mathcal{K}_i \tag{8.26}$$

$$\sum_{i \in \mathcal{I}} x_{ijk} \leq S_{kj}, \qquad \forall k \in \mathcal{K},\ \forall j \in \mathcal{J}_k \tag{8.27}$$

$$\sum_{i \in X_j} \sum_{k \in \mathcal{K}_i} c_{ijk} x_{ijk} = \sum_{r \in \mathcal{R}_j}, \qquad \forall_{jr} \in \mathcal{J} \tag{8.28}$$

$$v_{jr} \leq u_{jr} y_{jr}, \qquad \forall j \in \mathcal{J},\ \forall r \in \mathcal{R}_j \tag{8.29}$$

$$v_{j,r+1} \leq u_{jr} y_{j,r+1}, \qquad \forall j \in \mathcal{J};\ r = \{1, 2, \ldots r_j - 1\} \tag{8.30}$$

$$\sum_{r \in \mathcal{R}_j} y_{jr} = 1, \qquad \forall j \in \mathcal{J} \tag{8.31}$$

$$y_{jr} \in \{0, 1\} \text{ and } v_{jr} \geq 0, \qquad \forall \in \mathcal{J},\ \forall \in \mathcal{R}_j \tag{8.32}$$

$$x_{ijk} \geq 0, \qquad \forall i \in \mathcal{I},\ \forall j \in \mathcal{J}_i,\ \forall k \in \mathcal{K}_j \tag{8.33}$$

The objective function of (MS-MB-MS) minimizes the total procurement cost, the cost of defective items, and the cost of items missing their target delivery time. Constraints (8.26) represent the condition that the total demand of each product at each buyer will be satisfied. Constraints (8.27) ensure that the total number of products procured by each supplier to all plants is within the production and shipping capacity of that supplier. Constraints (8.28) determine the dollar amount of business awarded to supplier $j$. Constraints (8.29) and (8.30) link the purchase of the product with the business volume discount to the appropriate segment of the discount pricing schedule for each supplier. Constraints (8.31) ensure that only one discount bracket for each supplier's volume of business will apply. Finally, Constraints (8.32) and (8.33) ensure integrality and non-negativity on the decision variables.

Because the overall formulation is a mixed-integer linear program, this problem can be solved by any off-the-shelf software such as CPLEX. Note that this formulation, even though it evaluates the impact of quantity discounts, quality, and delivery performances, does not explicitly account for the inventory decisions. This would be one potential research area for the future.

## 8.3 Implications, Limitations, and Conclusions

In this chapter I provided an overview of several analytical models in integrated supplier selection and inventory models. We emphasized that our focus is on improving the performance of the supply chain by considering the total operational cost of logistics, rather than just the selecting of suppliers or optimizing the inventory costs at the buyers.

The increased importance of global sourcing not only underscores the importance of systematic and efficient procurement models and methods, but also highlights the importance of evaluating many dimensions of purchasing such as quality, service, and dependability, and their interactions with other logistics functions. One of the complexities of buying goods and services of foreign origin is the wide variability in characteristics among the producing countries. Although superior quality and a broader supply base favor global sourcing, there are potential problems, including long lead times, additional inventories, and a higher cost of doing business. We can only improve the global operations through more structured analytical models. This statement goes for any industry that favors outsourcing or global sourcing. The models presented in this chapter provide a good basis for more complicated models.

## References

Aissaoui, N., M. Haouari, and E. Hassini, (2007), Supplier selection and order lot size modeling: a review, *European Journal of Operational Research*, 34, 3516–3540.

Akinc, U. (1993), Selecting a set of vendors in a manufacturing environment, *Journal of Operations Management*, 11, 107–122.

Basnet, C., and J.M.Y. Leung, (2005), Inventory lot-sizing with supplier selection, *Computers and Operations Research*, 32, 1–14.

Bender, P.S., R.W. Brown, M.H. Isaac, and J.F. Shapiro, (1985), Improving purchasing productivity at IBM with a normative decision support system, *Interfaces*, 15, 106–115.

Benton, W.C., (1991), Quantity discount decisions under conditions of multiple items, multiple suppliers and resource limitation, *International Journal of Production Research*, 29, 1953–1961.

Benton, W.C., (2007), *Purhasing and Supply Management*, New York: McGraw-Hill Irwin.

Chaudhry, S.S., F.G. Forst, and J.L. Zydiac, (1993), Vendor selection with price breaks, *European Journal of Operational Research*, 70(1), 52–66.

Crama, Y., J.R. Pascual, and A. Torres, (2004), Optimal procurement decisions in the presence of total quantity discounts and alternative product recipes, *European Journal of Operational Research*, 159, 364–378.

Dahel, N., (2003), Vendor selection and order quantity allocation in volume discount environments, *Supply Chain Management: An International Journal*, 8(4), 335–342.

Dobler, D.W., L. Lee, and N. Burt, (1990), *Purchasing and Materials Management: Text and Cases*, New York: McGraw-Hill.

Federgruen, A., and Y.-S. Zheng, (1992), An efficient algorithm for computing an optimal $(r, q)$ policy in continuous review stochastic inventory systems, *Operations Research*, 40(4), 808–813.

Ghodsypour, S.H., and C. O'Brien, (1998), A decision support system for supplier selection using an integrated analytic hierarchy process and linear programming, *International Journal of Production Economics*, 56–57, 199–212.

Ghodsypour, S.H., and C. O'Brien, (2001), The total cost of logistics in supplier selection, under conditions of multiple sourcing, multiple criteria and capacity constraint, *International Journal of Production Economics*, 73, 15–27.

Jayaraman, V., R. Srivastava, and W.C. Benton, (1999), Supplier selection and order quantity allocation: a comprehensive model, *The Journal of Supply Chain Management*, 35, 50–58.

Keskin, B.B., S. Melouk, and I. Meyer, (2010a), A Simulation-Optimization Approach for Integrated Sourcing and Inventory Decisions, Computers and Operations Research, 37(9), 1648–1661.

Keskin, B.B., H. Üster, and S. Çetinkaya, (2010b), An Integrated Location/Inventory Optimization Approach to a Generalized Vendor Selection Problem, Computers and Operations Research, 37(12), 2182–2191.

Pan, A.C., (1989), Allocation of order quantity among suppliers, *Journal of Purchasing and Material Management*, 25(3), 36–39.

Rosenthal, E.C., J.L. Zydiak, and S.S. Chaudhry, (1995), Vendor selection with bundling, *Decision Science*, 26(1), 35–48.

Tempelmeier, H., (2002), A simple heuristic for dynamic order sizing and supplier selection with time varying data, *Production and Operations Management*, 11, 499–515.

# Chapter 9

# Inventory Optimization of Small Business Supply Chains with Stochastic Demand

Kathleen Campbell, Gerard Campagna, Anthony Costanzo, and Christopher Matthews

## Contents



## 9.1 Introduction

In an ever-changing world, small businesses have been seen to start, flourish, and then very often lose their competitive edge and become obsolete. In very small businesses, this cycle seems even more abbreviated as bigger businesses can see what is popular, incorporate it into their daily sales at cheaper prices, and leave the small business to suffer. Today more than ever, small businesses need to think and act creatively in order to compete. As technology becomes more available, supply chains in the world continue to integrate, and life cycles of products shorten, it is imperative that small businesses integrate technology, statistics, and a working knowledge of the supply chain into their daily operations. This chapter discusses a seasonal small business, Jake's Water Ice. Located in the beach town of Ocean City, New Jersey, Jake's is open only 13 weeks each year. Finding a way to maximize the resources available while minimizing both inventory and excess staff might help keep this small business running long into the future. The methodologies used are multiple linear regression and simulation. But the results give an idea of how a little time spent properly collecting, organizing, simulating, and analyzing data can help a company run efficiently. As for any good study, as analysis ends it is important to take note of what future data can be collected to help further the process.

## 9.2 Literature Review

In today's business climate, it is imperative that companies remain competitive to stay viable and show a reasonable profit. The current trend for any business to maintain successful operations is for it to utilize current technologies. In June 1994, the Boeing 777 was built with computer-aided design tools that were linked among 2,200 computers. They were able to design and even simulate the assembly of the plane, which was something that was never before seen (Wheatley 2008). Some 15 years later, the game has significantly changed. Some of the new technologies that are currently used include the use of simulation software to more effectively and efficiently manage and make business decisions, such as changing operational locations, switching transportation modes, and changing the way inventory is kept to keep up with shorter life cycles. Companies have often used simulation software in the past to imitate particular business processes. Today they are using this software in new and innovative ways. Much of this innovation can be seen in recent developments in supply chain, inventory optimization, and demand planning simulation software.

Supply chain, inventory, and demand simulation software are three different business processes that actually influence one another. The *supply chain* can be defined as having items flowing from one stage of supply to the next, both within the business and outside, in a seamless fashion (Kanaka 2008). The supply chain is more of a high-level optimization. It has a varied focus, including decisions such as warehouse location, truck usage, and choice of supplier. *Inventory optimization* is more focused on just how much inventory a company should maintain to make the most profit. Both supply chain and inventory software need an accurate forecast of demand to be implemented properly and that is why we begin with *demand planning* software.

"While no one is attempting to achieve the consultant's dream of one unit produced at the factory for every item scanned at the store," successful and more accurate demand planning is allowing companies to become more innovative and better planned than ever before (Casper 2008, 19). Demand software is not complicated in how it works. There are three engines: (1) a stat engine that collects historical data, (2) a collaboration engine to take input from people who interact with customers, and (3) a demand engine to help companies model and shape demand through promotions. These three engines can provide companies with real-time forecasts, which enable them to manage day-to-day operations with greater accuracy, but they can also be used for long-term planning. Accurate forecasting allows companies to easily identify market signals that allow them to cope with fluctuations in the market and also provide value to the demand planning system's abilities to simulate scenarios and perform what-if projections (Casper 2008, 29). Knowing what is going to happen beforehand is probably the most powerful tool a business can possibly possess.

Demand planning is the first step in a successful supply chain simulation. The supply chain is a business process that has been consistently studied and tested in business. The tools used to improve the supply chain have grown over time, but today there is an entirely new outlook and set of tools to improve it. Many companies have used enterprise resource planning (ERP), which provides global visibility of information from within a company and across the supply chain. ERP is a tool that has been used to improve business processes, but "Supply Chain Management software can consider all the relevant constraints in production, manufacturing, distribution, transportation, and warehousing simultaneously, while also making real-time adjustments to constraints" (Chang and Makatsoris 2008, 25). The traditional view of supply chain management was a series of decisions made in silos, that is, that each department made goals and strategies without any integration between the suppliers, manufacturers, distributors, retailers, and costumers. Decisions were much too non-integrated among all departments involved in the supply chain. "It's a linked problem that has rarely been linked" (Trebilcock 2008, 47). The new and improved view of supply chain incorporates everything from inventory in warehouses to marketing new products. All parts of the process are streamlined to develop a complete supply chain simulation. Traditional supply chain software merely takes a demand forecast to determine the right location and/or the right amount of inventory. The new supply chain wave of software takes all variables into account. With network design and optimization software, companies can determine the best locations for factories, warehouses, and transportation networks" (Trebilcock 2008). The most powerful part of supply chain analysis is the ability to conduct a what-if analysis and see how each changed factor affects the business. Supply chain software can answer questions regarding supplier performance, increased profitability, and the cost of inventory and delivery performance when expanding inventory (Chang and Makatsoris 2008).

Supply chain simulation is not something that can be accomplished overnight. One must first understand the supply and planning process, collecting excessive amounts of data including manufacturing process data and inventory and stock levels. One must then design a workflow diagram and compare performance measures and targets (Chang and Makatsoris 2008). There have been many companies that have been able to develop successful solutions to supply chain problems using this model.

One of these companies is Daimler Auto. Daimler Auto used simulation software to streamline its material handling business with its assembly line planning. Previously, the analysis of material flow, plant layouts, and material zones were afterthoughts to main assembly operations. By simulating the material process within the assembly context, planners could see the robustness of the assumptions they had made about things such as speeds, choke points, and handling capacities. The simulation "effectively closes the loop between providing a validation process to stress test plans against conditions that they might experience in the real world" (Wheatley 2008, 52).

Supply chain software covers many different aspects of the business, including inventory optimization; however, inventory management should also be considered in its own capacity before inserting it into supply chain analysis. The concept of zero inventory management has been around for a long time. Zero inventory management has the goal of trying "to reduce inventory to a minimum and enhances profit margins by reducing the need for warehousing and expenses related to it" (Kanaka 2008, 33). Companies have attempted to do this through the use of inventory management software. Inventory optimization software has helped companies monitor, manage, and optimize strategies to decide what to make, and what to buy and from whom. It also helps companies determine what to carry and in what form (Kanaka 2008). Inventory optimization software begins with accurate forecasts of demand and the availability of inventory throughout the supply chain. After these are forecasted, the software can be used for many useful decisions. While running, the software can take into consideration how much time passes between placing an order and having a product in a distribution center (DC). It can also balance the costs of carrying goods versus acquiring goods under different service requirements, and even model inventory levels to determine the rules for reordering and where to place that inventory (Trebilcock 2008). Companies are able to use this kind of software because of the great technology developments in business. Radio frequency identification (RFID) technology has allowed companies to collect data and increase efficiency by tracking the movement of items across systems while locating malfunctioning equipment and allowing companies to more quickly and accurately diagnose solutions (Kanaka 2008). The technologies of demand planning along with supply chain and inventory optimization solutions have been mostly a trend of larger companies due to the costs and difficulty of implementing them; however, more recently, these technologies have been implemented by companies of all shapes and sizes.

In the not too distant past, small businesses avoided implementing major supply-chain technologies and processes because upgrading could potentially be too costly, time consuming, and resource draining. Every decision for a small business is a big decision that can greatly affect the business as a whole. Big businesses usually implement such processes or technologies with relative ease because they have entire departments and teams to aid in the implementation and also the availability of financial resources to switch direction if things do not go as planned.

Small businesses are now starting to catch up with the implementation of supply chain technologies and processes. AMR Research, a research and advisory company for supply chain and IT professionals, found that "in 2008 mid-market companies will be 'aggressive' in buying supply-chain management software, due to continued pressure to reduce manufacturing costs and to help customers reduce their own costs" (Rafter 2009). Small to mid-sized businesses are realizing that regardless of their size, their customers expect them to comply with their demands and orders. This is causing these small to mid-sized businesses to adjust their practices and in most cases invest in new supply chain technologies and processes.

Available technologies for supply chain management are plentiful, ranging from simple Microsoft software to advanced logistical software. Considering that each company is different and their needs are unique, software such as Microsoft Excel can often satisfy most of a company's needs. On the other hand, companies that have more in-depth needs would use software such as UPS Trade Direct Cross Border, a product of UPS Supply Chain Solutions (Malykhina 2004). One example of a small business is Ikor Industries, a manufacturer of specialized industrial filtration products that has global clients and needs a supply chain veteran like UPS to help them in the distribution of their products. For a small business like Ikor Industries, too much time would be spent trying to learn and understand all the international laws as well as trying to keep track of all logistical information. For small businesses that do not know where to turn in order to explore their options for supply chain technologies and processes, special interest groups such as the Supply-Chain Council (Washington, D.C.), Supply Chain Management Review (an online industry magazine to network information), and the Supply Chain Management Research Center (research center located at the University of Alabama) can be utilized (Rafter 2009).

"Without good supply chain management, a company may lack access to vital information and the deficit can stop production from being as fast or efficient as possible" (Rafter 2009). In an article titled "Value of Information in a Capacitated Supply Chain," Choudhury et al. (2008, 117) point out that information technology "has a significant impact on the supply chain performance." It has revolutionized the way in which suppliers and businesses interact, as well as businesses and their customers. Utilizing relative supply chain data, such as inventory levels, forecasted data, or sales trends, companies can reduce cycle times, reduce inventory costs, or even improve customer satisfaction by reducing errors in their orders. To evaluate the flow of information in a supply chain over time, the authors used a discrete event simulation "because of the dynamic and stochastic nature" of the fictional demand. It is the demand variability, they stress, that is the most widely recognized factor in influencing supply chain performance. They conclude that as "the end item demand variance increases [or just demand, for short], the benefit due to the availability of real time inventory information increases" (Choudhury et al. 2008, 126). In other words, demand is dynamic because of many factors such as seasonal demand, the economy, and the weather, and the more demand fluctuates, the more real-time information, regarding inventory levels, shipment information, or customer information, is beneficial for a company (Hassini 2009).

Seasonal demand, which is dynamic, is a difficult thing to predict but it is worth the effort because it directly affects inventory costs and labor costs. An important part of managing a business' supply chain is identifying whether or not seasonal demand is an actual force that affects the company or if the seasonal demand is a result of business practices. This false sense of seasonal demand could be caused by businesses over-compensating for what they believe is a pattern in demand for their products. For example, a business can "maintain minimal levels of capacity . . . and operate the plant at maximum production levels" in order to satisfy current demand and obtain a level of inventory that would satisfy the supposed demand (Bradley and Arntzen 1999). In other words, businesses should produce what is currently needed while inventory levels are obtained for what will be required later. This self-induced pattern is driven by business practices and not by actual seasonal changes, which affect sales and inventory (Bradley and Arntzen 1999, 795).

To have an optimal performance during seasonal demand, a balance between capacity utilization and inventory levels must be obtained. This is especially difficult because the investment decisions for capacity utilization and inventory are made in separate places within companies. Capacity utilization decisions, such as purchasing new equipment, are normally made by upper management (strategic decisions) for many reasons, including the cost of purchasing new equipment. Although this style is typical, it does create one obvious problem: These major decisions take a lot of time and can cause delays in an investment of new equipment, which directly affects production levels. On the other hand, inventory-level decisions (operational decisions) are made by low-level managers who are not frequently involved with capacity utilization decisions.

In an article by Bradley and Arntzen (1999, 800), the authors create an aggregate planning model to combine these two concepts. They conclude by stating, "Variability *in demand* necessitates additional investment *in capacity*, or inventory, or both." Managers should "build inventories for which demand is certain . . . especially if that demand is going to happen soon" (Bradley and Arntzen 1999, 801). Having high levels of inventory during uncertain times of demand would most likely be costly to a company, both in terms of inventory holding costs and production costs. This issue must be considered, especially for small businesses that do not have storage capacity for inventory. Bradley and Arntzen conclude that managers "may choose not to build all the forecasted requirements for a *seasonal* product *in* advance. Instead, a portion of the forecasted *demand* could be produced at a point *in* time *closer* to the realization of *demand* when the level of *demand* is more certain" (Bradley and Arntzen 1999). In other words, managers build up inventory when it is close to the business' prime season. Doing so too far in advance can be costly. One way of avoiding this is proper documentation of sales and inventory, which would help managers make inventory decisions more intelligently.

Supply chain management and optimization are very complex and intricate issues that continue to grow both in importance and acceptance, and are unique to each company that adopts these ideals. New technologies are critical to obtaining an optimal supply chain for businesses. There are, however, obstacles that complicate matters. Inventory levels and seasonal or stochastic demand heavily influence decision making for supply chain management and optimization.

Stochastic seasonal demand and inventory optimization have a crucial impact and can be deciding factors as to whether a small business is profitable or it loses money. Stochastic seasonal demand can lead to various difficulties within a small business and can make inventory optimization difficult—and for a small business, very costly. "The heuristics used in business practice are shown to cost an average of 30% above optimal policy costs. A superior heuristic is constructed utilizing an analytic approximation for optimal policies that costs an average of 2% over optimal policy costs" (Metters 1997, 1017). Many small businesses do not have the resources to spend on these heuristics, especially considering, as the article indicates, that they would be spending up to 30 percent more than needed for the heuristics to begin with. The main goal of using optimization for the small business would be to resolve the "problems with determining the amounts to be produced in each period so as to minimize the discounted costs of production, inventory storage and lost sales" (Metters 1997, 1017). This is the exact goal we set out to accomplish using various simulations, modeling, and a few multiple linear regressions.

Any seasonal business, in our case Jake's Water Ice located in a shore town on the coast of New Jersey, has the challenge of determining how the issue of inventory optimization should be utilized. The inventory needs to be optimized not only for the weekly demands, but also the seasonal demand as a whole, in this case 13 weeks in the summer. This business needs to optimize the necessary amount of supplies, water ice, ice cream, and other products, taking into consideration the fluctuations of customers throughout the week and throughout the high and low seasons. The business must also consider that once the summer season comes to an end, remaining supplies are minimized. Several years of recorded data were available for review and analysis to build a model as well as forecast sales.

"There are two basic types of systems that are used when demand is stochastic. The first is the continuous review system in which the stock level is constantly monitored and an order for a fixed quantity is placed whenever inventory falls to a specific threshold level. The second is the periodic review system in which the inventory level is monitored at specific intervals followed by an order up to a pre-specified level" (Eynan and Kropp 2002, 1020). In this case the continuous review system is used and an order is placed when inventory falls below a specific level. The vast majority of supplies as well as water ice ingredients are purchased through one supplier, and there is a minimal monetary order required for delivery. Inventory is taken, and orders are put in on Tuesday, which is followed by a delivery on Thursday. The other two suppliers supply the hard and soft ice cream, and each runs on a different schedule.

The economy is another reason why inventory optimization is essential for today's small businesses. To ensure maximized profit, an accurate inventory optimization plan needs to be implemented, but it seems as if most businesses are not pursuing this. "It is surprising how many businesses have not made 'inventory optimization' a top priority" (Mulani 2009, 23). "Companies don't fully consider the range of benefits associated with superior inventory management." By properly managing its inventory, a business can lower carrying costs, improve asset utilization, raise customer service levels (essential in the food business), reduce operating costs, improve supplier relationships, and increase revenue and gross margin. All of these are great reasons why any business big or small should strive for inventory optimization.

To optimize inventory investment, a business needs to account for all of the variability in the supply chain. There are many different factors in a seasonal business that can determine a demand for a given time period. A water ice shop's sales can be directly impacted by the time of year, weather, competitive or friendly promotional activities within walking distance, as well as various other factors. This is what makes forecasting the demand so difficult, and can directly impact the inventory for a given time period. It is important that the forecasts are as accurate as possible because some of the inventory in the shop (ice cream, water ice) is volatile and can only be stocked for a certain period of time. Additionally, consideration must also be given to the fact that there are certain days that an inventory invoice must be placed to ensure timely delivery. This is important because a product shortage can negatively impact the business and drive away potential customers. The provided data is very useful for forecasting sales and inventory for specific periods (e.g., holiday or time of season), but certain factors such as weather must be factored differently.

## 9.3 Trend Analysis and Linear Regression

To successfully put a simulation model together for the water ice business and supply chain process, demand must first be forecasted. It is very easy for one to short change this part of the process, but forecasting demand is an essential and pivotal part. The simulation of this process, depends on accurately forecasted demand; without it, the results of the simulation are meaningless numbers. This is why we spent much time and effort precisely forecasting demand for a highly seasonal water ice store, using Jake's Water Ice in Ocean City, New Jersey, for our model.

The technique used for this forecast was multiple linear regression analysis. The data used was sales data for the water ice store for the summers of 2006, 2007, and 2008. The sales data for all three years began Memorial Day weekend and ended in the first week of September. The store maintains weekend hours from Memorial Day weekend through most of June. While there is an obvious demand to be open during Memorial Day weekend, there is such low demand thereafter that opening weekdays would be a liability to the company as costs would far exceed revenues. Prior to completing the regression, we assessed the data using a moving average that highlighted the need to clean the data. The date each year did not correspond to the same day of the week. Therefore, the data for every summer had to be mapped to the appropriate day of the week for comparison purposes. If July 15th fell on a Thursday one year and on a Friday the next year, the much heavier sales volume on

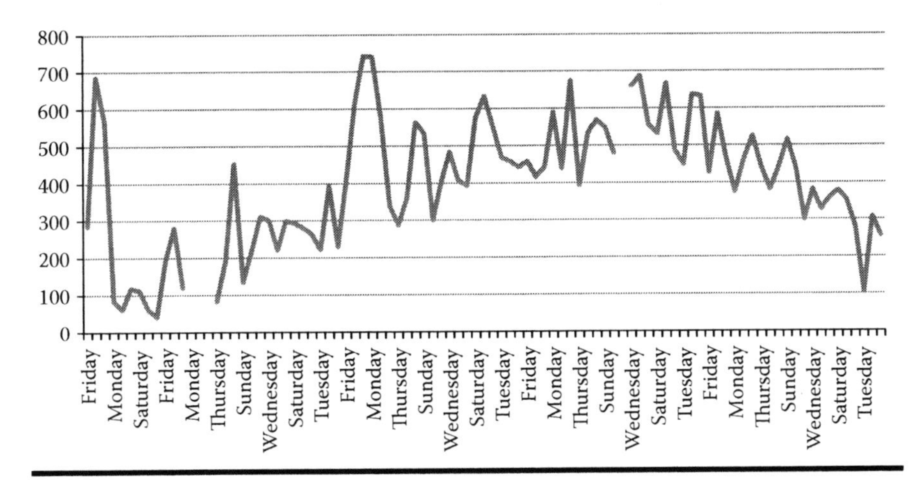


**Figure 9.1 Trend of sales for 2006 summer season.**

Friday would throw off the forecasted number using averaging. We began with the Friday of Memorial Day weekend and then grouped the data by day of the week, ignoring the actual dates themselves. The summer is approximately 13 weeks long; therefore, we have 13 weeks of data grouped by day of week, all using Memorial Day weekend as the starting point.

First, the yearly trends were observed. As can be seen in Figure 9.1, in 2006 there are gaps where the store was closed during specific time periods, perhaps only open on weekends. Also, there is a spike that represents Memorial Day and then sales remain low and begin an overall rise later in the summer, until there is a peak followed by a gradual decline. To determine if this was just a 1-year trend, it becomes necessary to look also at the other data collected.

While looking at the trend for 2007 (Figure 9.2), there are some small differences but the overall trend for the summer seems to have remained intact. The main patterns of increase and decrease are very similar. However, there isn't the same end of summer gradual decline. This can be due to the fact that the store did not open at all in September 2008 and therefore the end-of-season trend stops with Labor Day weekend. The overall sales did increase, as the highest income attained in 2006 was $800 and in 2007 it was $900.

Finally, in comparing 2008 (Figure 9.3) to the previous two summers, the center of the trend looks very similar to both 2006 and 2007. The increase and then gradual decrease matches 2006. However, there is a spikey set of data points at the beginning that lasted longer into the summer of 2008 than in any other summer. When asked, the manager assessed that this was affected by the decrease in weekly rentals, leading to homeowners spending long weekends at the shore later into the prime season and causing very busy weekends and very slow weekdays.

With these trends in mind, we began making a list of assumptions and questions that we needed to contemplate before beginning the regression. We started by talking

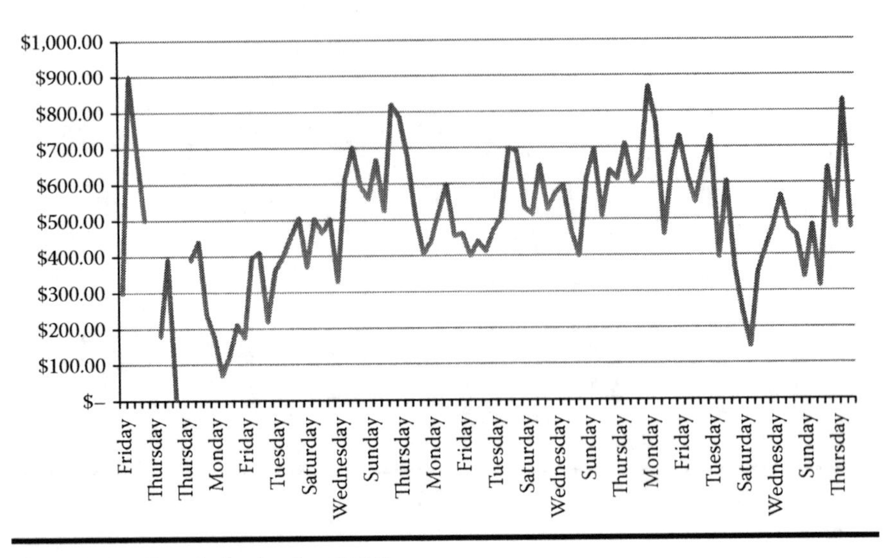


**Figure 9.2 Trend of sales for 2007 summer season.**

to the water ice staff about what factors throughout the summer really have an effect on sales. One of the main factors we decided to test was holidays; noticing the huge spike at the beginning of summer, we assumed that all summer holidays would have the same effect. Also, we assumed another factor that would have a great impact on sales was the particular day of the week. Obviously Saturday and Sunday would presumably have higher sales than the weekdays. Therefore, a regression variable was created to separate every day of the week to see if we could find any significance with

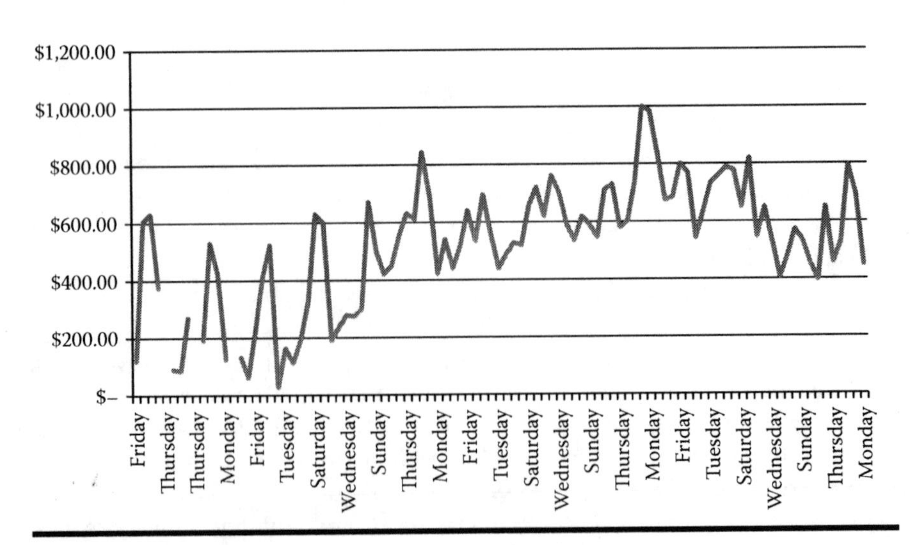


**Figure 9.3 Trend of sales for 2008 summer season.**

the weekend days. Looking at Figures 9.1 through 9.3, there is evidence that the weekends definitely have an effect on sales. Looking at Figure 9.1, there is evidence of increases and decreases in sales that follows a basic 7-day pattern. However, there also seems to be a seasonal pattern. Further review of Figures 9.2 and 9.3 shows that the trends are similar although not exactly the same. One can easily discern some recurring patterns throughout the 3 years that, when asked, the employees attributed to the shore rental season.

The last factor we considered in our first regression was the impact that a progression in years had on sales. Our first assumption was that there would be no particular impact from year to year on sales. We were under the assumption that sales would be flat from year to year. We wanted to test our assumption before performing the regression so we graphed the data from 2006 through 2008 on one graph to attempt to identify any sloping pattern over the 3-year span. After examining Figure 9.4, we actually did identify an increasing pattern in the sales data over the 3 years. With a correlation of only .37, we did not think it was extremely significant but it is definitely something that could impact sales and that needed to be considered in our regression and sales forecast.

Therefore, our independent variables for the initial regression are year, day of week, week of summer, and whether or not it was a holiday. Our dependent variable was the sales for Jake's Water Ice. While the results were mixed, we did find some very significant factors. The first was that year did, in fact, have a significant impact on sales because the $P$-value for this was fairly low. Holiday, as expected, did have a

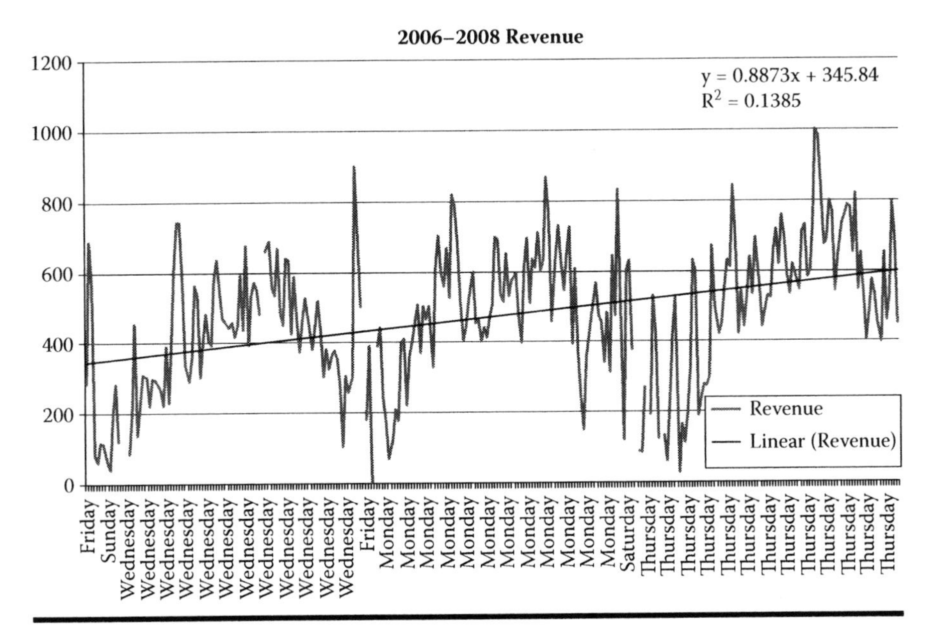


**Figure 9.4 Trend for sales using summers 2006 through 2008.**

**Table 9.1 Multiple Regression Run: All Variables**

| *Regression Statistics* | |
|---|---|
| Multiple R | 0.509019 |
| R Square | 0.2591 |
| Adjusted R | |
| Square | 0.236901 |
| Standard Error | 171.2911 |
| Observations | 276 |

ANOVA

| | *df* | *SS* | *MS* | *F* | *Significance F* |
|---|---|---|---|---|---|
| Regression | 8 | 2739610.3 | 342451.3 | 11.6715689 | 3.15442E-14 |
| Residual | 267 | 7833950.4 | 29340.64 | | |
| Total | 275 | 10573561 | | | |

| | *Coefficients* | *Standard Error* | *t Stat* | *P-value* | |
|---|---|---|---|---|---|
| Intercept | 280.8542 | 34.994404 | 8.02569 | 3.2186E-14 | |
| Week | 26.09339 | 2.8356284 | 9.201976 | 1.0511E-17 | |
| Monday | −81.8232 | 38.124939 | 2.146185 | 0.03275956 | |
| Tuesday | −77.2678 | 40.029378 | 1.930277 | 0.05463156 | |
| Wednesday | −48.7127 | 39.34164 | 1.238198 | 0.21673018 | |
| Thursday | −73.7596 | 38.375927 | 1.922028 | 0.05566628 | |
| Friday | −64.9713 | 36.754134 | 1.767729 | 0.0782484 | |
| Saturday | 21.06239 | 36.733829 | 0.573378 | 0.56687109 | |
| Holiday | 216.4835 | 49.548383 | 4.369134 | 1.7873E-05 | |

significantly low $P$-value and explained much of the variation on the days that were deemed holidays. Particular weeks of the summer also had an increasing relationship and had a significant effect on sales.

Table 9.1 shows that together these factors explained a lot of the variation of sales throughout the summer, as can be seen by looking at the coefficient of determination, or $R^2_{adj} = 0.509$. This tells us that 50.9 percent of the variation in sales can be explained by the factors we designated. Only some of the days of the week proved significant. Notice from the output that Sunday was used as the reference variable to the other days of the week. The results showed that Monday, Tuesday, Thursday, and Friday all had significantly lower sales when compared to Sunday using a 10 percent significance level. Wednesday and Saturday, however, were not significant when compared to Sunday, as can be seen by their high $P$-values. Saturday makes sense as it is similar to Sunday, in that is the weekend. However, there seemed to be no real reason as to why Wednesday was not significant as compared to Sunday. After contacting an employee, we discovered that this was because Sunday and Wednesday are Family Trivia nights at the water ice store. Wednesday on its own might have been significant but the trivia made it very comparable to Sunday sales. The important thing we noted was that the model would fit for now but if the trivia night were ever to change, the forecast for Wednesday would most likely become inaccurate.

While this data revealed some significance, we decided to look for other factors in the data that might increase our overall correlation while deleting the insignificant factors. There was more work to be done with the particular day of the week. Now that we were deleting Saturday and Wednesday, the other days could be compared to a combination of the three missing days. After taking out these high $P$-valued items, we were able to lower the degrees of freedom of our model while barely losing any of our explained variation, the $R^2_{adj}$, which is a strong determining factor of a good model. We did improve our model by taking out the higher $P$-values, but we still felt that our model was insufficiently explaining the variation in sales throughout the summer.

Therefore, through further research with the water ice store staff and reexamining the individual graphs for the 3 years, we discovered that there was definitely a clear pattern of seasonality. The graph shows that the sales data starts out low, then increases until mid-summer, followed by an end-of-summer decrease. After talking with staff we concluded the first 4 weeks of summer to be low season, the next 6 weeks after July 4th inclusive to be high season, and then the final 3 weeks of summer to be low season. This seems to be a good representation of seasonality when looking at Figures 9.1 through 9.3. After factoring a high and low season into our model and examining if they had a significant impact on sales, we were very pleased with the results, as seen in Table 9.2.

The percent of sales variation explained by our independent factors jumped from 50.9 percent to 74 percent, which is a significant increase. This implies a very strong correlation, .86, between the independent variables and sales. All the items in our regression have very low $P$-values, all but one of which are less than a 5 percent significance level and all are below 10 percent, which was chosen to be our level of significance. High season versus low season was a great input to our model as it really increased the amount of variation while adding an extremely low $P$-value to our model. Meanwhile, the standard error was relatively small, with a value of about 144. Overall, this is a pretty effective, if not perfect model.

**Table 9.2 Multiple Regression (Days of Week, Week #, Holiday)**

| *Regression Statistics* | |
|---|---|
| Multiple R | 0.740822606 |
| R Square | 0.548818134 |
| Adjusted R Square | 0.533552582 |
| Standard Error | 133.9200588 |
| Observations | 276 |

ANOVA

| | *df* | *SS* | *MS* | *F* | *Significance F* |
|---|---|---|---|---|---|
| Regression | 9 | 5802961.856 | 644773.5 | 35.95141133 | 3.30698E-41 |
| Residual | 266 | 4770598.85 | 17934.58 | | |
| Total | 275 | 10573560.71 | | | |

| | *Coefficients* | *Standard Error* | *t Stat* | *P-value* | |
|---|---|---|---|---|---|
| Intercept | 82.76051698 | 30.75049067 | 2.691356 | 0.007567213 | |
| Year | 57.504778 | 9.916603284 | 5.798838 | 1.88856E-08 | |
| Week | 23.8713584 | 2.205798054 | 10.8221 | 7.3944E-23 | |
| Monday | −60.32430875 | 25.06570545 | −2.40665 | 0.016781781 | |
| Tuesday | −67.45947157 | 26.58912623 | −2.53711 | 0.011749084 | |
| Thursday | −55.70553895 | 25.06317153 | −2.22261 | 0.027083408 | |
| Friday | −46.39049032 | 24.03133579 | −1.93042 | 0.054618155 | |
| Holiday | 124.1115973 | 48.17914159 | 2.576044 | 0.010533931 | |
| Holiday *Weekend | 335.2469713 | 72.72557369 | 4.609754 | 6.26478E-06 | |
| High Season | 186.5108832 | 16.38873354 | 11.38043 | 1.035E-24 | |

Although we are happy with our model for forecasting future sales and demand, it is imperative to also point out areas that could be studied to possibly give an even better future analysis. One main factor that we really would have liked to incorporate into our model is weather. It is especially obvious that weather would clearly have an impact on sales. While some data had been collected, it was not on a daily or even weekly basis and therefore we were unable to incorporate it into our model. We would assume that good weather is going to have a significant impact on sales as compared to a rainy and cold day. However, without the data, we are currently unable to back up our assumption. We believe that adding weather would have absorbed much more of the variation out of the current model and lead to a more accurate demand forecast. Although we were not able to incorporate this factor into our analysis, the model above is more than sufficient for further analysis of this type of seasonal business.

## 9.4 Simulation

To compete in the current economic downturn, it is imperative that all small businesses pay close attention to the workings of their supply chain. Therefore, the next area that we focused on was the supply chain of Jake's Water Ice. It is the supply chain that is the conduit of potential sales and growth as well as possible reductions in cost and wasted material. A better understanding of this supply chain can be beneficial by reducing cavalier spending and making better-educated estimates of purchases.

The supply chain of the water ice shop starts much like any other business, with suppliers. The three main suppliers used in this particular supply chain are Balford Farms, Island Ice Cream, and Gil's Wholesale. Together they provide the water ice shop with the essential tools and ingredients to make and sell hard and soft ice cream and water ice, which is made on site daily. Ordering from these suppliers depends on demand as well as certain restrictions set in place by both limited storage capacity and, of course, the suppliers themselves. For example, Balford Farms will not accept an order that is less than $150. There is also a limitation of delivery days, which in the summer is on Mondays and Fridays. So ordering items and managing inventory must be approached in a strategic way in order to find that equilibrium of orders and inventory without ever running out during the season. Meanwhile, the refrigeration space is limited, and there is a short shelf life on all milk-based products. This is one of the reasons why we run our simulation: to better understand that relationship between ordering goods, demand, and inventory.

After the suppliers deliver the goods to Jake's Water Ice shop, the items must be stored in inventory, which again must be considered because inventory space is limited. The water ice is then created on site and the soft ice cream is mixed as well. The hard ice cream comes ready to serve. During this step of the supply chain, elements such as labor and processing times must be considered. Because this water ice shop is located at the beach, the demand is very seasonal and therefore labor, inventory, and sales change as the seasonal rentals change.

The final step in the supply chain process for Jake's Water Ice shop is the actual sales of the goods to customers. The average transaction results in $5.20 in sales. Using our regression model we tested many elements that affect sales, including days of the week, holidays, and low season versus high season. Because this is a small water ice shop and the sales are seasonal, every step in the supply chain must be strategically considered because every decision affects the business and its bottom line. We employed simulation software to generate a hypothetical summer using our knowledge of the regression statistics and the supply chain.

We used Extend LT version 6.0.3 by Imagine That! Inc. This software is used to simulate a wide variety of scenarios, including transportation, manufacturing, and processing such as banking or customer service. These models are created when a series of blocks are interconnected in order to show a fully inclusive model. There are many different blocks in this software in order to give the user a large assortment of possibilities. Following the supply chain model, we used several blocks, including labor pool (to simulate the labor needed), transaction blocks (simulating the actual sales), import blocks (used to simulate the demand), and many others. Also, another useful concept that we incorporated is blocks that calculate graphs, charts, and summary statistics. We used these blocks to get a better understanding of our simulation and breakdown of each element.

In Figure 9.5, one can see the actual model created to simulate a typical order flow at Jake's Water Ice. The flow of the model is from left to right with the "global time units" or the units used throughout the simulation measured in minutes. Minutes

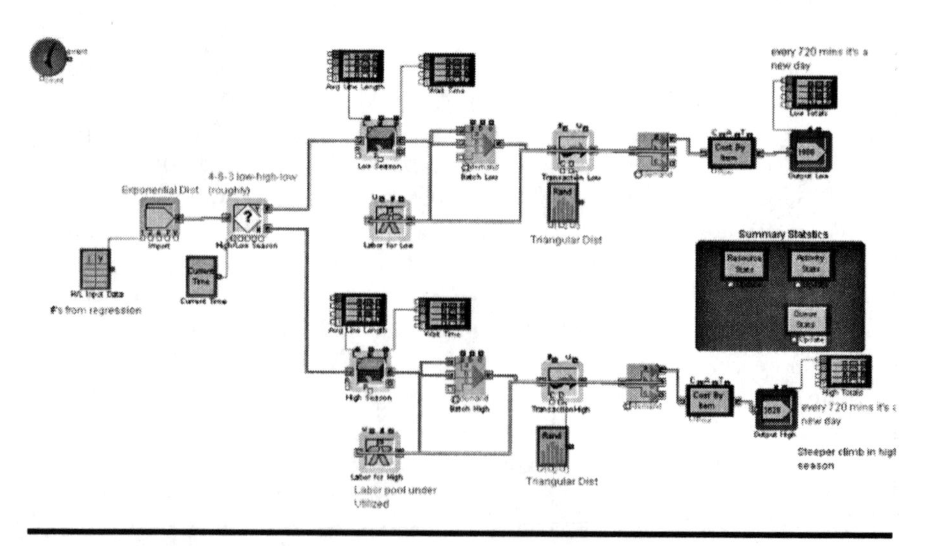


**Figure 9.5 Business simulation of typical customer order as modeled over one full season.**

are used because it is easier to interpret the results based on customer frequency without inflating the numbers or having fractional purchases. While we could have used hours, it was concluded that the results would be easier to understand and interpret in minute form.

Beginning with the far left part of the model, the Input Data block is the location where we input the average number of customers per 60 minutes for both high season and low season. We estimated these numbers from our calculations using the same historical data collected to create the regression model. The number of customer arrivals is based on an exponential distribution, with the average being different depending on whether it is high season or low season. While this is believed to be the best distribution of successes over a specific time frame, if a different distribution is considered, it would need to be altered in the second block (the Import block).

Next is the Decision block. In this block we tie in a Current Time block in order to simulate the actual minutes that are occurring during the simulation. For our purpose, within this block we created an equation to determine whether the current time is low season or high season. This crucial block then routes the simulation according to the current season. High season begins the weekend preceding or inclusive of July 4th and lasts roughly 6 weeks thereafter. We recognize that, in general, the pattern is 4 low-season weeks, followed by 6 high-season weeks, and then 3 more low-season weeks. This pattern will hold true for most summers. This Decision block can determine this breakdown with the equation we created.

After being sorted by the Decision block, the simulation then heads to our next block, the Stack block. The Stack block is indicated on the graph by low season and high season. This is a housing area for the customer and their demand. It can signify the actual line that occurs during service. For instance, at 8:30 P.M. when there is a set of customers lined up outside the store, the computer recognizes that multiple sales could be happening at once and the customers need to be stacked. Meanwhile, at slower times, stacking is feasible and must be accounted for using this block.

Next we needed to create a Batch block, which can be used to demonstrate customer demand. As each customer has a specific food choice, we will consider each order per customer to be a batch. It is vital that the Batch block be connected with the Labor block as they work concurrently. Because there is no automation in a business of this size, the one-to-one relationship between customer and employee must be maintained and accounted for.

Following this is the Transaction block, which signifies the interaction between the laborer and the consumer. The transaction has then occurred and records the exchange of money for product. The transaction time is created by the Triangular Distribution block located underneath it. We used a triangular distribution because, on average, the transaction times range from 1 to 4 minutes, while the assumption is that the "most likely" scenario would put the transaction time around 2 minutes.

As you progress forward, we un-batched the relationship between labor and the customer, signifying the end of the transaction. When running the simulation we can see the labor leaving the Un-batching block and returning back to the labor pool where they will remain until they are needed again for another transaction. When there are multiple customers, it is assumed that after a typical transaction, the laborer (or person working in the store) takes the next customer in line. However, when there are single customers, the employee will wait, while doing other work tasks, until the next customer arrives.

Located next in the model is a statistical block that calculates estimates for the cost per transaction, what time a revenue was incurred, and how many units were consumed. When collected properly, this can be very useful for assessing future sales. Further analysis might lead to an even better breakdown of the data to help simulate monthly, then weekly, and eventually daily or hourly sales, which would enable smarter staffing choices as well as informed purchasing and flavor availability decisions.

At the end of our model is the Output block, which tallies the number of complete transactions during the simulation. As you can see from this simulation that was run, the number of transactions during low season was 1,988 and during high season 3,628. These values change slightly each time the simulation is run, but they do show that during high season the number of transactions is almost double that during low season. It also gives a great estimate for expectations over the course of a season.

The other blocks present in the model are strictly tools that help with interpreting and understanding the data, whether graphically or integrally. Attached to both of the Stack blocks and Output blocks are Plotter/Discrete Event blocks that graphically show a certain piece of the data. Figure 9.6 provides an example of the average wait time for a customer during low season, while in Figure 9.7 one can see the same graph from the high-season perspective.

Within Figure 9.6, one can see that the $x$-axis represents the minutes of operation over the summer. Recalling that the slow season is the first 4 weeks and last 3 weeks of the summer, then the center of the picture is zero, representing that this is not the low season but the high season. The $y$-axis is the wait time, which ranges from 0 to

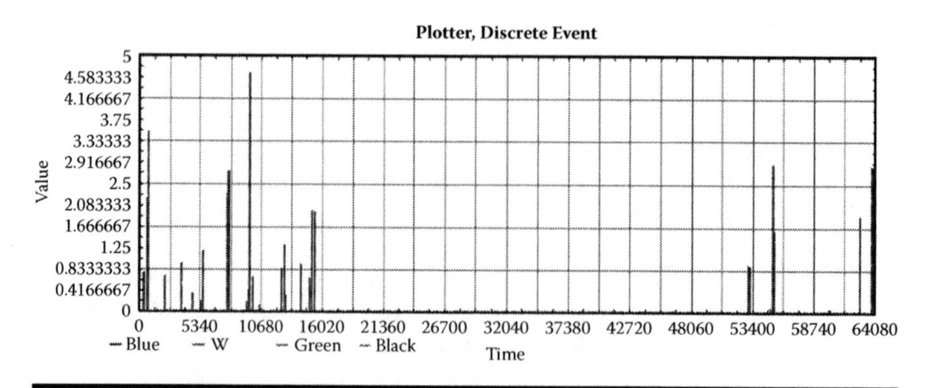


**Figure 9.6 Average wait time for a customer. Graph looks only at low season.**

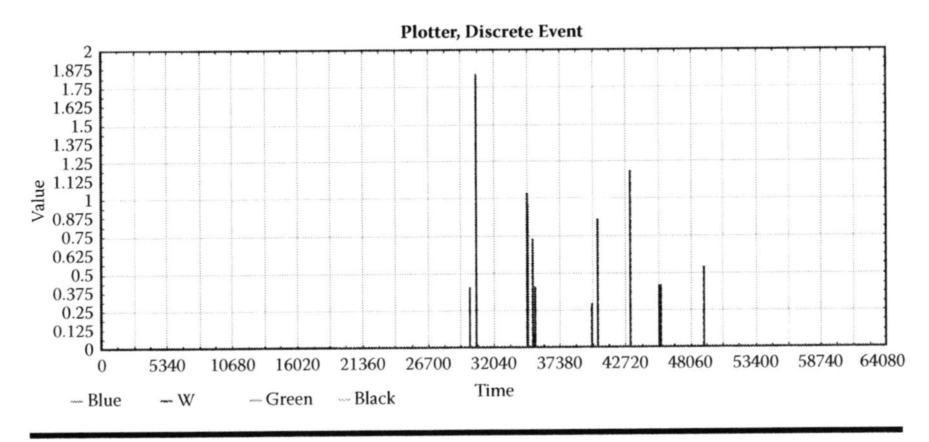


**Figure 9.7 Average wait time for a customer. Graph looks only at high season.**

about 4.5 minutes in Figure 9.6. This might help the employer decide to continue with current staffing choices as there was never an exceptionally long wait time. Figure 9.7 is the same graph for high season, where the wait time range is less than half that of the low season while the *x*-axis is held constant.

Another example is the representation of the number of customers, which is associated with sales transactions for low season and can be seen in Figure 9.8. Note that there is a plateau in the middle of the graph, which again represents the time frame of high season. This figure shows the first 4 weeks and the last 3 weeks in the summer as the customer transactions accumulate. It shows the increasing total of transactions during the hypothetical summer for low season. A similar graph of high season would have a very steep slope as compared to the almost 45-degree angle seen in most of Figure 9.8. Overall, this information could be looked at and broken

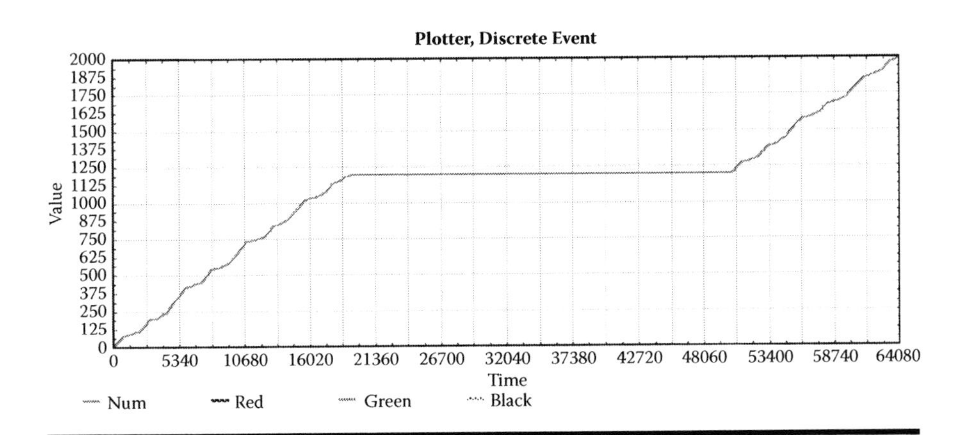


**Figure 9.8 Cumulative customer orders during slower season.**

**Table 9.3 Simulation Results**

| | |
|---|---|
| Avg sale/customer | $5.20 |
| Avg transaction time for all summer | 2.33 min |
| Avg sales for low season (7 wks) | $10,337.60 |
| Avg sales/wk for low season | $1,476.80 |
| Avg sales for high season (6 wks) | $18,865.60 |
| Avg sales/wk for high season | $3,144.27 |

down weekly to assess the highest increasing slopes, again helping management discern employee scheduling as well as when to have a more varied inventory. For example, one would notice that there is a very steep slope for time (in minutes) from 4,005 to 5,340. A more in-depth analysis should be done in this time frame to see what the cause is and how that can be used to perhaps increase customers in the intervals surrounding this time or better use inventory and employees to minimize cost.

Finally, the Activity Stats, Queue Stats, and Resource Stats blocks give us a summary of collected data during the simulation. In the Resource Stats block we see that all of the labor was utilized during low season, but surprisingly the labor was *underutilized* during high season. This is conclusive evidence that Jake's Water Ice shop is slightly overstaffed during the high season. Perhaps the manager can redistribute hours for employees to avoid excessive overlap, which could keep customers and employees happy while reducing payroll. In the Queue Stats block we see that the maximum wait time for a customer in low season was 4.6 minutes and only 1.8 minutes in high season (see Figure 9.5). This has a direct relationship with the number of employees working at that given time. The customer who waited 4.6 minutes was there during low season, and the shop was most likely employing only one or two employees and therefore could not process all the customers' requests in a timely fashion. However, it would be interesting to see whether this, in fact, had any effect on the customer's reaction and choice of dessert in the future or whether allowing a slightly longer wait time might help reduce costs during the high season.

Running the simulation obtained some notable results, which can be seen in Table 9.3. These figures are great estimates that will allow the water ice shop to get a feel for the projections of sales for the 2009 season and also for the factors that they may wish to keep an eye on for future simulations. Note that the results in Table 9.3 are subject to change every time the simulation is run and therefore should be used for estimation and not exact expectations. What we have is one possible solution to the simulation. Our results could have been different if we had changed some of the elements inside the simulation, such as changing the labor pool totals, altering the distributions in the simulation, changing our view of the sales (optimistic approach versus pessimistic approach), changing process times, etc.

## 9.5 Inventory Optimization

As stated in the literature review, seasonal demand, which is dynamic, is a difficult thing to predict but it is worth the effort because it directly affects inventory costs and labor costs. Therefore, although we ran into various problems due to the many variables to be taken into account, effort was put into trying to create an inventory system that might work for Jake's Water Ice. The first step was to recognize that there are three separate supply chains happening that do not overlap, and that each one must be considered in order to accurately estimate the inventory over time to create a good simulation in the future.

The three main inventory systems that enable Jake's Water Ice to operate include a hard ice cream distributor, a soft ice cream distributor, and a water ice ingredient and all other supplies distributor. Each of these individual categories has its own suppliers and specific variables, such as order minimums and storage capacity maximums. There are limitations within specific delivery dates as well as deadlines by which inventory must be ordered. We hope that with the breakdown we have created and careful data collection, a forthcoming simulation will prove invaluable to this and possibly other small businesses.

First, we start with the variables and constraints of the hard ice cream. The hard ice cream comes directly from the Island Ice Cream Company, which supplies 2.5-gallon tubs of product as well as frozen pretzels. This company has set delivery dates of Tuesday and Friday, and has a $100 minimum purchase requirement. This means that if a particular flavor of ice cream is getting low or runs out, an order needs to be placed before the set delivery dates while also meeting the minimum purchase. On the capacity utilization side, it is important to note that freezer space is very limited. There are two dip top freezers at the store that have enough space to allow for 16 different flavors of ice cream for serving. The only storage space at the proper temperature setting is four spots underneath the ice cream in each freezer allowing for a total of eight tubs underneath, leaving the store with a maximum of 24 tubs in inventory at any one time.

This problem could be considered a lot size problem with dynamic demand. Lot size problems are generally separated into two categories: periodic and continuous. Due to the fact that the store is seasonal and open for about 13 weeks, the orders can be seen as discrete and a periodic review will be our focus. There is a fixed cost, or fuel surcharge, for each delivery ($A_t$), the variable cost per order ($C_t$), and the size of the lot procured for each time period ($Q_t$). The starting season has no ice cream remaining from a previous time period, no ice cream should remain at the end of the season, and there must always be some ice cream available to customers while the store is open. There is also an inventory cost ($h_t$) to carry stock from time $t$ up

**Table 9.4 Island Ice Cream Inventory**

| *Week* | *1* | *2* | *3* | *4* | *5* | *6* | *7* | *8* | *9* |
|---|---|---|---|---|---|---|---|---|---|
| Expected Demand ($D_t$) | 90 | 10 | 12 | 14 | 10 | 15 | 25 | 55 | 100 |
| Fixed Cost ($A_t$) | 3 | 0 | 0 | 3 | 0 | 0 | 0 | 3 | 3 |
| Variable Unit Cost ($C_t$) | 180 | 0 | 0 | 90 | 0 | 0 | 0 | 90 | 120 |
| Unit Inventory Cost ($h_t$) | 15 | 10 | 10 | 15 | 15 | 15 | 15 | 15 | 20 |

until time $t + 1$. To estimate the size of the orders, which we again separate into low season and high season, we will want to minimize $Q_1$, $Q_{2,\dots}$, $Q_n$ in order to minimize purchasing costs and inventory costs over the $N$ time periods. Note that with only 13 weeks and two delivery days per week, the maximum value of $N$ is 26. $D_t$ represents the demand expectations for the company. A good simulation model would need to take all of these factors into consideration.

Looking at Table 9.4, one can see that there is a need for an initial build-up of hard ice cream inventory. While only a few flavors are needed in the first few weeks, the refrigeration unit must be filled in order to make sure that temperatures remain constant, and also it is pointless to run a refrigeration unit that holds eight tubs for only two tubs of ice cream. Therefore, the Memorial Day weekend purchase is a larger purchase because there will be an immediate high demand followed by a very slow interval. The next purchase will not be necessary for about 2 weeks, and this will remain constant for all low-season time periods. Because of the minimum requirement, it is not possible to replace single tubs, and one must wait until multiple items are low before ordering—hence the large time frame between purchases. On the other hand, once high season begins, both freezer units are started and orders will be placed twice a week as the inventory will move much faster and the major limitation becomes capacity for storage as well as making sure that no major flavors, such as vanilla, ever run out.

While this is a good initial table for development, individual orders would need to be monitored more closely in order to create an accurate simulation that could be used per customer order. The model that would need to be developed would have to take into account all these variables to help the store manager determine how much of each flavor needs to be ordered, and what dates to order each flavor, all the while making sure the minimum purchase is being consistently met. Although we had daily sales totals, we did not have information on how much of each individual flavor was being sold daily, weekly, or even monthly. Without this important piece of data, it was difficult to develop a working model that could accurately optimize the inventory order process because we could not determine what the demand was for each individual flavor.

Next we consider soft ice cream. This comes from a company called Balford Farms, which not only provides the soft ice cream mix, but also supplies whipped

**Table 9.5 Balford Farms Inventory**

| *Week* | *1* | *2* | *3* | *4* | *5* | *6* | *7* | *8* | *9* |
|---|---|---|---|---|---|---|---|---|---|
| Expected Demand ($D_t$) | 250 | 45 | 45 | 45 | 60 | 60 | 60 | 200 | 300 |
| Fixed Cost ($A_t$) | 4.5 | 4.5 | 4.5 | 4.5 | 4.5 | 4.5 | 4.5 | 4.5 | 4.5 |
| Variable Unit Cost ($C_t$) | 160 | 75 | 80 | 70 | 70 | 100 | 120 | 160 | 200 |
| Unit Inventory Cost ($h_t$) | 30 | 20 | 20 | 20 | 30 | 30 | 30 | 30 | 30 |

cream, milk, and iced tea for the store. The soft ice cream mix is purchased in lot sizes of 4 gallons, which is referred to as a case. The set delivery days are Monday and Friday, and Balford Farms has a two-case minimum purchase requirement. The only soft ice cream flavors served at this location are vanilla and chocolate. It is imperative to account for the fact that milk has a very short shelf life. Therefore, new product must be purchased weekly, regardless of sales, because the expiration dates make old items unusable. On the capacity side, there was only a maximum available refrigerated space to store about six cases without getting creative. In the high season, the two delivery days often put a strain on the capacity issue. Most milk companies used to deliver three or four times a week. However, to tighten their internal supply chains, deliveries are now made on only two days, making it very difficult for a small seasonal business to gauge inventory versus capacity. Again, once the individual data is collected, a simulation model for soft ice cream will be invaluable to the company.

Looking at Table 9.5, one can see that it is not the same inventory model as for the hard ice cream (see Table 9.4). Once the machine is running, there must always be product available for the machine. Also, the machine is cleaned often and product running throughout the machine must be thrown away. Also, due to expiration dates, it is rare that one can go very long between orders. In addition, because soft ice cream is used for cones as well as gelati and sundaes, there is a higher overall demand. The fuel surcharge is higher and the inventory cost is increased due to lost product. The representation is for low season, which is very similar to high season. The main difference would be that the unit inventory cost would decrease because much less product is wasted in proportion to product purchased in the high season. The machine is cleaned at regular intervals regardless of season. Because in low season there are less than half as many purchases between cleanings, the proportion of wasted product is much lower in high season.

Finally, all other supplies are purchased from Gil's Wholesale, which provides every other item the store needs, including spoons, all the various cups and sizes, lids, sugar, water ice ingredients, fudge, dip top, candy, jimmies, cleaning supplies, buckets, spatulas, and virtually every other item the business uses on a daily basis. This company is based in Philadelphia and provides delivery on Thursday with a minimum purchase requirement of $350. All orders less than that amount must be picked up directly at the store, which is an inconvenience. While there are two freezers on site dedicated to water ice that can store a maximum of 27 gallons, this factor doesn't really have an effect on the ordering because there are many other variables that could initiate a purchase (such as an order of cups, cones, or toppings). Therefore, the optimization for this particular part of the inventory is different. This is again a periodic review with a dynamic demand. There are still only thirteen weeks over the summer and, hence, thirteen possible delivery dates. Generally, unlike the ice cream, everything ordered from Gil's has a much longer shelf life and there is more room for the storage of these items throughout the store. Because only the ingredients to make the water ice are purchased and no refrigeration is needed, the capacity for storage is much larger.

Therefore, the manager keeps lists that the opening employee must check off each day. The opening employee is in charge of refilling all the supplies from cups to spoons, to toppings. The employees go through a list of every item that is in inventory and are asked to check off whichever they believe to be running low. The manager utilizes the list to determine if, in fact, anything is close to running out and adds that to the Tuesday order from Gil's, which will be delivered on Thursday. When done properly, this is an efficient way to make sure that not only are the orders large enough to meet the minimum requirements, but also to ensure that all supplies are replenished before they run out.

Although we are using this as a periodic review with a dynamic demand, there are some large differences. First, the initial supplies do not necessarily have to reset to zero each summer. Cups, spoons, and napkins can be stored and used regardless of how long they have been in inventory. Therefore, the ending inventory might also not be zero. In addition, the purchase of some of the products is harder to calculate because for every ice cream, gelati, or sundae that is made, a cup, a spoon, and some napkins are also used. Because of this, the simulation model that should be done once more data is collected may prove to be more continuous in nature and it may be harder to capture what is used per customer order.

Because of our lack of the specific individual orders beyond what was collected for the customer arrival simulation, more data is needed to provide an accurate inventory optimization. One method that could be used to obtain this data would be to track it at the register. Have the employees enter into the register exactly what the customer ordered, going as far as to include toppings, type of product (hard ice cream, soft ice cream, water ice), and exact flavor. You could then extract this data from the register every day or every week and enter it into a database. By doing this, we would know exactly what the most popular flavors are, how much of each flavor is being sold, when specific flavors sell more, etc. Because this might be infeasible, another method might be to track the current inventory and how quickly it turns over. Using one of these methods of collection, we would then be able to run a merged model with our first simulation giving an accurate representation of the entire business process, as well as allowing us to properly develop an inventory optimization model.

The simulation we did do of the high and low season waiting times and process times did give us a few conclusions for the inventory. First, from that model, we were able to determine that the number of customers during high season was almost double the number of customers during low season. There were shorter wait times and much more demand, meaning that there would be many more orders for supplies and product during this period. This would mean that the manager of the store would need to either order more product at the times he placed orders, or place orders more frequently. Another point we discovered from the demand simulation was that Saturday, Sunday, and Wednesday were the busiest days for the store. This can help the manager determine when to order more product based on how much inventory is available leading up to these high-demand days. Finally, had we had specific weather conditions, we could have also made some additional conclusions about both the regression and the inventory optimization.

Had we had the data we needed—specifically the amount of each individual product sold—we would have then been able to create a merged simulation model with our first demand model. This model would then show the entire business process from when customers come to the counter to when they pay and leave with their orders, updating the inventory model along the way. Being able to capture and interpret this information would make the supply chain run more efficiently. The model would connect at a point where the employee creates the individual product for the customer, also known as the batch block in our simulation. Our inventory model would then get updated, and the repository box in the inventory model would get lower. This process would be run for both the high and low seasons, and repository boxes would be created for the three main products: hard ice cream, soft ice cream, and water ice supplies.

As the customers come to buy product, the individual repository box that is affected would be updated to reflect the product inventory change. We would then be able to set a certain restock point for these boxes to help tell us approximately what dates and how quickly they reached the predetermined point during each season, indicating when the manager would need to order each individual product. This could be as specific as determining exactly when the manager would need to order a specific flavor, or as broad as just determining when to order one of the main product categories (i.e., hard ice cream, soft ice cream, water ice) and the manager could then determine the flavor by simply looking at what is in stock.

The process of forecasting demand, using the projected forecast to simulate a business process, and creating an inventory optimization model from the simulation output can be a very powerful business tool. One of the major pitfalls we experienced throughout our attempt to complete this process for the water ice business was data collection. Before beginning this project, we assumed that we had ample data to complete a detailed inventory model for the water ice business. Once we started along the process, however, we realized that the data we had would not be sufficient at that time to put in place the kind of inventory model we set out to create. In the future, data collection really needs to be the center of our focus. We plan to discuss a new data collection plan with the water ice store staff that focuses on being much more detailed regarding the type of daily data being collected than in the previous years. As discussed earlier in this chapter, we definitely believe that collecting weather data for each day the store is open would really have improved our regression model and therefore increased the accuracy of the demand forecast. The most important factor going forward, however, is an accurate estimate of each product (hard ice cream, water ice, soft ice cream, etc.) that is being sold each day. It was impossible for us to put together a specific laid-out inventory plan when this data did not exist.

The small water ice business, like most small companies, faces new challenges and competition yearly. Therefore, in an effort to remain profitable, simulating a summer and creating an inventory optimization plan can really give a small business like this a unique competitive advantage. There is an endless list of benefits to being able to run as close to zero inventory as possible. Our goal for the water ice company would be for an optimal flow of inventory throughout the summer where no inventory is being kept stagnant, and freshly delivered product is at the fingertips of the customers.

## References

Bradley, J.R., and B.C. Arntzen. (1999), The simultaneous planning of production, capacity and inventory in seasonal demand environments, *Operations Research,* 47(6), 795–806.

Casper, C. (2008), Planning comes of age, *Food Logistics,* 101, 19–24.

Chang, Y., and Makatsoris, H. (2008), Supply chain modeling using simulation, *International Journal of Simulation Modeling,* 2(1), 24–30.

Choudhury, B., Agarwal, Y.K., Singh, K.N., and Bandyopadhyay, D.K. (2008), Value of information in a capacitated supply chain, *INFOR: Information Systems and Operational Research,* 46(2), 117–127.

Eynan, A., and Kropp, D. (1998), Periodic review and joint replenishment in stochastic demand environments, *IEEE Transactions,* 30(11), 1025–1034.

Fraser, I., and Moosa, I. (2002), Demand estimation in the presence of stochastic trend and seasonality: the case of meat demand in the United Kingdom, *American Journal of Agricultural Economics,* 84, 83–89.

Hassini, E. (2009) Supply chain optimization: current practices and overview of emerging research opportunities, *INFOR: Information Systems and Operational Research,* 46(2), 93–96.

Kanaka, B. (2008). Optimization in supply chain: zero inventory approach, *Icfai University Journal of Supply Chain Management,* V(3), 33–42.

Malykhina, E. (2004, October 18), Small business' big supply chains, *Information Week,* pp. 90–92.

Metters, R. (1997), Production planning with stochastic seasonal demand and capacitated production, *IIE Transactions,* 29(11), 1017–1029.

Mulani, N. (2009, February 1), The case for inventory optimization, *Logistics Management,* (February): www.logisticsmgmt.com/article/CA6635273.html

Rafter, M.V. (2009), Let's Get Visible: Supply Chain Technology, http://technology.inc.com (accessed October 15, 2009).

Trebilcock, B. (2008), A new supply chain by design, *Modern Materials Handling,* 63(13),47–50.

Wheatley, M. (2008), Simulating the supply chain, *Automotive Logistics,* 2, 52–55.

# OPTIMIZATION MODELS OF SUPPLY CHAIN PROBLEMS

# Chapter 10

# A Dynamic Programming Approach to the Stochastic Truckload Routing Problem

Virginia M. Miori

## Contents

## 10.1 Introduction

The evolution of the truckload routing problem combines a typical assignment problem with a restricted probabilistic traveling salesman problem. Although a significant problem in the trucking industry, it has attracted much less attention in the literature than its less-than-truckload counterpart, the vehicle routing problem.

This chapter first presents related literature and then follows with the triplet formulation of the problem. The progression from the original formulation to one ideally suited for dynamic programming is presented along with the analysis of sample problems and the complexity of the algorithm.

## 10.2 Literature

Arunapuram et al. (2003) noted that truckload carriers were faced with a difficult problem. They presented a new branch and bound algorithm for solving an integer programming formulation of this vehicle routing problem with full truckloads (VRPFL). Time window and waiting cost constraints were also represented in the problem formulation. The resulting efficiency of the method was due to a column generation scheme that exploited the special structure of the problem. The column generation was used to solve the linear relaxation problems that arose at the nodes. The objective in solving the VRPFL was to find feasible routes that minimized cost. They noted that minimizing the total cost was equivalent to minimizing the cost of traveling empty.

Chu (2004) presented a heuristic that applied to both the VRP and the TRP (less-than-truckload versus truckload applications). The focus of the article was a heuristic technique, based on a mathematical model, used for solution generation for a private fleet of vehicles. Outside carriers that provided LTL service could be employed in the solution. Chu addressed the problem of routing a fixed number of trucks with limited capacity from a depot to customers. The objective was the minimization of total cost.

The TL-LTL heuristic developed by Chu was designed to not only build routes, but also to select the appropriate mode. The first step was the selection of customers to be served by mode. LTL customers were stripped out and routes built using the classic Clarke and Wright algorithm (1964). The final step was swapping customers between and within routes. Although Chu addressed the TRP, its primary focus was on the LTL mode and, in fact, it replicated a typical "cluster first–route second" approach.

Gronalt et al. (2003) dealt with the TRP with time window constraints. The approach they utilized was a savings-based approach. The objective was to minimize empty vehicle movements because the empty movements used the same number of vehicle resources as loaded movements, but generated no revenue. They provided an exact formulation of the problem and then calculated a lower bound on the solution based on a relaxed formulation using network flows. They further presented four different savings-based heuristics for the problem. Their results showed that these methods found good solutions quickly. A sensitivity analysis followed that determined the impact of time window tightness on the performance of the heuristics.

There were generally two different problem classes. The first was concerned with dynamic carrier allocation (e.g., train carriers, pallets) to different locations over time. The second class of problems dealt with the pickup and delivery of orders by a fleet of vehicles. In Gronalt et al. (2003), the authors dealt with a problem that belonged to the second class. Customers placed orders with a logistics service provider, necessitating shipments between two locations. The provider served these orders from a number of distribution centers. The shipments occurred at different echelons of the distribution network. The first was between pickup locations and the nearest distribution center. The second type fell between distribution centers. The last was between a distribution center and the order destination. Shipments between a distribution center and a customer were typically less-than-truckload shipments. Those shipments that took place between distribution centers were typically truckload shipments.

Gronalt et al.(2003) strived for the goal of minimizing empty vehicle movements for the truckload portion of the shipments. The results were also extended to a multi-depot problem. The algorithms included the savings algorithm; the opportunity savings algorithm, which incorporated opportunity costs into the calculation of savings values; and the simultaneous savings approach, which accepted a number of savings combinations at each iteration that satisfied a given condition.

The TRP has also been addressed in the literature as a multi-vehicle truckload pickup and delivery problem. Yang et al. (2000) presented a problem in which every job's arrival time, duration, and deadline are known. They considered the most general cost structure and employed alternate linear coefficients to represent different cost emphases on trucks' empty travel distances, jobs with delayed completion times, and the rejection of jobs. Note again that this problem did not require all potential loads to be serviced. The problem was presented first as an offline problem with cost minimization as its objective. A rolling horizon and real-time policies were then introduced.

The authors examined a specific stochastic and dynamic vehicle routing problem. They devised heuristic stationary policies to be used by the dispatcher for each situation, at each decision epoch. The policies were based on observations of the offline problem and were intended to preempt empty movements. If a vehicle was already en route to a job, it could be dispatched to serve another job based on these rules. A series of simulations showed that the policies developed were very efficient.

## 10.3 Triplet Formulation

The formulation of the TRPTW model utilizes a triplet concept rather than a lane concept (Miori 2006). The TRPTW invariably results in a loaded lane movement followed by an empty lane movement; a triplet ties these movements together.

The triplet creates a logical connection between the associated loaded and empty movements, although the formulation does allow for two loaded movements as well. The empty movements represent a stochastic element of the TRP; triplet notation eliminates this.

A route is composed of a series of triplets to be serviced by a single vehicle. A triplet may be composed of a loaded and empty movement, two loaded movements, or two empty movements. We must begin and end a route at the same domicile. Multiple domiciles may be present in a single instance of the TRPTW model. Time conditions are placed on the delivery of freight to the customers. The remaining conditions are imposed by the Department of Transportation (DOT). Restrictions are placed on the number of driving hours in a given week and the number of work hours in a given week.

The TRPTW can be stated as a restricted probabilistic TSP. We customize the standard TSP constraints for our triplet formulation and supplement with load, schedule and triplet feasibility, vehicle utilization, and DOT regulations. Let

$N =$ Number of nodes
$V =$ Number of vehicles
$w_i =$ Total allowed transit hours per route

The time window specifications are particular to the location of each node. These require the subscript $i$, where $i = 1, \dots, N$. $D_i$ is a model parameter that is calculated as the solution is generated. It falls within the delivery time window and is based on route starting times and transit times.

$D_i =$ Departure time from node $i$
$[e_i, l_i] =$ Early and late arrival time window for node $i$

The demand for service and the travel time parameters are provided for each lane pair. Lanes are designated $ij$ where $I = 1, \dots, N$ and $j = 1, \dots, N$.

$y_{ij} = 1$ if node pair $ij$ represents a loaded movement, 0 otherwise
$t_{ij} =$ Travel time between nodes $i$ and $j$

The costs per triplet per vehicle and the decision variables employ designations of location, subscripts, $i$, $j$, and $k$ where $i = 1, \dots, N$; $j = 1, \dots, N$; $k = 1, \dots, N$. In addition, the vehicle must be specified for the decision variables, subscript $v$ where $v = 1, \dots, V$.

$C_{ijk} =$ Cost to service triplet $ijk$
$x^v_{ijk} = 1$ if triplet $ijk$ is served by vehicle $v$, 0 otherwise

The objective is stated as a cost minimization function. The complete formulation follows:

$$Minimize \sum_i \sum_j \sum_k c_{ijk} x^v_{ijk} \tag{10.1}$$

subject to:

$$\sum_v \sum_k x^v_{ijk} + x^v_{kij} \geq 1 \quad \forall y_{ij} = 1 \tag{10.2}$$

$$\sum_{ijlmv} (x^v_{ijk} - x^v_{klm}) = 0 \quad \forall k, v \tag{10.3}$$

$$D_i + t_{ij} \leq D_j \quad \forall i, j \tag{10.4}$$

$$e_i \leq D_i \leq l_i \quad \forall i, j \tag{10.5}$$

$$\sum_{ijk} (t_{ij} + t_{jk}) x^v_{ijk} \leq H \quad \forall v \tag{10.6}$$

The decision variables must take on binary values:

$$x^v_{ijk} \in \{0, 1\} \tag{10.7}$$

## 10.4 A Penalty Method for the TRP

In this section we present a Lagrangian relaxation of the TRP formulation. This prepares us for the restatement as a Markov process and the Dynamic Programming formulation. We introduce a Lagrange multiplier for each constraint to be relaxed and form a linear combination of the constraints with their corresponding Lagrange multipliers as coefficients.

$$minimize \sum_{ijk} \left[ c_{ijk} x^v_{ijk} + \lambda_{1a} \sum_{ijk} [\max(0, (t_{ij} + t_{jk}) x^v_{ijk} - H)] \right]$$

$$+\lambda_b \left( \sum_{ijk} \left[ \max\left(0, H - (t_{ij+t_{jk}}) x^v_{ijk}\right) \right] + \lambda_2 \left( \sum_{vk} \left[ y_{ij} \left( x^v_{ijk} + x^v_{kij} \right) - 1 \right] \right) \right) \tag{10.8}$$

subject to:

$$\sum_{ijlmv} \left( x^v_{ijk} - x^v_{klm} \right) = 0 \quad \forall k \tag{10.9}$$

$$D_i + t_{ij} \leq D_j \quad \forall i, j$$

$$e_i \leq D_i \leq l_i \quad \forall i \tag{10.10}$$

$$x_{ijk}^{v} \in \{0, 1\} \tag{10.11}$$

In this case, only two of the constraints will be relaxed, that is, the DOT hours of service and the load satisfaction constraints. The remaining constraints will be satisfied through later development of transition probabilities for the dynamic programming formulation and solution.

The DOT hours of service constraint is weighted by $\lambda_1$. The load satisfaction constraint is weighted by $\lambda_2$.

The Lagrangian relaxation provides an excellent way to relax constraints. By placing them in the objective function and applying the Lagrange multipliers, we create a penalty function. The solution of a relaxation of this type does, however, optimize over the decision variables as well as the Lagrange multipliers. In this application, we instead treat the multipliers as penalties within the penalty function. Therefore, alternate values of $\lambda_1$ and $\lambda_2$ will be evaluated, but not treated as variables to be optimized.

## 10.5 Restatement as a Markov Process

The formulation above uses traditional transportation problem notation, which is extended to handle the additional requirements imposed by the TRP. As noted, a number of solution techniques have been used to address the problem. This chapter seeks to solve the TRP using dynamic programming and therefore requires changes in how the problem is approached and in the problem notation.

We move from the traditional view of the problem to a tree structure as shown in Figure 10.1. The originating node of the tree represents the domicile from which all vehicles commence their routes. The same node also represents the ultimate destination of each route. All successive levels represent the addition of a node, which may be a lane or a triplet, to each vehicle's route. Upon selecting a node at each level, examination of the best node to select at the next level can take place. The expected depth of any tree will be determined from the average miles accumulated by the nodes as the total DOT allowable miles divided by the average miles per node.

Dynamic programming solutions depend on our ability to evaluate the cost-to-go of the tree. Given the stochastic nature of demand in the TRP, at best we will be able to calculate the expected cost-to-go. To do so requires probabilistic information. These probabilities are determined from a count data model that predicts the number of loads available and produces the likelihood of the occurrence of particular loads at that level, where the level corresponds to a time window.

A key component of the probabilities, no matter the method of determination, is consideration of only the current level. We do not have to consider the probabilities associated with the selection of nodes in the path leading to the current level. This

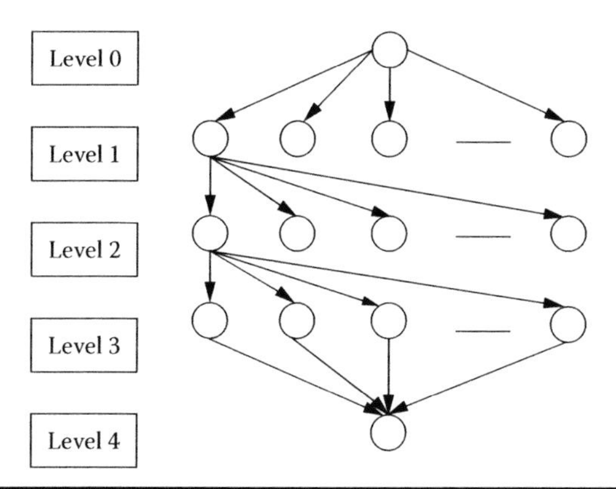


**Figure 10.1 General tree structure.**

important property, also known as the Markov property (Equation 10.1), allows us to view the tree as a finite Markov process.

$$P(X_{k+1} = x_{k+1} | X_o = x_o, \ldots, X_k = x_k) = P(X_{k+1} = x_{k=1} | X_k = x_k) \tag{10.12}$$

In our finite Markov process, transitions between nodes at successive levels are captured in a transition probability matrix. In a homogeneous Markov process, a single stochastic matrix $\mathbf{P}$ represents transitions at any level. We require instead a time nonhomogeneous Markov process due to variation in transition probabilities for each level of our tree. Therefore, we employ a sequence of stochastic transition probability matrices. The superscript in the notation signifies the level of the dynamic program (tree structure) at which this matrix applies:

$$\mathbf{P}^{(k)}, k = 1, \ldots, N \tag{10.13}$$

The determination of the transition probability matrices includes a number of factors, among them the capture of the time window information. Triplets are served at particular levels of the tree based on the need to deliver loads during the corresponding time frame. Therefore, the probability of a triplet being served at a particular level of the tree will increase when the time window matches the level of the tree and decreases otherwise. The inclusion of time window information in the transition probabilities ensures that these constraints will be satisfied. Recall that our relaxed TRP was left with only these constraints, resulting in the satisfaction of all constraints in the TRP.

We can now carry the notation and the new view of the problem into a dynamic programming formulation. The transition probability matrices will play a key role in the evaluation of the expected cost-to-go in the dynamic program.

## 10.6 Dynamic Programming Formulation

Prior to the presentation of the dynamic programming formulation, we must define the necessary functions and notation. We begin with a general notation that reflects the problem size and node designation:

$A(k+1, k)$ = Available nodes at level $k+1$; depends on node chosen at level $k$
$n_{k+1}$ = Cardinality of $A(k+1, k)$ (the number of nodes available at level $k+1$, which depends on the node selected at level $k$)
$s(k)$ = Node (triplet) chosen at level $k$

The transition probability matrices may be referenced in general terms or on a node-specific basis:

$\mathbf{P}^{(k)}$ = Transition probability matrix at level $k$
$\mathbf{P}^{(k)}_{i,j}$ = Transition probability of loaded movement from node $i$ to node $j$ at level $k$

The costs, distances, and demand associated with the individual problem to be solved using the dynamic program may refer to an entire node (triplet), although in some cases it must be specified for individual stops within a triplet.

$\mathbf{C}_{s(k),j}$ = Cost of adding node $j$ after the node chosen at level $k$
$\mathbf{E}_{q,r}$ = Cost of moving from stop $q$ to stop $r$
$\mathbf{F}_{q,r}$ = Distance between stop $q$ and stop $r$
$L_{q,r}$ = Loads available between stop $q$ and stop $r$
$f(\mathrm{i}, k)$ = Expected cost-to-go if we choose node $i$, stage $k$
$f(1, 0) = 0$

Several values are determined arithmetically for (and within) the dynamic program:

$T$ = Total load requirement for the routing scenario
$D(k-)$ = Max(DOT maximum hours – actual hours,0) at level $k$
$D(k+)$ = Max(actual hours – DOT maximum hours,0) at level $k$
$U(i, k)$ = Number of unserved loads if we choose node $i$ at level $k$
$E_{\delta(s(k)),1}$ = Return cost to domicile

Notation breaking down nodes into stops is also necessary in evaluating the dynamic program:

$(u, v)$ = A lane represented by two vertices: origin and destination
$(u, v, w)$ = A triplet represented by three vertices: origin, intermediate, and destination
$\alpha(s(k))$ = Initial stop in the node chosen at level $k$ in the triplet representation

$\beta(s(k))$ = Intermediate stop in the node chosen at level $k$ in the triplet representation

$\delta(s(k))$ = Final stop in the node chosen at level $k$ in the triplet representation

In the dynamic programming recursion, we are selecting the node z at the next level, and the penalty values (Lagrange multipliers $\lambda_{1a}$, $\lambda_{1b}$, and $\lambda_2$). The node z minimizes the sum of the cost of that node and the cost-to-go. The unserved cost is calculated as an expected value. The probability of a load (or loads) is applied to the number of loads subsequently left unserved if this load is carried. The probability that no loads are served is then applied to the total number of unserved loads given that none are served. $\lambda_{1a}$, $\lambda_{1b}$, and $\lambda_2$ are the penalties on the relaxed constraints: DOT hours of service restriction and load satisfaction.

All routes start from a single domicile or terminal (T) and return to that same domicile at the completion of the route. The construction of the routes requires that the domicile location is coincident with origin 1 of the first triplet selected. Successive triplet selections require origin 2 of the existing triplet to coincide with origin 1 of the next triplet in sequence. This method of selection continues through to the triplet selected at the penultimate level of the tree. The route is subsequently completed by the return to the domicile.

The full recursion follows:

$$f(s(k), k) = \min{}_{\mathrm{T}}(z \in (k+1, k), \lambda_1\lambda_2)\{C^{((k))}_{(s(k),z)} + \dashv E_{(s(k),Z))} + \gamma_{1a}(D_{(k-)}) \\ + \lambda_{ib}(L\mathbf{P}^{((k))}_{s(k)z})\lambda_2 U(s(k), k) + (1 - \mathbf{P}^{((k))}_{s(k)z})\lambda_2 U(1, k) + f(z, k+1)\} \tag{10.14}$$

The mileage matrix ($\mathbf{M}_l$) for triplets is generated from the mileage matrix for individual stops. The distance from the initial stop of the triplet to the intermediate stop is added to the mileage from the intermediate stop to the final stop in the triplet. The cost is accumulated in the same fashion. The product of the cost-per-mile for each leg of the triplet and the mileage for each leg of the triplet are calculated:

$$\mathbf{M}_l = \mathbf{F}_{\alpha(l),\beta(l)} + \mathbf{F}_{\beta(l),\delta(l)} \tag{10.15}$$

$$\mathbf{C}_l = \mathbf{E}_{\alpha(l),\beta(l)} \cdot \mathbf{F}_{\alpha(l),\beta(l)} + \mathbf{E}_{\beta(l),\sigma(l)} \cdot \mathbf{F}_{\beta(l),\delta(l)} \tag{10.16}$$

Recall that the original formulation of the TRP contains a constraint that reflects the DOT limitation on total miles traveled in a route. This constraint is relaxed and included in the recursion with a penalty term:

$$D(k) = \left| \sum_{g=1}^{k} [\mathbf{M}_{s(g)+\mathbf{F}_{s(g-1)s(g)}}]^{-m} \right| \tag{10.17}$$

The original formulation also contains a constraint that guarantees service of all loads. An additional term should also be included to penalize unserved loads:

$$U(s(k), k) = \max \left\{ 0, T - \sum_{g=0}^{k} [\mathbf{L}_{x(s(g)),\beta(s(g))} + \mathbf{L}_{\beta(s(g)),}\ \delta(s(g))] \right\} \quad (10.18)$$

The connection cost results from any discontinuity in the route generated. Because all triplets are required to share a stop, the last stop of triplet n corresponds to the first stop of triplet n + 1.

A discontinuity can only occur between the final stop in the last triplet and the domicile. In this instance, the cost and distance between the final stop in the departing node and the domicile is determined for inclusion in the recursion ($E_{\delta(s(k)),1}$), and in fact produces the stopping condition in the recursion.

## 10.7 Dynamic Programming Model Results

Prior to examining the results of the dynamic programming model, it was important to establish a baseline. To accomplish this, the dynamic programming code was run on sample data set 1, in a deterministic simulation mode. All probabilities associated with any level of demand included the case of zero demand and were elevated to the value of 1.0, indicating that the demand was no longer uncertain. The probabilities for the zero demand case could also be set to 0.0 because the expectation requires the product of the probability and demand. In either case, the resulting value is zero.

The outcome of the deterministic dynamic programming run was then compared to the outcome of the optimization with deterministic demand. We turned again to AMPL (Fourer et al. 2003) and employed the NEOS (Czyzyk et al. 1998, Gropp and Moré 1997), and feaspump (Bertacco et al. 2005, Dolan 2001, Fischetti et al. 2005) solver to model the deterministic demand case. The feaspump solver was initially developed to solve binary, or 0-1, linear programming problems, much like the problems we solved. It was then extended to solve general mixed-integer linear programming problems.

After establishing the baseline, we could then move on to a comparison of the dynamic programming results to a typical dispatcher produced solution. The dynamic programming model was then run on a series of stochastic scenarios. We first reflected adjustments to penalty parameters and then, having selected appropriate penalties, performed runs using additional sample data sets.

Early in the testing process it became clear that the unserved load penalty should take on values with an order of magnitude of $10,000 to ensure that every unit of demand was served, a critical constraint in this model. Other approaches to the TRPTW have allowed the carrier to "cherry pick" the demand that they served (Chu 2004, Frantzaskakis and Powell 1990, Powell 1996, Yang et al. 2000). In doing so, they actually eliminated the source of greatest complexity, building routes to fulfill unappealing demand requirements. This model was predicated on ultimately fulfilling the need to serve all demand.

The next penalty to consider was that applied to insufficient use of time, or alternatively, insufficient utilization of vehicles. Because the dynamic programming model was already biased toward producing longer routes due to the nature of the recursion, the penalty on using insufficient time did not have a strong impact. It was therefore left at a value of zero.

The remaining penalty, which when applied to excess time, has proven to have several very interesting impacts on the routes produced. The first and foremost was that of restricting the routes in length in order to meet Department of Transportation Hours of Service Rules.

If these rules are violated, the offending carrier must pay a fine. No restrictions are placed on the drivers' ability to do their job. It stands to reason therefore that carriers would be willing to incur fines in certain routing scenarios. The fines are established by the states and range between $500 and $2000 per instance (Federal Motor Carrier Safety Administration website, http://www.fmcsa.dot.gov/). The order of magnitude of the excess time penalty directly reflects these fines.

The issue of time restrictions presented an opportunity afforded by the development of routes in this way. The maximum number of hours driving is not a single specific number, but is actually made up of a series of driving scenarios and the hours allowed for each (http://www.fmcsa.dot.gov/). The carriers can interpret them and apply them in whatever manner is most advantageous to them. In addition, the ability to vary the excess hour penalty serves as a useful tool for dispatchers. Each adjustment to the penalty allows the production of additional route scenarios, and dispatchers can select the one that more fully meets their needs on any given day. This would then provide a basis for negotiation of work rules with the driver union.

Upon completion of the runs for the base case, routes were generated for four additional data sets. These data sets varied in terms of number of locations served, geographic dispersion of locations, amount of demand, and probability of demand instances. Each run was followed by the analysis of results and data to characterize the attributes of the problem setting. Anomalies within each setting were identified to provide greater insight into continued research.

### 10.7.1 Sample 1

The Sample 1 data set, as already mentioned, is a simulation of truckload demand. The domicile is in the most central location. Five locations were oriented on a grid that represented typical truckload distances and directions. This data set became the baseline analysis data and was therefore evaluated using a number of different techniques. A manual procedure was performed to create routes as they would be in a typical dispatch solution with deterministic demand. This implies the delivery of

**Table 10.1 Manual Lane Routes: Typical Dispatch Solution**

| *Lane Route* | *Hours* | *Cost* ($) |
|---|---|---|
| 1: (1,1) (1,4) (4,1) (1,1) | 32.34 | 1,846 |
| 2: (1,1) (1,2) (2,1) (1,1) | 16.48 | 1,121 |
| 3: (1,2) (2,1) (1,3) (3,1) | 48.72 | 3,031 |
| 4: (1,3) (3,2) (2,4) (4,1) | 68.06 | 4,215 |
| 5: (1,3) (3,5) (5,1) (1,1) | 51.32 | 3,187 |
| 6: (1,4) (4,3) (3,1) (1,1) | 60.24 | 3,601 |
| 7: (1,5) (5,1) (1,5) (5,1) | 40.00 | 2,420 |
| 8: (1,5) (5,4) (4,5) (5,1) | 36.96 | 2,283 |
| 9: (1,1) (1,3) (3,2) (2,1) | 36.02 | 2,290 |
| 10: (1,1) (1,1) (1,3) (3,1) | 32.24 | 1,910 |
| 11: (1,1) (1,2) (2,3) (3,1) | 36.02 | 2,239 |
| 12: (1,1) (1,2) (2,5) (5,1) | 36.36 | 2,342 |
| 13: (1,1) (1,3) (3,4) (4,1) | 60.24 | 3,668 |
| 14: (1,1) (1,4) (4,2) (2,1) | 48.52 | 2,965 |
| 15: (1,1) (1,5) (5,2) (2,1) | 36.36 | 2,317 |
| 16: (1,1) (1,5) (5,4) (4,1) | 34.60 | 2,068 |
| 17: (1,5) (5,3) (3,1) (1,1) | 51.42 | 3,058 |
| 18: (1,3) (3,2) (2,1) (1,1) | 36.02 | 2,290 |
| *Totals* | 761.92 | 46,852 |

only one truckload unit of demand with a possible backhaul to cover an additional unit of demand. The manual procedure produced the routes presented in Table 10.1.

A mixed-integer linear programming solution for the deterministic demand scenario was generated using AMPL and the NEOS *feaspump* solver. Table 10.2 contains the routes produced using deterministic demand within a mixed-integer linear program (MILP).

We require a deterministic run of the dynamic programming model to provide a strong comparison to the optimal solution provided by the MILP. Because the

**Table 10.2 *Feaspump* Integer Linear Programming Solution**

| *Triplet Route* | *Hours* | *Cost* ($) |
|---|---|---|
| 1: (1,1,1) (1,1,1) (1,1,5) (5,2,3) (3,3,1) | 55.90 | 3,435 |
| 2: (1,1,1) (1,1,1) (1,1,4) (1,2,1) (1,1,1) | 48.52 | 2,965 |
| 3: (1,1,3) (3,2,1) (1,1,2) (2,2,2) (2,1,1) | 52.50 | 3,411 |
| 4: (1,1,1) (1,1,2) (2,4,1) (1,1,1) (1,1,1) | 48.52 | 3,045 |
| 5: (1,1,1) (1,1,1) (1,2,2) (2,5,4) (4,4,1) | 50.96 | 3,200 |
| 6: (1,1,1) (1,4,3) (3,2,1) (1,1,1) (1,1,1) | 64.02 | 3,982 |
| 7: (1,3,3) (3,5,1) (1,1,1) (1,1,1) (1,1,1) | 51.42 | 3,187 |
| 8: (1,1,5) (5,5,1) (1,1,1) (1,5,1) (1,1,1) | 40.00 | 2,420 |
| 9: (1,1,1) (1,1,1) (1,1,1) (1,3,4) (4,4,1) | 60.24 | 3,668 |
| 10: (1,1,1) (1,1,5) (5,1,3) (3,1,1) (1,1,1) | 52.24 | 3,120 |
| 11: (1,1,5) (5,5,1) (1,1,1) (1,1,1) (1,1,1) | 20.00 | 1,210 |
| 12: (1,3,3) (3,3,3) (3,3,1) (1,1,1) (1,1,1) | 32.24 | 1,910 |
| 13: (1,1,1) (1,5,4) (4,5,3) (3,3,1) (1,1,1) | 68.38 | 4,131 |
| *Totals* | 644.94 | 39,685 |

dynamic program is the implementation of a penalty method, the constraints on the number of hours traveled and load servicing are no longer viewed as hard constraints. In addition, the penalty parameters are not optimized; instead, we run scenarios testing penalty values to determine those that are most effective. To match the requirements of the baseline comparison, we set the penalty on excess hours to be sufficiently restrictive. At a penalty of $2000, we can see that the costs from the deterministic run of the dynamic programming model deviated less than 2.6 percent from the MILP. The hours were within 3.1 percent of the optimal. The total cost of the deterministic solutions differs from the optimal solution due to the impact of the non-optimal penalties and the ability of the dynamic program to create routes that exceed 70 hours in duration. The results of this run are presented in Table 10.3.

The results of the stochastic scenarios with various levels of penalty are shown in Tables 10.4, 10.5, and 10.6. The penalty values for the excess hours range are $700, $600, and $500, respectively. Each decrease in penalty value demonstrated a reduction in the number of routes required to fulfill all demand.

**Table 10.3 Sample 1 Deterministic Routes: $2000 Penalty per Hour**

| *Triplet Route* | *Hours* | *Cost ($)* |
|---|---|---|
| 1: (1,1,1) (1,4,1) (1,3,2) (2,1,3) (1,1,1) | 84.38 | 5,103 |
| 2: (1,1,1) (1,5,1) (1,5,3) (3,2,3) (1,1,1) | 78.62 | 4,823 |
| 3: (1,1,1) (1,2,5) (5,4,5) (5,4,2) (1,1,1) | 75.96 | 4,816 |
| 4: (1,1,1) (1,1,1) (1,2,1) (1,2,5) (5,1,1) | 52.84 | 3,463 |
| 5: (1,1,1) (1,1,1) (1,1,1) (1,3,4) (4,1,1) | 60.24 | 3,668 |
| 6: (1,1,1) (1,3,2) (2,4,4) (4,5,2) (1,1,1) | 78.54 | 4,960 |
| 7: (1,1,1) (1,2,1) (1,4,1) (1,1,1) (1,1,1) | 48.72 | 2,966 |
| 8: (1,1,1) (1,5,4) (4,4,4) (4,1,1) (1,1,1) | 34.60 | 2,068 |
| 9: (1,1,1) (1,4,3) (3,3,3) (3,1,1) (1,1,1) | 60.24 | 3,602 |
| 10: (1,1,1) (1,3,5) (5,1,1) (1,1,1) (1,1,1) | 51.42 | 3,188 |
| *Totals* | 625.56 | 38,656 |

**Table 10.4 Sample 1 Stochastic Routes: $700 Penalty per Hour**

| *Triplet Route* | *Hours* | *Cost ($)* |
|---|---|---|
| 1: (1,1,1) (1,2,3) (3,2,1) (1,1,3) (3,1,1) | 72.04 | 4,529 |
| 2: (1,1,1) (1,3,2) (2,1,3) (3,2,3) (1,1,1) | 75.46 | 4,756 |
| 3: (1,1,1) (1,2,1) (1,5,3) (3,4,2) (1,1,1) | 103.94 | 6,479 |
| 4: (1,1,1) (1,1,1) (1,1,2) (2,5,2) (2,1,1) | 52.72 | 3,449 |
| 5: (1,1,1) (1,3,5) (5,1,1) (1,1,1) (1,1,1) | 51.42 | 3,187 |
| 6: (1,1,1) (1,4,1) (1,1,1) (1,5,4) (4,1,1) | 66.84 | 3,914 |
| 7: (1,1,1) (1,5,1) (1,2,4) (4,1,1) (1,1,1) | 68.52 | 4,255 |
| 8: (1,1,1) (1,1,2) (2,4,1) (1,1,1) (1,1,1) | 48.52 | 3,045 |
| 9: (1,1,1) (1,5,4) (4,5,3) (3,1,1) (1,1,1) | 68.38 | 4,131 |
| *Totals* | 607.84 | 37,745 |

**Table 10.5 Sample 1 Stochastic Routes: $600 Penalty per Hour**

| *Triplet Route* | *Hours* | *Cost ($)* |
|---|---|---|
| 1: (1,1,1) (1,2,3) (3,2,1) (1,1,3) (3,1,1) | 72.04 | 4,529 |
| 2: (1,1,1) (1,3,2) (2,1,3) (3,2,3) (1,1,1) | 75.46 | 4,756 |
| 3: (1,1,1) (1,2,1) (1,5,3) (3,4,2) (1,1,1) | 103.94 | 6,479 |
| 4: (1,1,1) (1,1,1) (1,1,2) (2,5,2) (2,1,1) | 52.72 | 3,449 |
| 5: (1,1,1) (1,3,5) (5,1,1) (1,1,1) (1,1,1) | 51.42 | 3,187 |
| 6: (1,1,1) (1,4,1) (1,4,1) (1,5,4) (4,1,1) | 99.08 | 5,760 |
| 7: (1,1,1) (1,5,1) (1,2,4) (4,1,1) (1,1,1) | 68.52 | 4,255 |
| 8: (1,1,1) (1,5,4) (4,5,3) (3,1,1) (1,1,1) | 68.38 | 4,131 |
| *Totals* | 563.24 | 34,907 |

We see that varying levels of penalty associated with excess hours of travel produced routes with lengths that exceeded the maximum hours specified. In an applied setting, we turn this to our distinct advantage. The dispatcher is left with the option of creating longer routes. Prior to the drastic increase in fuel prices, this was not as critical, but in the current economic climate, it is in the best interest of transportation companies to extend the length of the routes. Companies will be in a better financial position by paying the cost of driver layovers in order to avoid excess fuel usage.

**Table 10.6 Sample 1 Stochastic Routes: $500 Penalty per Hour**

| *Triplet Route* | *Hours* | *Cost ($)* |
|---|---|---|
| 1: (1,1,1) (1,2,1) (1,1,5) (5,1,3) (1,1,1) | 52.60 | 3,598 |
| 2: (1,1,1) (1,3,5) (5,2,1) (1,2,3) (3,1,1) | 103.80 | 6,533 |
| 3: (1,1,1) (1,3,2) (2,4,5) (5,4,2) (1,1,1) | 93.08 | 5,830 |
| 4: (1,1,1) (1,4,3) (3,3,2) (2,5,2) (2,1,1) | 100.26 | 6,311 |
| 5: (1,1,1) (1,5,4) (4,1,3) (3,3,4) (1,1,1) | 78.72 | 4,841 |
| 6: (1,1,1) (1,5,1) (1,5,3) (3,1,1) (1,1,1) | 71.42 | 4,268 |
| *Totals* | 499.86 | 31,081 |

The location of the domicile has not only facilitated the creation of efficient routes, but has also given way to many route permutations that are also effective. This may be easily seen when examining the results from the multiple penalty levels under consideration. At each level, we were able to create similar routes that reflected different priorities within the route building.

The dynamic programming model will always attempt to create fewer routes if possible, despite the lack of a stated objective. Any decrease in the number of routes suggests the overt consideration of capital costs in the dynamic programming model. As capital costs are accumulated, the model may then form a financial bias against additional routes.

The stochastic scenario with a penalty of $500 per excess hour, presented in Table 10.6, showed that all loads were still served with a decrease of four routes in the solution. These differences result from the application of likelihoods to the demand and the reduction of the penalty on excess hours. The dynamic programming code evaluated alternate paths based on the expected cost-to-go.

Note that the longest route in the solution where the penalty of $500 is 103.80 hours (Table 10.6). A dispatcher can easily accommodate a route of this length by scheduling a single layover, thus extending the route by less than 2 days while accruing significant savings in fuel costs. The flexibility in delivery schedule for truckload movements is a very important factor in allowing the dispatcher to take advantage of cost reductions accrued through longer routes. A new constraint in the formulation is suggested by this analysis. This constraint would trade off additional fuel costs against layover costs, and would further assist the dispatcher in assessing the value of longer routes.

### *10.7.2 Sample 2*

The Sample 2 data set has ten locations to be routed. These sites are drawn from actual truckload demand recorded between these locations. The probabilities associated with the demand were generated in the count data regression analysis. The cities contained in this sample are found in Table 10.7. Harrisburg, Pennsylvania, is the domicile in this sample. Note that it falls well within the implied boundaries of the region. The routes generated and their associated hours of service and cost are presented in Table 10.8.

The route costs are substantially higher in this data set and in the data sets that follow. The increase heavily reflects the recent increase in fuel costs. Carriers have added surcharges to allow them to recover the increases and, in turn, are negotiating new contracts that they hope will protect them on a longer-term basis.

The average route length is longer than that of the test data, but still well within manageable levels for the dispatcher. The location of the domicile also provided greater flexibility in the route-building process, resulting in efficiencies in the routes.

**Table 10.7 Sample 2 Locations**

| *Number* | *City* | *State* |
|---|---|---|
| 1: | Harrisburg | PA |
| 2: | Oklahoma City | OK |
| 3: | New Milford | CT |
| 4: | Mobile | AL |
| 5: | Greenville | MS |
| 6: | Summit Argo | IL |
| 7: | Philadelphia | PA |
| 8: | Cincinnati | OH |
| 9: | Memphis | TN |
| 10: | Albany | GA |

**Table 10.8 Sample 2 Stochastic Routes: $500 Penalty per Hour**

| *Triplet Route* | *Hours* | *Cost ($)* |
|---|---|---|
| 1: (1,1,1) (1,4,5) (5,6,2) (2,4,9) (9,1,1) | 98.96 | 10,160 |
| 2: (1,1,1) (1,9,7) (7,7,7) (7,9,5) (5,1,1) | 86.35 | 8,858 |
| 3: (1,1,1) (1,7,8) (8,8,8) (8,6,8) (8,1,1) | 34.86 | 3,634 |
| 4: (1,1,1) (1,3,1) (1,7,1) (1,3,6) (6,1,1) | 50.93 | 5,492 |
| 5: (1,1,1) (1,10,1) (1,3,8) (8,10,9) (9,1,1) | 97.08 | 9,860 |
| 6: (1,1,1) (1,6,9) (9,9,9) (9,2,7) (1,1,1) | 60.92 | 6,157 |
| 7: (1,1,1) (1,3,6) (6,9,9) (9,2,1) (1,1,1) | 69.17 | 7,326 |
| 8: (1,1,1) (1,10,10) (10,2,1) (1,3,4) (4,1,1) | 114.97 | 11,856 |
| 9: (1,1,1) (1,3,3) (3,4,9) (9,3,1) (1,1,1) | 67.38 | 7,050 |
| 10: (1,1,1) (1,8,4) (4,5,7) (7,4,6) (6,1,1) | 110.92 | 11,519 |
| *Totals* | 791.54 | 81,912 |

**Table 10.9 Sample 3 Locations**

| Number | City | State |
|---|---|---|
| 1: | Champaign | IL |
| 2: | Clackamas | OR |
| 3: | Mobile | AL |
| 4: | Mira Loma | CA |
| 5: | Mesquite | TX |
| 6: | Oklahoma City | OK |
| 7: | Tracy | CA |

### 10.7.3 Sample 3

The Sample 3 data set has seven locations to be routed, again drawn from actual truckload demand recorded between these locations. The probabilities were generated in the count data regression analysis. The cities contained in this sample are found in Table 10.9. The domicile was located in Champaign, Illinois. Upon examination of these locations, we saw that the region boundaries were essentially defined by the locations, and none of these cities could truly be considered to be centrally located. Therefore, the domicile in Sample 3 sits at the northeastern edge of the region defined. The routes generated and their associated hours of service and cost are presented in Table 10.10.

Four of the routes generated exceed the hours of service target, but are easily managed within a slightly longer route window. The length of the routes reflected the geographic dispersion of the locations and their distance from the domicile. It is unlikely that a run of Sample 3 data could result in routes of any shorter distance.

### 10.7.4 Sample 4

The Sample 4 data set has six locations to be routed. The cities contained in this sample are found in Table 10.11. The domicile in Sample 4 is Summit Argo, Illinois. The domicile fell at the northwestern edge of the region being served. The routes generated and their associated hours of service and cost are presented in Table 10.12.

The routes generated for this data set are significantly shorter than the previous data sets considered. Upon investigation, we find that this is a function of geographic dispersion and location of the domicile. If the domicile is not nestled neatly into the center of the locations to be serviced, the routes become more complex to define.

**Table 10.10 Sample 3 Stochastic Routes: $500 Penalty per Hour**

| *Triplet Route* | *Hours* | *Cost ($)* |
|---|---|---|
| 1: (1,1,1) (1,3,5) (5,6,7) (7,2,7) (7,1,1) | 94.11 | 10,160 |
| 2: (1,1,1) (1,5,1) (1,6,1) (1,5,1) (1,1,1) | 65.42 | 8,858 |
| 3: (1,1,1) (1,6,1) (1,6,5) (5,6,1) (1,1,1) | 46.40 | 3,634 |
| 4: (1,1,1) (1,5,6) (6,1,3) (3,1,3) (1,1,1) | 59.76 | 5,492 |
| 5: (1,1,1) (1,6,4) (4,7,4) (4,1,4) (1,1,1) | 93.79 | 9,860 |
| 6: (1,1,1) (1,3,1) (1,5,3) (3,1,6) (1,1,1) | 64.55 | 6,157 |
| 7: (1,1,1) (1,5,3) (3,5,4) (4,1,7) (1,1,1) | 105.57 | 7,326 |
| 8: (1,1,1) (1,6,5) (5,6,4) (4,7,4) (4,1,1) | 72.79 | 11,856 |
| 9: (1,1,1) (1,1,1) (1,1,1) (1,5,3) (3,1,1) | 31.38 | 7,050 |
| 10: (1,1,1) (1,1,1) (1,1,1) (1,6,3) (3,1,1) | 33.37 | 11,519 |
| *Totals* | 667.14 | 72,123 |

The domicile in this data set is Summit Argo, Illinois. While not completely separate from the other locations, it is certainly at the edge of the geographic region over which the loads are carried. In addition, as the network of locations becomes wider, and the hours of travel increase, even our lower penalty on excess hours becomes a significant deterrence. The end result is the production of more routes that travel shorter distances.

**Table 10.11 Sample 4 Locations**

| *Number* | *City* | *State* |
|---|---|---|
| 1: | Summit Argo | IL |
| 2: | Greenville | MS |
| 3: | Mobile | AL |
| 4: | Jacksonville | IL |
| 5: | Harrisburg | PA |
| 6: | New Milford | CT |

**Table 10.12 Sample 4 Stochastic Routes: $500 Penalty per Hour**

| *Triplet Route* | *Hours* | *Cost* ($) |
|---|---|---|
| 1: (1,1,1) (1,4,2) (2,1,4) (4,3,1) (1,1,1) | 48.94 | 5,356 |
| 2: (1,1,1) (1,4,5) (5,6,1) (1,4,1) (1,1,1) | 34.19 | 3,676 |
| 3: (1,1,1) (1,4,3) (3,4,5) (5,6,2) (1,1,1) | 61.52 | 6,586 |
| 4: (1,1,1) (1,4,6) (6,4,6) (6,1,2) (1,1,1) | 70.30 | 7,501 |
| 5: (1,1,1) (1,2,1) (1,4,3) (3,4,2) (2,1,1) | 64.58 | 6,951 |
| 6: (1,1,1) (1,6,4) (4,1,4) (4,6,5) (5,1,1) | 60.80 | 6,521 |
| 7: (1,1,1) (1,3,2) (2,3,2) (2,4,5) (1,1,1) | 49.28 | 5,363 |
| 8: (1,1,1) (1,6,3) (3,1,6) (6,3,1) (1,1,1) | 85.70 | 9,468 |
| 9: (1,1,1) (1,6,2) (2,4,2) (2,6,4) (1,1,1 ) | 81.25 | 8,782 |
| 10: (1,1,1) (1,1,4) (4,6,3) (3,1,1) (1,1,1) | 49.94 | 5,377 |
| *Totals* | 606.50 | 65,582 |

This characteristic of the dynamic programming model also works well within the dispatcher mind-set. The decreased penalty provides visibility into additional possibilities that may still be applied to the routing operation. Dispatchers will be willing to extend route durations, although not beyond some level that they eventually deem acceptable.

### *10.7.5 Sample 5*

The Sample 5 data set has nine locations to be routed. The cities contained in this sample are found in Table 10.13. Summit Argo, Illinois, is again the domicile but in this case it fell in a more central location to the region. It was toward the north but was centrally located when considering the eastern and western directions. The routes generated and their associated hours of service and cost are presented in Table 10.14.

The routes generated for this final data set are all in excess of our target of 70 hours. This solution again provides a basis for renegotiation of work rules. With the size of the region and the orientation of Summit Argo, Illinois, on the northern edge of the region, it creates a complex and formidable route-building task. Compromising on work rules in order to keep a transportation company profitable is a viable and important alternative.

**Table 10.13 Sample 5 Locations**

| *Number* | *City* | *State* |
|---|---|---|
| 1: | Summit Argo | IL |
| 2: | Baltimore | MD |
| 3: | Champaign | IL |
| 4: | Harrisburg | PA |
| 5: | Greenville | MS |
| 6: | Mobile | AL |
| 7: | Mesquite | TX |
| 8: | New Milford | CT |
| 9: | Oklahoma City | OK |

**Table 10.14 Sample 5 Stochastic Routes: $500 Penalty per Hour**

| *Triplet Route* | *Hours* | *Cost ($)* |
|---|---|---|
| 1: (1,1,1) (1,7,3) (3,6,5) (5,7,9) (9,1,1) | 86.71 | 9,469 |
| 2: (1,1,1) (1,6,5) (5,1,9) (9,7,3) (3,1,1) | 80.49 | 8,696 |
| 3: (1,1,1) (1,6,3) (3,4,3) (3,5,1) (1,1,1) | 86.93 | 9,294 |
| 4: (1,1,1) (1,3,9) (9,8,4) (4,8,2) (2,1,1) | 72.99 | 7,991 |
| 5: (1,1,1) (1,8,3) (3,8,1) (1,9,6) (6,1,1) | 125.99 | 13,048 |
| 6: (1,1,1) (1,9,3) (3,9,3) (3,8,1) (1,1,1) | 94.32 | 10,211 |
| 7: (1,1,1) (1,8,4) (4,8,3) (3,6,1) (1,1,1) | 82.65 | 8,747 |
| 8: (1,1,1) (1,3,8) (8,7,9) (9,8,4) (4,1,1) | 109.64 | 11,721 |
| 9: (1,1,1) (1,3,6) (6,1,3) (3,7,1) (1,1,1) | 75.78 | 8,258 |
| 10: (1,1,1) (1,7,9) (9,7,8) (8,3,1) (1,1,1) | 84.38 | 9,045 |
| *Totals* | 899.88 | 96,481 |

## 10.8 Efficiency of Solution Method

The dynamic programming solution methodology provides significant efficiencies over the enumerative approach. Before discussing the number of numerical evaluations required in each of these, we first introduce additional appropriate notation.

$N$ = Number of geographic locations
$L$ = Number of levels in the decision tree
$U_i$ = Number of unserved loads at level $i$

The number of unserved loads at each level may be represented as a polynomial expression of the number of geographic locations, $N$ where $N \geq 3$. For $N < 3$, the problem reduces to a trivial case. This general expression, with $c$'s representing constant terms, is presented in Equation (10.19):

$$U_i = N^2 - c_1 N^2 - c_2 N - c_2 \tag{10.19}$$

The number of evaluations required to enumerate the routes within the TRPTW follows in Equation (10.20). At each level, we calculate the cost of the current triplet and the recursive cost-to-go from that triplet. For every triplet, a single calculation is performed for every change in the number of unserved loads, which is in fact a change in the state. The maximum number of states is equivalent to the number of unserved loads at each level. Applying the unserved loads shown in Equation (10.19) then leads us to an order of magnitude for the number of calculations of $N^{11}$:

$$N^2(U_1 U_2 N^2 + U_2 N^5) \tag{10.20}$$

The dynamic programming model leverages off of our ability to store calculations for the associated state spaces. This reduces the number of calculations required. The worst-case scenario is shown in Equation (10.21). Because each triplet can end in only one of $N$ possible locations, the costs-to-go need only be calculated for the $N$ final stops within the triplet. Applying the total number of unserved loads, as represented in Equation (10.19), to each level then results in a magnitude for the number of calculations of $N^6$:

$$\sum_{i=1}^{L-1} U_i N^i \tag{10.21}$$

The best-case scenario for the dynamic programming model is shown in Equation (10.22). In this case the number of unserved loads is reduced to one at each level. This can mean that there is simply one unserved load, or it may also indicate that the state remains the same across all triplets examined at each level. In other words, the number and the nature of the unserved loads remain the same, which means just a single calculation is required for each triplet. This results in a magnitude of $N^3$ for the number of calculations.

$$\sum_{i=1}^{L-1} N^i \tag{10.22}$$

The magnitude of greatest interest to us is neither the best case nor the worst case, but rather the average case. Because at most two loads can be fulfilled by each triplet, this will reflect a range of three possible states for each triplet at each level. In smaller problems ($N \geq 3$, but still close to 3), the average number of calculations required will have a magnitude of $N^4$. As the problem size increases ($N \to \infty$), the magnitude will approach the best case of $N^3$.

## 10.9 Conclusions

We have been able to draw a number of interesting conclusions from the analysis undertaken in this research. Not only does the model directly address the question of truckload routes, it also has significant implications for other aspects of truckload operations.

The routes generated by the dynamic programming model tended to be longer than 70 hours. Recall that the constraint on hours of service was relaxed, and the excess hours were penalized. The penalty amount is commensurate with the fine carriers might be willing to incur in the interest of reducing other costs. Rather than exceeding the hours of service on a routine basis, we recommend that the carriers alter the work rules under which they operate. Rather than incurring fines, the routes may be extended with the simple addition of a driver layover. In the past, this practice has been discouraged due to the higher priority of quality of life for the drivers. In the face of massive increases in fuel prices, however, the profitability of the carriers is threatened and their negotiating position with the unions is significantly enhanced. The results of the model provide an excellent basis for this renegotiation.

We also noted that the number of routes generated decreased with decreases in the penalty on excess hours. This naturally leads to fleet size evaluations. We reflect possible changes in the work rules, while running various penalty scenarios to determine the appropriate fleet size.

Location analysis may also be built into this solution technique. We can examine the location of the domicile relative to the locations served. This allows us to run multiple scenarios to examine two location alternatives. The first is the possibility of shifting domicile locations in order to better serve the customers assigned to that domicile. The second location consideration is the realignment of customers to domiciles.

## 10.10 Future Research

Through the analysis procedure, two additional interesting constraints came to light. The first of these addressed the addition of capital costs to the model data. As all who are interested in supply chain management can attest, we often have to look beyond our specific problem in order to find a truly efficient solution. In this case, we should look at the cost of owning or leasing additional equipment. Being able to reduce the number of routes is in itself a worthwhile achievement, but being able to additionally reduce the associated fleet size is a greater achievement.

The second constraint is one that came to light in view of the massive increases in fuel costs. Like all other union shops, transportation companies have been subject to restrictions imposed by union rules and the need to drastically increase the quality of life of their drivers. Long-haul truckload drivers, those who carry the demand discussed in this research, in particular face unappealing work conditions. An additional constraint may be added that trades off layover cost against the fuel cost of returning home within the applicable number of hours. This analysis would pave the way for revised work rules that could offer the drivers longer runs, but also longer breaks between runs. The cumulative time on the road over the course of a month would be the same but the quality of life at home would greatly increase.

These two constraints lead the way to a number of other constraints that would allow dispatchers to integrate more of the decision making into the model. Examples of such constraints are consideration of seniority, customers with strict time windows, and customers with particularly soft time windows.

The existing model may also be used to assess a number of other characteristics of the transportation environment. These include proposed changes in the dispatch methodology, renegotiation of work rules, reassessment of domicile location, fleet size analysis, and assignment of delivery locations to domiciles.

## References

Arunapuram, S., Mathur, K., and Solow, D. (2003), Vehicle routing and scheduling with full truckloads, *Transportation Science,* 37(2), 170–182.

Bertacco, L., Fischette, F., and Lodi, A. (2005), A feasibility pump heuristic for general mixed-integer problems, http://neos.mcs.anl.gov/neos

Chu, C.-W. (2004), A heuristic algorithm for the truckload and less-than-truckload problem, *European Journal of Operational Research,* 1–11.

Clarke, G., and Wright, J.W. (1964), Scheduling of vehicles from a central depot to a number of delivery points, *Operations Research,* 12, 568–581.

Czyzyk, J., Mesnier, M., and Moré, J. (1998), The NEOS Server, *IEEE Journal on Computational Science and Engineering,* 5, 68–75.

Federal Motor Carrier Safety Administration website (2006), http://www.fmcsa.dot.gov/

Dolan, E. (2001), The NEOS Server 4.0 Administrative Guide, Technical Memorandum ANL/MCS-TM-250, Mathematics and Computer Science Division, Argonne National Laboratory, Argonne, IL.

Fischetti, M., Glover, F., and Lodi, A. (2005), The feasibility pump, *Mathematical Programming A*, 104, 91–104.

Fourer, R., Gay, D.M., and Kernighan, B.W. (2003), *AMPL: A Modeling Language for Mathematical Programming, second edition*, Pacific Grove, CA: Duxbury Thomsom Brooks/Cole.

Frantzaskakis, L.F., and Powell, W.B. (1990), A successive linear approximation procedure for stochastic, dynamic vehicle allocation problems, *Transportation Science*, 24(1), 40–57.

Gronalt, M., Hartl, R.F., and Reimann, M. (2003), New savings based algorithms for time constrained pickup and delivery of full truckloads, *European Journal of Operational Research*, 151, 520–535.

Gropp, W., and Moré, J. (1997), Optimization environments and the NEOS Server, *Approximation Theory and Optimization*, Cambridge: Cambridge University Press, pp. 167–182.

Miori, V.M. (2006), Dynamic Programming Applied to a New Formulation of the Stochastic Truckload Routing Problem, Doctoral thesis, Philadelphia, PA: Drexel University.

Powell, W.B. (1996), A stochastic formulation of the dynamic assignment problem, with an application to truckload motor carriers, *Transportation Science*, 30(3), 195–219.

NEOS Solvers (2006). http://neos.mcs.anl.gov/neos/solvers

Yang, J., Jaillet, P., and Mahmassani, H.S. (2000), Study of a Real-time Multi-vehicle Truckload Pickup-and-Delivery Problem, Research funded by NSF, Grant DMI-9713682.

*Chapter 11*

# Modeling Data Envelopment Analysis (DEA) Efficient Location/Allocation Decisions

Ronald K. Klimberg, Samuel J. Ratick, Vinay Tavva, Sasanka Vuyyuru, and Daniel Mrazik

## Contents



## 11.1 Introduction

The initial focus of mathematical programming models for facility location was based upon the spatial availability of a server to fulfill demand [for a review of covering models, see ReVelle (1987) and Schilling et al. (1993)]. Toregas et al. (1971) formulated the Location Set Covering Problem (LSCP) that required coverage of all demand points within some time standard. A derivative of this was the $p$-center problem, which minimized the maximum distance separating any demand point from its nearest facility (Hakimi 1964, ReVelle and Swain 1970). Both of these models considered demand, server capability, and service time to be homogeneous. Budgetary constraints, which might make full coverage impossible, led to the development of the Maximal Covering Location Problem (MCLP) (Church and ReVelle 1974, White and Case 1974). Non-uniform distribution of demand nodes was incorporated into these partial covering models to develop the best deployment of a given number of facilities.

The Capacitated Facility Location Problem (CPLP) extends the classical transportation problem by choosing among a set of potential sites for locating supply facilities those that minimize cost—defined as the sum of transportation costs and fixed costs of opening facilities given a limited supply capabilities (Daskin 1995). The formulation for the CPLP model is:

**Model: Capacitated Facility Location Problem (CPLP)**

$$\text{MIN} \sum_{k=1}^{K} \sum_{l=1}^{L} c_{kl} b_{kl} + \sum_{k=1}^{K} F_k y_k \tag{11.1}$$

s.t.

$$\sum_{k=1}^{K} x_{kl} \geq 1 \quad \forall l \tag{11.2}$$

$$x_{kl} \leq y_k \quad \forall k, l \tag{11.3}$$

$$\sum_{k=1}^{K} b_{kl} = \text{dem}_l \quad \forall l \tag{11.4}$$

$$b_{kl} \leq \text{MIN}[\text{dem}_l, \text{Cap}_k] y_k \quad \forall k, l \tag{11.5}$$

$$y_k, x_{kl} = 0, 1$$

$$b_{kl} \geq 0$$

(*Note:* We use the standard DEA notation, as seen later. As such, we have had to modify the standard location notation.)

$k = 1, \ldots, K$ index of facility locations
$l = 1, \ldots, L$ index of demand locations

where:

- Parameters:
  $c_{kl}$ = Cost of shipping one unit of demand from facility $k$ to demand $l$
  $\text{dem}_l$ = Number of units of demand at $l$

$F_k$ = Fixed cost of opening/using facility $k$
$Cap_k$ = Capacity of facility $k$

- Decision Variables:

$$x_{kl} = \begin{cases} 1 & \text{if facility } k \text{ serves demand } l \\ 0 & \text{o/w} \end{cases}$$

$$y_k = \begin{cases} 1 & \text{if facility } k \text{ "opened" (i.e., used)} \\ 0 & 0 \text{ o/w} \end{cases}$$

$b_{kl}$ = Number of units shipped from facility $k$ to demand location $l$

The Objective Function (11.1) is calculated as the sum of total transportation and fixed opening costs. The transportation costs are calculated as the product of the per-unit transportation costs and the amount shipped from facility $k$ to demand $l$ (i.e., $b_{kl}$). The constraints represented in (11.2) ensure that every demand is covered/satisfied, and the constraints in (11.3) ensure that only open facilities can supply demands. Constraint (11.4) ensures that all demands are satisfied by shipments from open facilities. Constraints in (11.5) ensure that the amount shipped from facility $k$ to demand $l$, $b_{kl}$, is either less than the demand requirement at $l$ ($dem_l$) or the supply at $k$ ($Cap_k$), whichever is smaller; if facility $k$ is not opened, then $b_{kl}$ is forced to zero.

The review of Current et al. (1990) of 45 facility location papers demonstrated that these problems are inherently multi-objective. They classified the most common objectives into four categories: cost minimization, demand oriented, profit maximization, and environmental concern. The cost minimization objective is the traditional objective of most facility location models. Demand-oriented objectives focus on measuring the "closeness" of the facilities, where "closeness" can be measured in terms of coverage or response time. In a previous paper (Klimberg and Ratick 2008), we postulated that a good location pattern is one that not only optimizes the spatial interaction among facilities and the demands they serve, such as the above-mentioned models and objectives, but also optimizes the performance (efficiency) of those facilities at the chosen locations. We used a linear programming approach called data envelopment analysis (DEA)—embedded within location models—to evaluate the efficiency of the facilities.

## 11.2 Data Envelopment Analysis (DEA)

DEA, which has been used for about 30 years, was developed to address the common problem of measuring organizational efficiency when many different types of measures of inputs and outputs are available (Charnes et al. 1978). DEA utilizes linear programming to produce measures of the relative efficiency of comparable units that employ multiple inputs and outputs. It requires only that the selected inputs and outputs be quantifiable. The technique can analyze these multiple inputs and outputs in their natural physical units without reducing or transforming them into some common measurement (such as dollars). DEA takes into account these multiple inputs and outputs to produce a single aggregate measure of the relative efficiency of each comparable unit. DEA defines efficiency as the ratio of weighted outputs to weighted inputs:

$$E = \text{Efficiency} = \frac{\text{Weighted sum of outputs}}{\text{Weighted sum of inputs}}$$

The more output produced for a given amount of resources, the more efficient (i.e., less wasteful) the operation.

The comparable decision-making units (DMUs) that employ multiple inputs and outputs are the components of the organization being evaluated, (e.g., hospitals—in the healthcare system, departments—in a hospital/company, or individuals, in a department). DEA has been applied in many diverse areas, including hospitals, healthcare organizations, physicians, pharmacies, drug reimbursement, armed forces, criminal courts, schools, university departments, banks, electric utilities, strip mining, manufacturing productivity, railroad property evaluation (Emrouznedjad et al. 2008, Klimberg and Kern 1992, Seiford 1996, Seiford and Thrall 1990, Tavares 2002). Further, relative spatial efficiency has been measured using DEA with the spatial variables of total travel distance and number uncovered (Athanassopoulos and Storbeck 1995, Desai and Storbeck 1990, Desai et al. 1995). DEA uniquely evaluates all the DMUs and all their inputs and outputs simultaneously, and conservatively identifies the sets of relatively efficient and relatively inefficient DMUs. Thus, the solution of a DEA model provides a manager with a summary with comparable DMUs grouped together and ranked by relative efficiency.

The Charnes et al. (1978) DEA model is a linear fractional program that compares the ratio of weighted outputs to weighed inputs. The efficiency of the r-th decision-making unit (DMU), $w_r$, is obtained by solving the following linear fractional formulation:

**Model: Charnes et al. Linear Fractional Data Envelopment Analysis Model (DEA1)**

$$\text{MAX } w_r = \frac{\sum_{j=1}^{J} u_j O_{jr}}{\sum_{i=1}^{I} v_i I_{ir}} \tag{11.6}$$

s.t.

$$\frac{\sum_{j=1}^{J} u_j O_{jk}}{\sum_{i=1}^{I} v_i I_{ik}} \leq 1 \quad \forall k \tag{11.7}$$

$$u_j \geq 0 \quad \forall j,$$

$$v_i \geq 0 \quad \forall i$$

where:

$i = 1, \ldots, I$ Inputs used at DMU
$j = 1, \ldots, J$ Outputs produced at DMU
$k = 1, \ldots, r, \ldots, K$ DMUs

- Parameters:
  $O_{jk}$ = Amount of the $j$-th output for the $k$-th DMU
  $I_{ik}$ = Amount of the $i$-th input for the $k$-th DMU
- Decision variables:
  $u_j$ = Weight assigned to the $j$-th output
  $v_i$ = Weight assigned to the $i$-th input

This formulation finds the set of weights, $u_j$ and $v_i$, that maximize the efficiency, $w_r$, of DMU$r$ Equation (11.6). The constraints in Equation (11.7) require the ratio of the sum of weighted outputs to the sum of weighted inputs (that is, the efficiency of each DMU, including the r-th DMU) to be no larger than 1. A similar DEA formulation is solved sequentially for each DMU.

This initial DEA1 model is a linear fractional program and is not a truly linear program and therefore cannot be solved using the simplex method of linear programming. With a few simple modifications, Charnes et al. (1978) transformed this initial formulation into a linear program. First, they assume that the denominator of the DMU being evaluated is equal to 1: $\sum_{i=1}^{I} v_i I_{ir} = 1$.

The other modification is to multiply both sides of the equation for Constraint (11.7) by the sum of the weighted inputs. This change linearizes Constraint (11.7) to yield the following:

$$\sum_{j=1}^{J} u_j O_{jk} \leq \sum_{i=1}^{I} v_i I_{ik} \quad \forall k$$

Additionally, a special case called weakly efficient causes the DEA model to be modified in practice. A particular DMU is weakly efficient if it is actually inefficient, but when applying the DEA model, one or more of its weights ($u_j$ or $v_i$) are equal to zero and it is incorrectly considered efficient. To avoid such situations we add the restriction that the weights are required to be greater than some small number $\varepsilon$, where $\varepsilon$ is a small infinitesimal value. The positivity of the weights $u_j$ and $v_i$ guarantees that a weakly efficient DMU would not be found efficient. The modified linear DEA1 formulation therefore becomes:

**Model: Charnes et al. Modified Data Envelopment Analysis Model (DEA2)**

$$\text{MAX } w_r = \sum_{j=1}^{J} u_j O_{jr} \tag{11.8}$$

st.

$$\sum_{i=1}^{I} v_i I_{ir} = 1 \quad (11.9)$$

$$\sum_{j=1}^{J} u_j O_{jk} - \sum_{i=1}^{I} v_i I_{ik} \leq 0 \quad \forall k \quad (11.10)$$

$$u_j, \ v_i \geq \varepsilon \quad \forall j, i$$

where:

$\varepsilon$ = A small infinitesimal value

## 11.3 Combined DEA/Location Model

Typical DEA consists of subsequently solving a slightly modified linear program for each DMU. However, to link DEA to a facility location model, the DEA model should be modified such that the DEA efficiencies of all the DMUs are calculated in one linear program. One approach by Thomas et al. (2002) combines facility location and DEA by iteratively executing the location/DEA for all combinations. In each iteration, DEA efficiencies are simultaneously solved for all DMUs and given a predetermined location.

The model we developed in Klimberg and Ratick (2008), which solves for the DEA value for each DMU simultaneously, is the following:

**Model: Simultaneous DEA (SDEA)**

$$\text{MAX} \sum_{r} (1 - d_r) = \sum_{r} w_r \quad (11.11)$$

st.

$$\sum_{i=1}^{I} v_{ri} I_{ir} = 1 \quad \forall r \quad (11.12)$$

$$\sum_{j=1}^{J} u_{rj} O_{jr} + d_r = 1 \quad \forall r \quad (11.13)$$

$$\sum_{j=1}^{J} u_{rj} O_{jk} - \sum_{i=1}^{I} v_{ri} I_{ik} \leq 0 \quad \forall k; \forall r; k \neq r \quad (11.14)$$

$$u_{rj}, \ v_{ri} \geq \varepsilon \quad \forall j, i, r$$

where:

- Decision variables:
  $u_{rj}$ = Weight assigned to the $j$-th output for DMU $r$
  $v_{ri}$ = Weight assigned to the $i$-th input for DMU $r$

To allow for the simultaneous solution of the DEA model for all DMUs, the Objective Function (11.11) now maximizes the sum of the efficiencies. The constraints in Equation (11.12) require the sum of DMU $r$'s weighted inputs to be equal to 1, and they are written for each DMU. The constraints in Equation (11.13) define efficiency as the sum of DMU $r$'s weighted outputs; these constraints are also written for each DMU. The constraints in Equation (11.9) require the sum of each set of weighted outputs to be less than the corresponding sum of weighted inputs [note that in Equation 11.14, $K$ sets of $(K - 1)$ constraints are written because the weights for each $r$ need to be tested with the input/output vectors for all other DMUs, where $K$ is the total number of DMUs being evaluated].

Unlike typical DEA in which the efficiency of each DMU is maximized sequentially, the above formulation maximizes the sum of the efficiencies for all the DMUs. In Klimberg and Ratick (2008) we embedded the above simultaneous DEA model into the capacitated facility location model (CPLP), and we assigned one DEA output variable to be the amount demanded. As a result, the DEA efficiencies are now dynamic, that is, the amount of output may change depending on how much is supplied and hence the DEA efficiencies will change. An additional aspect considered is that all potential locations are not necessarily opened while DEA considers the efficiencies of all locations. In our formulation, the efficiency of a location is equal to 0 if it is not used. The combination of the SDEA model with the CPLP results in the following capacitated adjustable simultaneous DEA/CPLP formulation (CASD) (Klimberg and Ratick 2008):

**Model: Capacitated Adjustable Simultaneous DEA/CPLP Model (CASD)**

$$\text{MAX} \sum_{k=1}^{K} \sum_{l=1}^{L} (1 - d_{kl}) \tag{11.15}$$

$$\text{MIN} \sum_{k=1}^{K} \sum_{l=1}^{L} c_{kl} b_{kl} + \sum_{k=1}^{K} F_k y_k \tag{11.16}$$

s.t.

$$\sum_{k=1}^{K} x_{kl} \geq 1 \quad \forall l \tag{11.17}$$

$$x_{kl} \leq y_k \quad \forall k, l \tag{11.18}$$

$$\sum_{k=1}^{K} \mathrm{b}_{kl} = \mathrm{dem}_l \quad \forall l \tag{11.19}$$

$$\mathrm{b}_{kl} \leq \mathrm{MIN}[\mathrm{dem}_l, \mathrm{O}_{mkl}]\mathrm{y}_k \quad \forall k, l \tag{11.20}$$

$$\sum_{i=1}^{I} \mathrm{v}_{kli}\mathrm{I}_{kli} = x_{kl} \quad \forall k, l \tag{11.21}$$

$$\sum_{j=1}^{J} \mathrm{u}_{klj}\mathrm{O}_{jkl} + \mathrm{d}_{kl} = \mathrm{x}_{kl} \quad \forall k, l \tag{11.22}$$

$$\sum_{j=1}^{J} \mathrm{u}_{klj}\mathrm{O}_{jrs} - \sum_{i=1}^{I} \mathrm{v}_{kli}\mathrm{I}_{irs} \leq 0 \quad \forall k, l; \forall r, s; (k \neq r \text{ and } l \neq s) \tag{11.23}$$

$$\mathrm{u}_{klj} \geq \varepsilon \mathrm{x}_{kl} \quad \forall k, l, j \tag{11.24}$$

$$\mathrm{v}_{kli} \geq \varepsilon \mathrm{x}_{kl} \quad \forall k, l, i \tag{11.25}$$

$$\mathrm{u}_{klj}\mathrm{O}_{jkl} \leq \mathrm{x}_{kl} \quad \forall k, l \tag{11.26}$$

$$\mathrm{b}_{kl} \geq \mathrm{x}_{kl} \quad \forall k, l \tag{11.27}$$

$$\mathrm{y}_k, \mathrm{x}_{kl} = 0, 1$$

$$\mathrm{b}_{kl}, \mathrm{u}_{klj}, \mathrm{v}_{kli} \geq 0$$

where:

$m$ = Index of the output that is used to satisfy demand in the CPLP

The first objective function in the above formulation, (11.15), and the corresponding DEA-related Constraints (11.21) through (11.23) are similar to the SDEA model, the difference being the additional index and the value $\mathrm{x}_{kl}$. If facility $k$ serves demand $l$ (i.e., $\mathrm{x}_{kl} = 1$), the corresponding input and output weights are required to be greater than $\varepsilon$ because of Constraints (11.24) and (11.25), respectively; the constraints in (11.26) require the weighted outputs to be less than 1, for all facilities, demands, and output types. On the other hand, if facility $k$ does not serve demand $l$ ($\mathrm{x}_{kl} = 0$), the constraints in Equations (11.24) and (11.25) require the input and output weights to be non-negative, and the constraints in Equations (11.21) and (11.26) force them to be equal to 0. The constraints in Equation (11.27) require at least one unit to be shipped from facility $k$ to demand $l$, if $\mathrm{x}_{kl} = 1$, allowing for that facility to be used in the calculation of DEA efficiency scores. Thus, if $\mathrm{x}_{kl} = 0$, the DEA input and output weights for facility $k$ are equal to 0, and that facility is not considered in the computation of the relative DEA efficiency scores for that location/allocation pattern. The second Objective Function (11.16) and Constraints (11.17) through (11.20) are related to the facility location part of the model. The second Objective Function (11.16) calculates the total cost (transportation and fixed opening costs) of supplying the demand in the system. The transportation costs are calculated as the product of the per unit transportation costs and the amount shipped from facility $k$ to demand $l$ (i.e., $b_{kl}$). Total fixed costs of opening facilities is obtained by summing over all facilities the product of the fixed costs ($F_k$) and the integer variable, $y_k$, which is 1 if that facility is chosen to be opened in the optimal solution. The constraints represented in (11.17) assure that every demand is satisfied, and the constraints in Equation (11.18) assure that only open facilities can supply demands. Constraint (11.19) assures that all demands are satisfied by shipments from open facilities. Constraints in Equation (11.20) ensure that the amount shipped from facility $k$ to demand $l$, $b_{kl}$, is either less than the demand requirement at $l$ ($dem_l$) or the supply at $k$ ($O_{mkl}$),whichever is smaller; if facility $k$ is not opened, then $b_{kl}$ is forced to zero. Normally, the supply at $k$ is the capacity of facility $k$, in this case, we use one of the outputs, $m$.

## 11.4 Example and Nonlinear Model

In Klimberg and Ratick (2008) we created and used an example with seven facilities serving fifteen demands, each with four inputs and three outputs, to test the DEA/capacitated facility location model (CASD). A series of multi-objective solutions for the CASD model were obtained by varying the relative weights on the minimum efficiency objective. Table 11.1 gives the objective function values for these solutions.

**Table 11.1 Solution Values for the CASD Model**

| *Objective Function Values* | | | | | |
|---|---|---|---|---|---|
| Total fixed costs | \$3,003 | \$2,739 | \$2,248 | \$2,316 | \$1,874 |
| Total transport costs | \$9,066 | \$10,367 | \$12,835 | \$16,662 | \$20,931 |
| Total costs | \$12,069 | \$13,106 | \$15,083 | \$18,978 | \$22,805 |
| Number of open facilities | 6 | 5 | 3 | 3 | 2 |
| Number of facility-demand links | 23 | 18 | 21 | 22 | 20 |
| Total sum of efficiency scores | 12.71 | 11.69 | 17.42 | 21.21 | 20.00 |
| Minimum efficiency score for solution | 0.2351 | 0.4586 | 0.6330 | 0.8297 | 1.0000 |

As the solutions in Table 11.1 illustrate, the simultaneous consideration of the spatial efficiency of transportation and fixed costs, together with the facility efficiency as measured by the DEA score, provides insight into the trade-offs between these objectives. To keep the CASD formulation linear, we allowed a fraction of one of the output variables to satisfy demand. In such a case, the fractional contribution of that output to the DEA score is not properly weighted in the DEA solution. If we keep the true DEA weight and allowed the amount of output to vary as given in the CASD formulation, it would have resulted in a nonlinear program. In this chapter we address that problem by extending the CASD formulation to allow all the output variables to varying—changing them from parameters to variables—and therefore allowing the DEA scores to dynamically change. The formulation for this capacitated adjustable, simultaneous dynamic DEA/CPLP model (CASD$^2$) is the following:

**Model: Capacitated Adjustable Simultaneous Dynamic DEA/CPLP Model (CASD$^2$)**

$$\text{MAX} \sum_{k=1}^{K} \sum_{l=1}^{L} (1 - \mathrm{d}_{kl}) \tag{11.28}$$

$$\text{MIN} \sum_{k=1}^{K} \sum_{l=1}^{L} \mathrm{c}_{kl} \mathrm{b}_{kl} + \sum_{k=1}^{K} \mathrm{F}_k \mathrm{y}_k \tag{11.29}$$

s.t.

$$\sum_{k=1}^{K} \mathrm{x}_{kl} \geq 1 \quad \forall l \tag{11.30}$$

$$\mathrm{x}_{kl} \leq \mathrm{y}_k \quad \forall k, l \tag{11.31}$$

$$\sum_{k=1}^{K} \mathrm{b}_{kl} = \mathrm{dem}_l \quad \forall l \tag{11.32}$$

$$\sum_{j=1}^{J} \mathrm{p}_{jkl} = \mathrm{b}_{kl} \quad \forall k, l \tag{11.33}$$

$$\mathrm{p}_{jkl} \leq \mathrm{O}_{jkl} \mathrm{y}_k \quad \forall k, l \tag{11.34}$$

$$\sum_{i=1}^{I} \mathrm{v}_{kli} \mathrm{I}_{kli} = x_{kl} \quad \forall k, l \tag{11.35}$$

$$\sum_{j=1}^{J} \mathrm{u}_{klj}\mathrm{p}_{jkl} + \mathrm{d}_{kl} = \mathrm{x}_{kl} \quad \forall k, l \tag{11.36}$$

$$\sum_{j=1}^{J} \mathrm{u}_{klj}\mathrm{p}_{jrs} - \sum_{i=1}^{I} \mathrm{v}_{kli}\mathrm{I}_{irs} \leq 0 \quad \forall k, l; \forall r, s; (k \neq r \text{ and } l \neq s) \tag{11.37}$$

$$\mathrm{u}_{klj} \geq \varepsilon \mathrm{x}_{kl} \quad \forall k, l, j \tag{11.38}$$

$$\mathrm{v}_{kli} \geq \varepsilon \mathrm{x}_{kl} \quad \forall k, l, i \tag{11.39}$$

$$\mathrm{u}_{klj}\mathrm{O}_{jkl} \leq \mathrm{x}_{kl} \quad \forall k, l \tag{11.40}$$

$$\mathrm{b}_{kl} \geq \mathrm{x}_{kl} \quad \forall k, l \tag{11.41}$$

$$\mathrm{y}_k, \mathrm{x}_{kl} = 0, 1$$

$$\mathrm{b}_{kl}, \mathrm{u}_{klj}, \mathrm{v}_{kli}, \mathrm{p_{jkl}} \geq 0$$

where:

- Decision variables:
  $\mathrm{p}_{jkl}$ = Number of units of output $j$ shipped from facility $k$ to demand location $l$

The CASD$^2$ is similar to the CASD formulation except:

1. In Constraints (11.36) and (11.37), the parameter $\mathrm{O}_{jkl}$ is replaced by the decision variable $\mathrm{p}_{jkl}$, thus making these constraints nonlinear.
2. Constraint (11.20) in the CASD formulation is removed from the CASD$^2$ formulation and Constraints (11.33) and (11.34) are added. Constraint (11.33) requires that all the output supplied to demand location $l$ must satisfy its demand. Constraint (11.34) forces $\mathrm{p}_{jkl}$ to equal zero if facility $k$ is not opened and further limits $\mathrm{p}_{jkl}$ to be less than $\mathrm{O}_{jkl}$ if facility $k$ is opened.

To solve the CASD$^2$ model, which is a mixed-integer nonlinear programming (MINLP) problem, we used AIMMS linear programming software and, in particular, the AIMMS outer approximation (AOA) algorithm. The AOA algorithm controls the interaction between a master MIP model (we used the CPLEX solver), and the NLP submodel (we used the SNOPT solver). On a 2.4-GHz Intel Core 2 Duo T7700 with 3GB of RAM laptop, it took about 10 to 20 minutes to run the model.

**Table 11.2 Solution Values for the CASD$^2$ Model**

| *Objective Function Values* | | | |
|---|---|---|---|
| Total fixed costs | $2,433 | $2,501 | $2,495 |
| Total transport costs | $7,786 | $7,786 | $7,786 |
| Total costs | $10,219 | $10,287 | $10,281 |
| Number of open facilities | 5 | 5 | 5 |
| Number of facility-demand links | 15 | 7 | 7 |
| Total sum of efficiency scores | 15.00 | 17.00 | 17.00 |
| Minimum efficiency score for solution | 1.00 | 1.00 | 1.00 |

Table 11.2 lists the multi-objective solutions from the CASD$^2$ model obtained by varying the weight on the minimum efficiency objective. In general, the costs are lower and now every DEA score for each facility $k$ and demand $l$ opened is 100 percent as compared to the linear CASD model. Because we are now allowing all three outputs to satisfy demand, there is so much excess output that demand can be easily satisfied, and the model reallocates the outputs in such a way to allow the DEA score for each open facility to be 100 percent. To address the excess output problem, we arbitrarily multiplied each demand location by 10. The multi-objective solutions to the CASD$^2$ model with the increased demands are listed in Table 11.3. There is a larger trade-off between costs and DEA efficiency illustrated in these solutions (although there is still an abundance of output).

**Table 11.3 Solution Values for the CASD$^2$ Model with the Demands Increase (Multiplied By 10)**

| *Objective Function Values* | | | | | |
|---|---|---|---|---|---|
| Total fixed costs | $2,991 | $3,026 | $3,113 | $3,163 | $3,200 |
| Total transport costs | $7,786 | $7,786 | $7,786 | $7,786 | $7,786 |
| Total costs | $10,777 | $10,812 | $10,899 | $10,949 | $10,986 |
| Number of open facilities | 5 | 5 | 5 | 5 | 5 |
| Number of facility-demand links | 32 | 32 | 35 | 36 | 37 |
| Total sum of efficiency scores | 30.80 | 30.88 | 34.74 | 35.99 | 37.00 |
| Minimum efficiency score for solution | 0.743 | 0.757 | 0.912 | 1.00 | 1.00 |

## 11.5 Conclusions

In this chapter we extended the capacitated adjustable simultaneous DEA/CPLP (CASD) model previously developed to consider the more realistic situation in which outputs from the facilities are decision variables. In such a case, the DEA scores are allowed to dynamically change and the model becomes nonlinear. In comparison with our earlier results, this change provides a wider range of choices in optimizing the combined DEA/Location model, allowing all facilities in each solution to obtain a DEA efficiency score of 1. Modifying the problem by increasing demands by an order of magnitude provided a greater range of trade-offs between the DEA (facility) efficiency objective and spatial efficiency measured in the location model objective.

The CASD$^2$ model provided a more realistic situation by allowing the outputs to vary. However, several issues still remain to be addressed. First, although all the outputs were allowed to change, realistically outputs are likely related to multiple demand patterns. As a result, the model should be expanded to address these multiple demands. In addition, the inputs should vary—in a production function-type relationship—with varying outputs. In this case, the outputs and inputs would in all likelihood change proportionally. Finally, using the AIMMS AOA algorithm, each of the nonlinear solutions was only shown to be locally optimal. Other solution procedures, such as genetic algorithms and tabu search, which have been shown to be proficient in solving nonlinear and integer problems, should be examined and compared to those obtained using the AIMMS procedure.

The results from the CASD$^2$ model developed in this chapter reinforces the idea that facility efficiency—as measured using DEA—should be incorporated into location modeling as a complementary objective to the optimal spatially efficient solutions provided by those models. The results of this type of spatial decision support system would provide important and perceptive information to decision makers.

## References

Athanassopoulos, A. D., and Storbeck, J.E. (1995), Non-parametric models for spatial efficiency, *The Journal of Productivity Analysis*, 6, 225–245.

Charnes A., Cooper, W.W., and Rhodes, E. (1978), Measuring efficiency of decision making units, *European Journal of Operational Research*, 2, 429–444.

Church, R.L., and ReVelle, C. (1974), The maximal covering location problems, *Papers of the Regional Science*, 32(1), 101–118.

Current, J., Min, H., and Schilling, D. (1990), Multiobjective analysis of facility location decisions, *European Journal of Operational Research*, 49, 295–307.

Daskin, M.S. (1995), *Network and Discrete Location: Models, Algorithms and Applications*, New York: John Wiley & Sons,.

Desai, A., and Storbeck, J.E. (1990), A data envelopment analysis for spatial efficiency, *Computers, Environment, and Urban Systems*, 14, 145–156.

Desai, A., Haynes, K., and Storbeck, J.E. (1995), A spatial efficiency for the support of location decision, in *Data Envelopment Analysis: Theory, Methodology and Applications,* edited by Charnes, A., Cooper, W.W., Lewin, A., and Seigord, L., Dordrecht: Kluwer Academic.

Emrouznedjad, A., Parker, B.R., and Tavares, G. (2008), Evaluation of research in efficiency and productivity: a survey and analysis of the first 30 years of scholarly literature in DEA, *Socio-Economic Planning Sciences,* 44, 151–157.

Hakimi, S.L. (1964), Optimal location of switching centers and the absolute centers and medians of a graph, *Operations Research,* 12, 450–459.

Klimberg, R.K., and Kern, D. (1992), Understanding Data Envelopment Analysis (DEA), Boston University School of Management Working Paper, pp. 92–44.

Klimberg, R.K., and Ratick, S.J. (2008), Modeling data envelopment analysis (DEA) efficient location/allocation decisions, *Computers & Operations Research,* 35, 457–474.

ReVelle, C. (1987), Urban public facility location, in *Handbook of Regional and Urban Economics,* Vol. II, edited by E.S. Mills, Elsevier, Amsterdam, pp. 1053–1096.

ReVelle, C., and Swain, R. (1970), Central facilities location, *Geographical Analysis,* 2, 30–42.

Schilling, D.A., Jayaraman, V., and Barkhi, R. (1993), A review of covering models in facility location, *Location Science,* 1, 25–55.

Seiford, L.M. (1996), Data envelopment analysis: the evaluation of the state of the art (1978–1995), *The Journal of Productivity Analysis,* 9, 99–137.

Seiford, L.M., and Thrall, R.M. (1990), Recent developments in DEA: the mathematical programming approach to frontier analysis, *Journal of Econometrics,* 46, 7–38.

Tavares, G., (2002), A bibliography of Data Envelopment Analysis (1978–2001), RUTCOR, Rutgers University; also available at http://rutcor.rutgers.edu/pub/rrr/reports2002.1_2002.pdf

Thomas, P., Chan, Y., Lehmkuhl, L., and Nixon, W. (2002), Obnoxious-facility location and data envelopment analysis: a combined distance-based formulation, *European Journal of Operational Research,* 141(3), 495–514.

Toregas, C., Swain, R., ReVelle, C., and Bergman, L. (1971), The location of emergency service facilities, *Operations Research,* 19(6), 1363–1373.

White, J., and Case, K. (1974), On covering problems and the central facility location problem, *Geographical Analysis,* 6(3), 281–293.

# Chapter 12

# Sourcing Models for End-of-Use Products in a Closed-Loop Supply Chain

Kishore K. Pochampally and Surendra M. Gupta

## Contents

## 12.1 Introduction

The current environmental trends threaten to radically alter the planet and many species upon it. According to the U.S. Environmental Protection Agency's *Municipal Solid Waste Fact Book*, 15 states have less than 10 years of landfill space left and 6 states have less than 5 years of landfill space left [10]. Increased consumption results in increased use of raw material and energy, thereby depleting the world's finite natural resources. Environmentally conscious manufacturing (ECM) involves planning, developing, and implementing manufacturing processes and technologies that minimize or eliminate hazardous waste and reduce scrap [15]. Effectively, ECM means saving natural resources, energy, clean air and water, landfill space, and money. Additional benefits of ECM include safer and cleaner facilities, lower future costs for disposal, reduced environmental and health risks, and improved product quality at lower cost and higher productivity.

While the traditional supply chain focuses on sourcing, manufacturing, and distributing products to consumers, retailers, and distribution outlets, a reverse supply chain aims to recover end-of-use products from consumers for recycling and disposal. The combination of traditional and reverse supply chains is called a closed-loop supply chain. Reverse and closed-loop supply chains are an integral part of ECM.

Strategic planning is one of the most challenging elements of managing a reverse/closed-loop supply chain [11]. To effectively satisfy drivers such as profitability, environmental regulations, and asset recovery, only the appropriate end-of-use products must be collected from the appropriate suppliers. To this end, this chapter addresses the following two crucial problems and proposes a quantitative decision-making model for each of them: (1) Which end-of-use products must be selected to reprocess? and (2) Which suppliers must the selected end-of-use products be purchased from? For the first problem, we propose a Linear Physical Programming [8] model wherein the selection criteria are presented in terms of ranges of different degrees of desirability. For the second problem, we express the goals in terms of certain performance indices and propose a model that combines Analytic Network Process [13] and Goal Programming [4]. The models are demonstrated via numerical examples.

The chapter is organized as follows: Section 12.2 gives introductions to the techniques used in this chapter, viz., Linear Physical Programming, Analytic Network Process, and Goal Programming; Sections 12.3 and 12.4 present the first and second problems, respectively; finally, Section 12.5 gives some conclusions.

## 12.2 Techniques Used

This section gives introductions to the following techniques used in this chapter: Linear Physical Programming (Section 12.2.1), Analytic Network Process (Section 12.2.2), and Goal Programming (Section 12.2.3).

### 12.2.1 Linear Physical Programming

The Linear Physical Programming (LPP) method [8] uses four distinct classes (1S, 2S, 3S, and 4S) to allow decision makers to expresses their preferences for the value of each criterion (for decision making) in a more detailed, quantitative, and qualitative way than when using a weight-based method like the Analytic Hierarchy Process [14]. These classes are defined as follows: smaller-is-better (1S), larger-is-better (2S), value-is-better (3S), and range-is-better (4S). Figure 12.1 depicts these different classes.

The value of the $p$-th criterion, $g_p$, for evaluating the alternative of interest, is categorized according to the preference ranges shown on the horizontal axis. Consider, for example, the case of Class 1S. The preference ranges are:

| | |
|---|---|
| Ideal range | $g_p \leq t_{p1}^+$ |
| Desirable range | $t_{p1}^+ \leq g_p \leq t_{p2}^+$ |
| Tolerable range | $t_{p2}^+ \leq g_p \leq t_{p3}^+$ |
| Undesirable range | $t_{p3}^+ \leq g_p \leq t_{p4}^+$ |
| Highly undesirable range | $t_{p4}^+ \leq g_p \leq t_{p5}^+$ |
| Unacceptable range | $g_p \geq t_{p5}^+$ |

The quantities $t_{p1}^+$ through $t_{p5}^+$ represent the physically meaningful values that quantify the preferences associated with the $p$-th generic criterion. Consider, for example, the cost criterion for Class 1S. The decision maker could specify a preference vector by identifying $t_{p1}^+$ through $t_{p5}^+$ in dollars as (10 20 30 40 50). Thus, an alternative having a cost of \$15 would lie in the Desirable range, an alternative with a cost of \$45 would lie in the Highly Undesirable range, and so on. We can accomplish this for a non-numerical criterion too, such as color, by (1) specifying a numerical preference structure and (2) quantitatively assigning each alternative a specific criterion value from within a preference range (e.g., Desirable, Tolerable).

The class function, $Z_p$, on the vertical axis in Figure 12.1 is used to map the criterion value, $g_p$, into a real, positive, and dimensionless parameter ($Z_p$ is, in fact, a piecewise linear function of $g_p$). Such a mapping ensures that different criteria values, with different physical meanings, are mapped to a common scale. Consider Class 1S again. If the value of a criterion, $g_p$, is in the Ideal range, then the value of

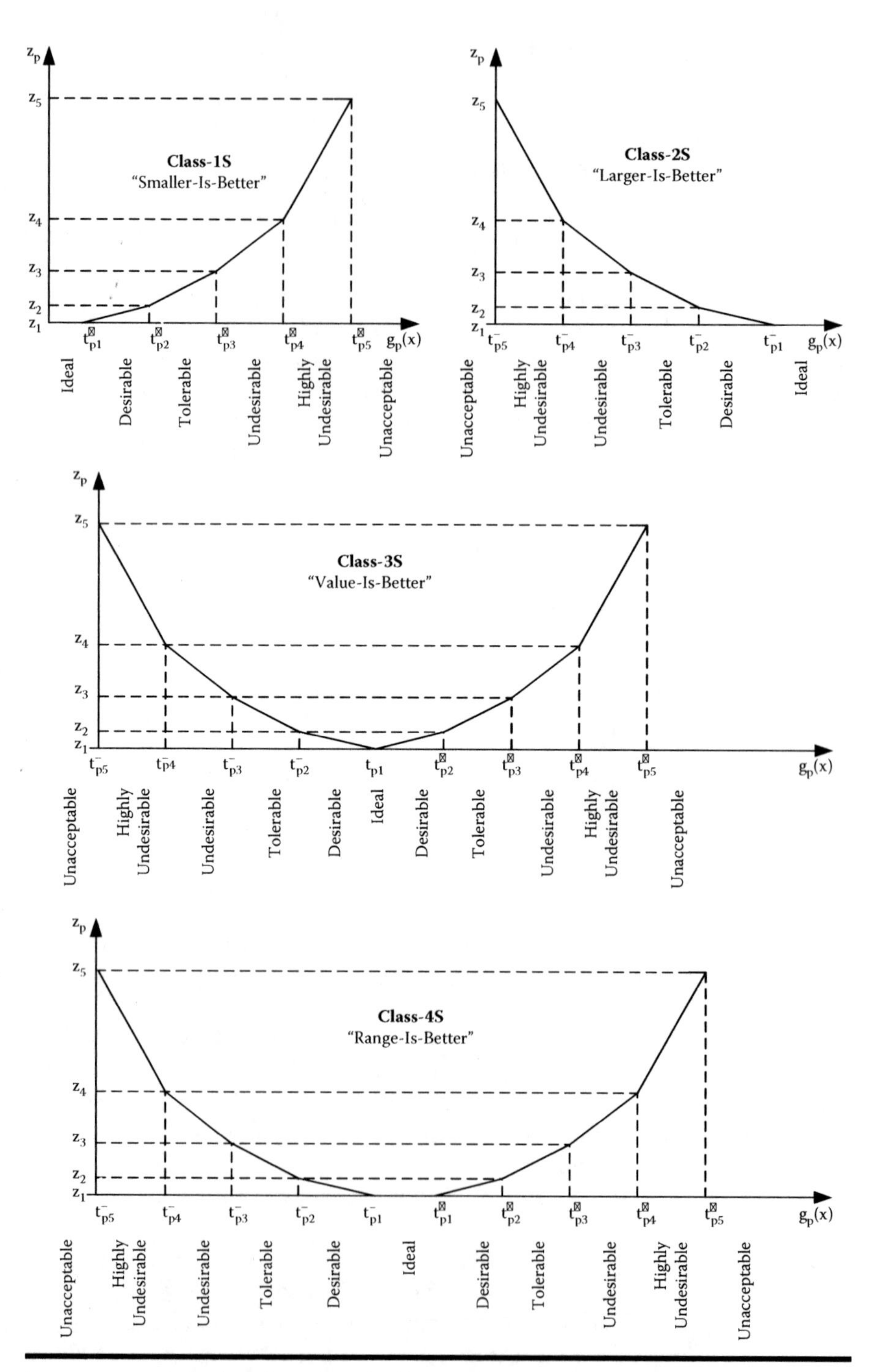


**Figure 12.1 Soft class functions for linear physical programming.**

the class function is small (in fact, zero), while if the value of the criterion is greater than $t_{p5}^{+}$, that is, in the Unacceptable range, then the value of the class function is very high. Class functions have several important properties, such as (1) they are non-negative, continuous, piecewise linear, and convex, and (2) the value of the class function, $Z_p$, at a given range intersection (say, *Desirable-Tolerable*) is the same for all class types.

Basically, ranking of the alternatives is performed in four steps:

*Step 1–Identify criteria for evaluating each of the alternatives.*
*Step 2–Specify preferences for each criterion, based on one of the four classes* (see Figure 12.1).
*Step 3–Calculate incremental weights.* Based on the preference structures for the different criteria, the LPP weight algorithm [8] determines incremental weights, $\Delta w_{pr}^{+}$ and $\Delta w_{pr}^{-}$ (used in Step 4), that represent the incremental slopes of the class functions, $Z_p$. Here, $r$ denotes the range intersection.
*Step 4–Calculate total score for each alternative.* The formula for the total score, $J$, of the alternative of interest is constructed as a weighted sum of deviations over all ranges ($r = 2$ to 5) and criteria ($p = 1$ to $P$), as follows:

$$J = \sum_{p=1}^{P} \sum_{r=2}^{5} (\Delta w_{pr}^{-} d_{pr}^{-} + \Delta w_{pr}^{+} d_{pr}^{+}) \tag{12.1}$$

where $P$ represents the total number of criteria (each belonging to one of the four classes in Figure 12.1), $\Delta w_{pr}^{+}$ and $\Delta w_{pr}^{-}$ are the incremental weights for the $p$-th criterion, and $d_{pr}^{+}$ and $d_{pr}^{-}$ represent the deviations of the $p$-th criterion value of the alternative of interest from the corresponding target values. An alternative with a lower total score is more desirable than one with a higher total score.

The most significant advantage of using LPP is that no weights need to be specified for the criteria for evaluation. The decision maker only needs to specify a preference structure for each criterion, which has more physical meaning than a physically meaningless weight that is arbitrarily assigned to the criterion.

Note that there are no decision variables in the above ranking procedure. LPP can also be used in a problem consisting of decision variables, by minimizing $J$ in Equation (12.1), and subjecting (if necessary) each criterion, $g_p$, to a constraint that falls into either one of the four classes (also called *soft* classes) in Figure 12.1 or one of the following four *hard* classes:

| | |
|---|---|
| Class 1H | Must be smaller (i.e., $g_p \leq t_{p,\max}$) |
| Class 2H | Must be larger (i.e., $g_p \geq t_{p,\min}$) |
| Class 3H | Must be equal (i.e., $g_p = t_{p.\text{val}}$) |
| Class 4H | Must be in range (i.e., $t_{p,\min} \leq g_p \leq t_{p,\max}$) |

### 12.2.2 Analytic Network Process

The Analytic Network Process (ANP) [13] allows for dependence within a set of criteria (inner dependence) as well as between sets of criteria (outer dependence). ANP allows for a complex relationship among decision levels and attributes, as it does not require a strict hierarchical structure. The looser network structure in ANP allows the representation of any decision problem irrespective of what criteria come first or what comes next. ANP is used in a wide variety of problem areas (for example, analyzing alternatives in reverse logistics for end-of-life computers [12], evaluating connection types in design for disassembly [3], and modeling the metrics of lean, agile, and leagile supply chains [1]). The steps involved in the ANP methodology are as follows:

*Step 1. Model development and problem formulation.* In this step, the decision problem is structured into its constituent components. The relevant criteria, the sub-criteria, and alternatives are chosen and are structured in the form of a control hierarchy.

*Step 2. Pairwise comparisons.* In this step the decision maker is asked to carry out a series of pairwise comparisons with respect to the scale shown in Table 12.1, where two main criteria are simultaneously compared with respect to the problem objective, two sub-criteria are simultaneously compared with respect to their main criteria, and pairwise comparisons are performed to address the interdependencies among the sub-criteria. The relative matrix of comparative importance values is then used to weigh the criteria using mathematical techniques like eigen vector, mean transformation, or row geometric mean.

*Step 3. Super matrix formulation.* The super matrix allows for resolution of interdependencies that exist among the sub-criteria. It is a partitioned matrix

**Table 12.1 Scale for Pairwise Judgments**

| *Comparative Importance* | *Definition* |
|---|---|
| 1 | Equally important |
| 3 | Moderately more important |
| 5 | Strongly important |
| 7 | Very strongly more important |
| 9 | Extremely more important |
| 2, 4, 6, 8 | Intermediate judgment values |

where each sub-matrix is composed of a set of relationships between and within the levels as represented by the decision maker's model. The super matrix M is made to converge to obtain a long-term stable set of weights. For convergence, M must be made column stochastic; that is done by raising M to the power of $2^{k+1}$, where $k$ is an arbitrarily large number.

*Step 4. Selection of the best alternative.* The selection of the best alternative depends on the "desirability index." The desirability index, *DI*, for alternative $i$ is defined as:

$$DI = \sum_{j-1}^{J} \sum_{k=1}^{K_j} P_j A_{kj}^{D} A_{kj}^{I} S_{ikj} \tag{12.2}$$

where $P_j$ is the relative importance weight of main criterion $j$, $A_{kj}^{D}$ is the relative importance weight for sub-criterion $k$ of main criterion $j$ for the dependency ($D$) relationships among sub-criteria, $A_{kj}^{I}$ is the stabilized relative importance weight (determined by the super matrix) for sub-criterion $k$ of main criterion $j$ for interdependency ($I$) relationships among sub-criteria, and $S_{ikj}$ is the relative impact of alternative $i$ on sub-criterion $k$ of main criterion $j$.

### 12.2.3 Goal Programming

Goal Programming [4] provides a way of tackling situations where a variety of objectives can be focused on simultaneously. It is generally applied to linear problems and deals with the achievement of specific targets/goals. The basic approach involves formulating an objective function for each objective and seeks a solution that minimizes the sum of the deviations of these objective functions from their respective goals. To this end, several criteria should be considered in the problem situation at hand. For each criterion, a target value is determined. Next, the deviation variables are introduced, which may be positive or negative (represented by $\rho_k$ and $\eta_k$, respectively). The negative deviation variable, $\eta_k$, represents the underachievement of the $k$-th goal. Similarly, the positive deviation variable, $\rho_k$, represents the overachievement of the $k$-th goal. Finally, for each criterion, the desire to overachieve (minimize $\eta_k$) or underachieve (minimize $\rho_k$), or satisfy the target value exactly (minimize $\rho_k + \eta_k$) is articulated [5].

Goal Programming problems can be categorized according to the type of mathematical programming model. Another categorization is according to how the goals compare in importance. In the case of preemptive goal programming, there is a hierarchy of priority levels for the goals, so the goals of primary importance receive first attention, and so forth. In the case of non-preemptive goal programming, all the goals are of roughly comparable importance [5].

## 12.3 Selection of End-of-Use Products

Section 12.3.1 gives the nomenclature used in the model. Section 12.3.2 presents the formulation of the Linear Physical Programming model, and Section 12.3.3 gives a numerical example to illustrate the model.

### 12.3.1 Nomenclature

$C_{df}$ Disposal cost factor (cost per unit weight)
$C_{dx}$ Disposal cost of product $x$
$C_r$ Reprocessing cost per unit time
$C_{rf}$ Recycling revenue factor (revenue per unit weight)
$C_{rpx}$ Total reprocessing cost of product $x$
$CC_x$ Collection cost of product $x$
$D_{xy}$ Disposal cost index of component $y$ in product $x$ (0 = lowest, 10 = highest)
$E_{xk}$ Sub-assembly $k$ in product $x$
$m_{xy}$ Probability of missing component $y$ in product $x$
$M_x$ Number of sub-assemblies in product $x$
$N_{xy}$ Multiplicity of component $y$ in product $x$
$p_{xy}$ Probability of breakage of component $y$ in product $x$
$P_{xy}$ Component $y$ in product $x$
$PRC_{xy}$ Percent of recyclable contents by weight in component $y$ of product $x$
$R_{rcx}$ Total recycling revenue of product $x$
$R_{rsx}$ Total resale revenue of product $x$
$R_{rsxy}$ Resale value of component $y$ in product $x$
$RC_{xy}$ Recycling revenue index of component $y$ in product $x$ (0 = lowest, 10 = highest)
$Root_x$ Root node of product $x$
$SU_x$ Supply of product $x$ per period
$T(E_{xk})$ Time to disassemble sub-assembly $k$ in product $x$
$T(Root_x)$ Time to disassemble $Root_x$
$W_{xy}$ Weight of component $y$ in product $x$
$x$ Product type
$X_{xy}$ Decision variable signifying selection of component $y$ to be retrieved from product $x$ for reuse ($X_{xy}$ = 1 for reuse, 0 for recycle)
$y$ Component type
$Z$ Overall profit

### 12.3.2 Model Formulation

The criteria considered in the model fall into either Class 1S or Class 2S.

#### 12.3.2.1 Class 1S Criteria (Smaller-Is-Better)

*Total collection cost per period* ($g_1$): $g_1$ of product $x$ is calculated by multiplying the supply of $x$ per period ($SU_x$) by the cost of collecting one product from consumers ($CC_x$):

$$g_1 = SU_x * CC_x \tag{12.3}$$

*Total reprocessing cost per period* ($g_2$): $g_2$ of product $x$ is calculated using the disassembly time of the root node [$T(Root_x)$], disassembly time of each sub-assembly [$T(E_{xk})$], supply of $x$ per period ($SU_x$) and the remanufacturing cost per unit time ($C_r$):

$$SU_x \left[ T(Root_x) + \sum_{k=1}^{M_j} T(E_{xk}) \right] C_r \tag{12.4}$$

*Total disposal cost per period* ($g_3$): $g_3$ of product $x$ is calculated by multiplying the component disposal cost by the number of units of components disposed, as follows:

$$\sum_y [SU_x \cdot DI_{xy} W_{xy}(1 - PRC_{xy})\{N_{xy}(1 - m_{xy}) - N_{xy}(1 - b_{xy} - m_{xy})\}] C_{df} \tag{12.5}$$

$DI_{xy}$ is the disposal cost index that varies in value from 1 to 10 representing the degree of nuisance created by the disposal of component $y$ of product $x$; $C_{df}$ is the disposal cost factor.

*Loss-of-sale cost* ($g_4$): $g_4$ of product $x$ represents the periodical worth of not meeting the demand on time. It can occur because of the unpredictability in the supply of end-of-use products. This can be obtained from an expert in the field.

*Worth of investment cost* ($g_5$): $g_5$ of product $x$ represents the periodical worth of the fixed cost of the production facility and the machinery required to reprocess. This can also be obtained from an expert in the field.

#### 12.3.2.2 Class 2S Criteria (Larger-Is-Better)

*Total reuse revenue per period* ($g_6$): $g_6$ of product $x$ is influenced by the supply of $x$ per period ($SU_x$), resale value of component $y$ ($RSR_{xy}$), multiplicity of component $y$($N_{xy}$), and breakage and missing probabilities of component $y$ ($p_{xy}$, $m_{xy}$). The reuse revenue equation can be written as follows:

$$\sum_y [SU_x \cdot RSRxy.N_{xy} \cdot (1 - m_{xy} - b_{xy})] \tag{12.6}$$

*Total recycling revenue per period* ($g_7$): $g_7$ of product $x$ is influenced by the supply of $x$ per period ($SU_x$), recycling revenue index of component $y$ ($RCRI_{xy}$),

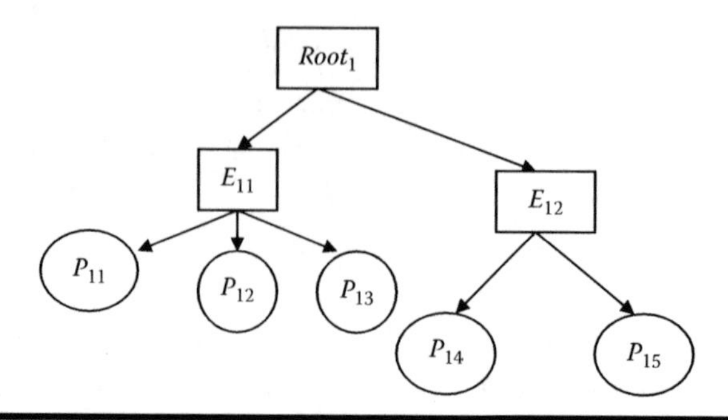


**Figure 12.2 Structure of Product 1.**

percentage of recyclable content in component $y$ ($PRC_{xy}$), multiplicity of component $y$ ($N_{xy}$), weight of component $y$ ($W_{xy}$), recycling revenue factor ($C_{rf}$), and breakage and missing probabilities of component $y(p_{xy}, m_{xy})$. The recycling revenue equation can be written as

$$\sum_y [SU_x \cdot RCRI_{xy} \cdot W_{xy} PRC_{xy} \cdot N_{xy} \cdot \{N_{xy}(1 - m_{xy}) - N_{xy}(1 - m_{xy} - b_{xy})\} C_{rf}] \tag{12.7}$$

## 12.3.3 Numerical Example

Consider three end-of-use products (1, 2, and 3) whose structures are shown in Figures 12.2, 12.3, and 12.4, respectively.

The data for the three products are given in Tables 12.2 through 12.4, respectively. The target values for the criteria are given in Table 12.5, and Table 12.6 shows the criteria values for each product. Table 12.7 shows the incremental weights obtained using the LPP weight algorithm [8].

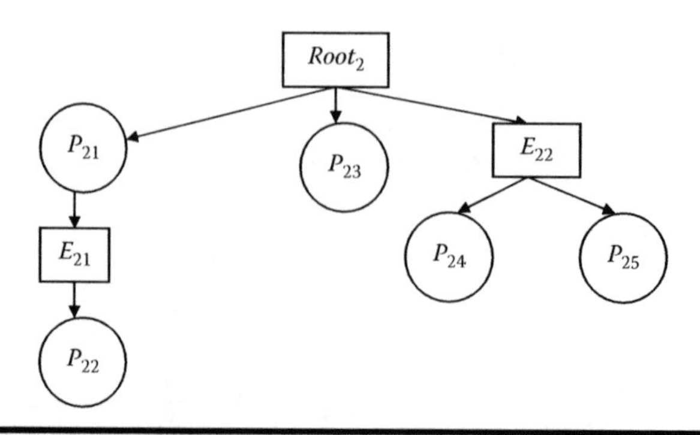


**Figure 12.3 Structure of Product 2.**

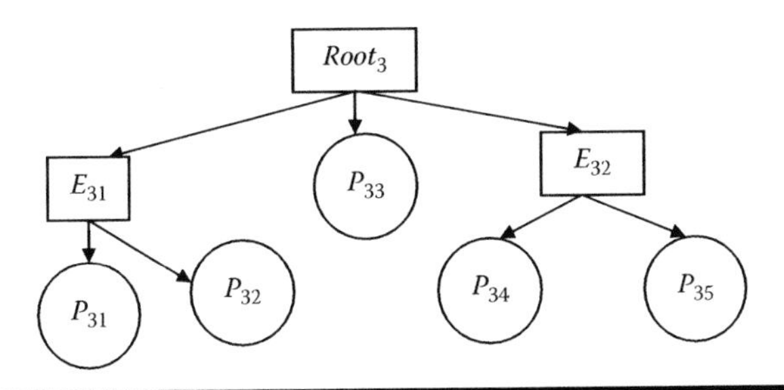


**Figure 12.4 Structure of Product 3.**

Tables 12.8, 12.9, and 12.10, show the deviations of criteria values from target values for the three products, respectively. For example, for criteria $g_1$ of product 1, deviation at $s = 2$ is the absolute value of the number obtained by subtracting the criteria value (i.e., 7.5; shown in bold in Table 12.6) from the target value (i.e., 10; shown in bold in Table 12.5).

The total score for each product is calculated using Equation (12.1) (using the incremental weights from the LPP algorithm and deviations from target values) and is shown in Table 12.11. Because alternatives with lower scores are more desirable than the ones with higher scores, product 1 is the best of the lot.

**Table 12.2 Data of Product 1**

| Part | $RSR_{xy}$ | $N_{xy}$ | $W_{1y}$ | $RCRI_{xy}$ | $PRC_{xy}$ | $DI_{xy}$ | $p_{xy}$ | $m_{xy}$ |
|---|---|---|---|---|---|---|---|---|
| $P_{11}$ | 6 | 3 | 3.5 | 5 | 0.65 | 6 | 0.1 | 0.3 |
| $P_{12}$ | 7 | 2 | 4.5 | 4 | 0.4 | 4 | 0.2 | 0.5 |
| $P_{13}$ | 5 | 4 | 5 | 5 | 0.3 | 4 | 0.4 | 0.1 |
| $P_{14}$ | 7.5 | 2 | 6 | 3 | 0.25 | 5 | 0.2 | 0.2 |
| $P_{15}$ | 5.5 | 1 | 5.5 | 2 | 0.45 | 1 | 0.5 | 0.0 |

**Table 12.3 Data of Product 2**

| Part | $RSR_{xy}$ | $N_{xy}$ | $W_{2y}$ | $RCRI_{xy}$ | $PRC_{xy}$ | $DI_{xy}$ | $p_{xy}$ | $m_{xy}$ |
|---|---|---|---|---|---|---|---|---|
| $P_{21}$ | 4 | 2 | 9 | 2 | 0.56 | 3 | 0.2 | 0.5 |
| $P_{22}$ | 6 | 5 | 3 | 1 | 0.5 | 6 | 0.2 | 0.2 |
| $P_{23}$ | 7 | 4 | 5 | 2 | 0.48 | 6 | 0.4 | 0.1 |
| $P_{24}$ | 2 | 2 | 6 | 3 | 0.2 | 1 | 0.1 | 0.1 |
| $P_{25}$ | 4.5 | 3 | 1 | 4 | 0.25 | 3 | 0.1 | 0.3 |

**Table 12.4 Data of Product 3**

| *Part* | $RSR_{xy}$ | $N_{xy}$ | $W_{3y}$ | $RCRI_{xy}$ | $PRC_{xy}$ | $DI_{xy}$ | $p_{xy}$ | $m_{xy}$ |
|---|---|---|---|---|---|---|---|---|
| $P_{31}$ | 4.5 | 3 | 8.5 | 5 | 0.4 | 5 | 0.1 | 0.3 |
| $P_{32}$ | 4 | 4 | 6 | 4 | 0.65 | 6 | 0.1 | 0.1 |
| $P_{33}$ | 5 | 6 | 4.1 | 2 | 0.2 | 3 | 0.2 | 0.0 |
| $P_{34}$ | 2 | 5 | 3.5 | 3 | 0.35 | 2 | 0.3 | 0.2 |
| $P_{35}$ | 3.5 | 1 | 2 | 1 | 0.25 | 7 | 0.5 | 0.5 |

**Table 12.5 Target Values of Criteria**

| *Criteria* | $t^+_{p1}$ | $t^+_{p2}$ | $t^+_{p3}$ | $t^+_{p4}$ | $t^+_{p5}$ |
|---|---|---|---|---|---|
| $g_1$ | 10 | 12 | 15 | 17.5 | 20 |
| $g_2$ | 3 | 5 | 9 | 10 | 15 |
| $g_3$ | 1 | 4 | 5.5 | 7.5 | 8 |
| $g_4$ | 2 | 5 | 7.5 | 9 | 10 |
| $g_5$ | 1 | 4 | 8 | 9 | 10 |
| | | | | | |
| Criteria | $t^-_{p1}$ | $t^-_{p2}$ | $t^-_{p3}$ | $t^-_{p4}$ | $t^-_{p5}$ |
| $g_6$ | 10 | 12.5 | 15 | 20 | 25.5 |
| $g_7$ | 5 | 13 | 17.5 | 20 | 25 |

**Table 12.6 Criteria Values for Each Product**

| *Criteria* | *Product 1* | *Product 2* | *Product 3* |
|---|---|---|---|
| $g_1$ | 7.5 | 7.5 | 10 |
| $g_2$ | 3.5 | 2.75 | 3 |
| $g_3$ | 2.97 | 1.6 | 1.8 |
| $g_4$ | 3.5 | 4 | 5 |
| $g_5$ | 2 | 2.5 | 2 |
| $g_6$ | 18.3 | 22.8 | 24.9 |
| $g_7$ | 18.4 | 15.7 | 19 |

**Table 12.7 Output of LPP Weight Algorithm**

| *Criteria* | $\Delta w^+_{p2}$ | $\Delta w^+_{p3}$ | $\Delta w^+_{p4}$ | $\Delta w^+_{p5}$ | $\Delta w^-_{p2}$ | $\Delta w^-_{p3}$ | $\Delta w^-_{p4}$ | $\Delta w^-_{p5}$ |
|---|---|---|---|---|---|---|---|---|
| $g_1$ | 0.05 | 0.0967 | 0.62773 | 2.63296 | – | – | – | – |
| $g_2$ | 0.05 | 0.076 | 2.41416 | 0.020321 | – | – | – | – |
| $g_3$ | 0.0333 | 0.3027 | 0.93408 | 24.33473 | – | – | – | – |
| $g_4$ | 0.0333 | 0.16827 | 1.49184 | 11.10897 | – | – | – | – |
| $g_5$ | 0.0333 | 0.09267 | 2.41416 | 10.26225 | – | – | – | – |
| $g_6$ | – | – | – | – | 0.04 | 0.0492 | 0.010258 | 0.122333 |
| $g_7$ | – | – | – | – | 0.0125 | 0.037056 | 0.14936 | 0.022875 |

**Table 12.8 Deviations of Criteria Values from Targets, for Product 1**

| *Criteria* | $s = 2$ | $s = 3$ | $s = 4$ | $s = 5$ |
|---|---|---|---|---|
| $g_1$ | **2.5** | 4.5 | 7.5 | 10 |
| $g_2$ | 0.5 | 1.5 | 5.5 | 6.5 |
| $g_3$ | 1.9 | 1.1 | 2.6 | 4.6 |
| $g_4$ | 1.5 | 1.5 | 4 | 5.5 |
| $g_5$ | 1 | 2 | 6 | 7 |
| $g_6$ | 6.7 | 1.7 | 3.3 | 5.8 |
| $g_7$ | 6.6 | 1.6 | 0.9 | 5.4 |

**Table 12.9 Deviations of Criteria Values from Targets, for Product 2**

| *Criteria* | $s = 2$ | $s = 3$ | $s = 4$ | $s = 5$ |
|---|---|---|---|---|
| $g_1$ | 2.5 | 4.5 | 7.5 | 10 |
| $g_2$ | 0.25 | 2.25 | 6.25 | 7.25 |
| $g_3$ | 0.6 | 2.4 | 3.9 | 5.9 |
| $g_4$ | 2 | 1 | 3.5 | 5 |
| $g_5$ | 1.5 | 1.5 | 5.5 | 6.5 |
| $g_6$ | 2.2 | 2.8 | 7.8 | 10.3 |
| $g_7$ | 9.3 | 4.3 | 1.8 | 2.7 |

**Table 12.10 Deviations of Criteria Values from Targets, for Product 3**

| *Criteria* | $s = 2$ | $s = 3$ | $s = 4$ | $s = 5$ |
|---|---|---|---|---|
| $g_1$ | 0 | 2 | 5 | 7.5 |
| $g_2$ | 0 | 2 | 6 | 7 |
| $g_3$ | 0.8 | 2.2 | 3.7 | 5.7 |
| $g_4$ | 3 | 0 | 2.5 | 4 |
| $g_5$ | 1 | 2 | 6 | 7 |
| $g_6$ | 0.1 | 4.9 | 9.9 | 12.4 |
| $g_7$ | 5.93 | 0.93 | 1.57 | 6.07 |

## 12.4 Selection of Suppliers

In this section, the Analytic Network Process (ANP) is used to calculate the performance indices (efficiency scores) of candidate collection centers, with respect to qualitative criteria taken from the perspective of a remanufacturing facility interested in buying end-of-use products from the collection centers. Then, Goal Programming is employed to determine the quantities of products to be transported from the candidate collection centers to the remanufacturing facility, while satisfying two important goals of the remanufacturing facility: maximize total value of purchase and minimize total cost of purchase. Sections 12.4.1 and 12.4.2 present the applications of ANP and Goal Programming, respectively.

### *12.4.1 Application of ANP*

The problem of evaluating the efficiencies of the candidate collection centers is framed as a four-level hierarchy (see Figure 12.5). The first level contains the objective of evaluation of the candidate collection centers; the second level consists of the main evaluation criteria taken from the perspective of a remanufacturing facility; the

**Table 12.11 Total Scores and Ranks of Products**

| *Product* | *Score* | *Rank* |
|---|---|---|
| 1 | 315.3144 | 1 |
| 2 | 339.1954 | 3 |
| 3 | 317.8653 | 2 |

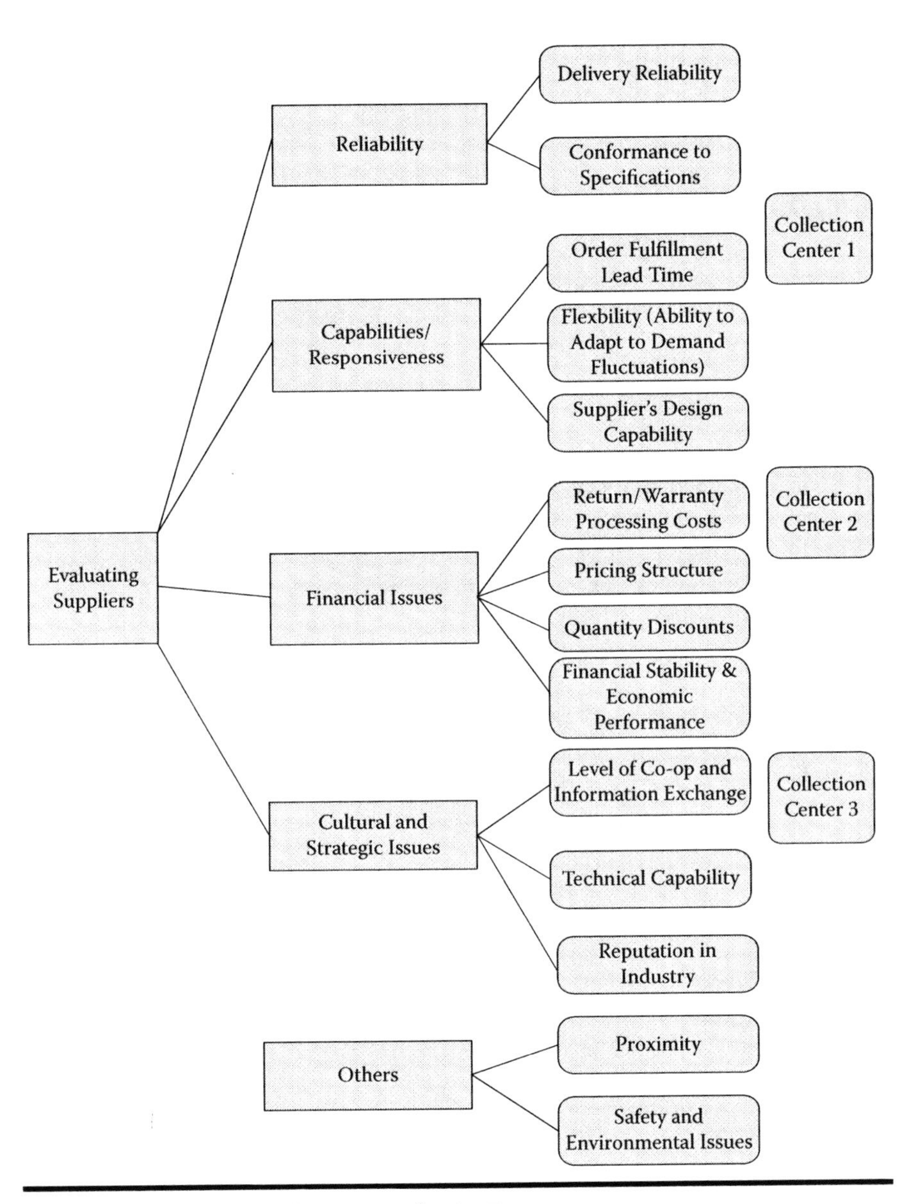


**Figure 12.5 Hierarchical Structure for ANP.**

third level contains the sub-criteria under each main criterion; and the fourth level contains the candidate collection centers. The main and sub-criteria considered are the following (see [2, 6, 7, 9, 16]):

- *Reliability.* This criterion relates to a collection center delivering the end-of-use products at the right time, at the right remanufacturing facility, in the right quantity, and in the promised condition. The sub-criteria considered under this main criterion are (a) delivery reliability, and (b) conformance to standards.
- *Capability/responsiveness.* This criterion reflects the velocity at which a collection center supplies the end-of-use products to the remanufacturing facility and the collection center's ability to adapt to sudden demand fluctuations. The sub-criteria considered here are order fulfillment lead time, flexibility in adapting to demand fluctuations, and design capabilities.
- *Financial issues.* This criterion reflects the costs and other financial aspects involved. The sub-criteria are return/warranty processing costs, pricing structure, quantity discounts, and financial stability and economic performance of the collection center.
- *Cultural and strategic issues.* This criterion consists of the following sub-criteria: level of cooperation and information exchange between the collection center and the remanufacturing facility; collection center's reputation in the industry; and collection center's technical capability (how knowledgeable the collection center is about the product).
- *Others.* This criterion considers miscellaneous aspects that are not considered in the other criteria. They are proximity of the collection center to the remanufacturing facility (it affects the transportation cost and the transit time), and safety and environmental aspects (because the collection center and the remanufacturing facility are closely involved, any safety issues with the collection center directly reflect on the remanufacturing facility's reputation; environmental aspects are concerned about the collection center's effort in pursuing environmental consciousness or a "green image").

It is assumed that there exist interdependencies among the sub-criteria on the third level in the hierarchy.

Three collection centers, S1, S2, and S3, are considered in the numerical example to illustrate the application of ANP. Table 12.12 shows the pairwise comparison matrix for the main criteria (second level in the hierarchy) and also the normalized eigen vector of the matrix. The elements of the normalized eigen vector are the impacts given to the main criteria with respect to the objective (first level in the hierarchy).

Tables 12.13 through 12.17 show the pairwise comparison matrices of sub-criteria with respect to their main criteria and also the corresponding normalized eigen vectors of the matrices.

**Table 12.12 Comparative Importance Values of Main Criteria**

| *Criteria* | *Reliability* | *Responsiveness* | *Financial Issues* | *Cultural and Strategic Issues* | *Others* | *Normalized Eigen Vector* |
|---|---|---|---|---|---|---|
| Reliability | 1 | 1/4 | 3 | 1 | 1/6 | 0.114 |
| Responsiveness | 4 | 1 | 1 | 5 | 3 | 0.431 |
| Financial Issues | 1/3 | 1/5 | 1 | 1/2 | 1/3 | 0.065 |
| Cultural and Strategic Issues | 1 | 1/3 | 2 | 1 | 1/3 | 0.102 |
| Others | 6 | 1/3 | 3 | 3 | 1 | 0.285 |

**Table 12.13 Comparative Importance Values of Sub-criteria under "Reliability"**

| *Sub-criteria* | *Delivery Reliability* | *Conformance to Specs* | *Norm. Eigen Vector* |
|---|---|---|---|
| Delivery Reliability (*DR*) | 1 | 1/3 | 0.25 |
| Conformance to Specs (*CS*) | 3 | 1 | 0.75 |

**Table 12.14 Comparative Importance Values of Sub-criteria under "Responsiveness"**

| *Sub-criteria* | *Order Fulfillment Lead Time (OFT)* | *Flexibility (F)* | *Design Capability (DC)* | *Norm. Eigen Vector* |
|---|---|---|---|---|
| Order Fulfillment Lead Time (*OFT*) | 1 | 3 | 4 | 0.623 |
| Flexibility (*F*) | 1/3 | 1 | 2 | 0.239 |
| Design Capability (*DC*) | 1/4 | 1/2 | 1 | 0.137 |

**Table 12.15 Comparative Importance Values of Sub-criteria under "Financial Issues"**

| *Sub-criteria* | *Returns/ Warranty Processing Costs (RC)* | *Pricing Structure (PS)* | *Qty. Discounts (QD)* | *Economic Performance and Financial Stability (EP)* | *Norm. Eigen Vector* |
|---|---|---|---|---|---|
| Returns/Warranty Processing Costs (*RC*) | 1 | 1/3 | 1/2 | 1/4 | 0.099 |
| Pricing Structure (*PS*) | 3 | 1 | 2 | 1 | 0.345 |
| Qty. Discounts (*QD*) | 2 | 1/2 | 1 | 1/2 | 0.185 |
| Economic Performance and Financial Stability (*EP*) | 4 | 1 | 2 | 1 | 0.37 |

**Table 12.16 Comparative Importance Values of Sub-criteria under "Cultural and Strategic Issues"**

| *Sub-criteria* | *Level of Cooperation and Information Exchange (Co-op)* | *Technical Capability (TC)* | *Reputation* | *Norm. Eigen Vector* |
|---|---|---|---|---|
| Level of Cooperation and Information Exchange (*Co-op*) | 1 | 5 | 6 | 0.722 |
| Technical Capability (*TC*) | 1/5 | 1 | 2 | 0.174 |
| Reputation (*R*) | 1/6 | 1/2 | 1 | 0.103 |

**Table 12.17 Comparative Importance Values of Sub-criteria under "Others"**

| *Sub-criteria* | *Proximity (P)* | *Safety and Environment (SE)* | *Norm. Eigen Vector* |
|---|---|---|---|
| Proximity (*P*) | 1 | 1/3 | 0.25 |
| Safety and Environment (*SE*) | 3 | 1 | 0.75 |

**Table 12.18 Matrix of Interdependencies (Super Matrix M)**

| | *DR* | *CS* | *OFT* | *F* | *DC* | *RC* | *PS* | *QD* | *EP* | *Co-op* | *TC* | *R* | *P* | *SE* |
|---|---|---|---|---|---|---|---|---|---|---|---|---|---|---|
| *DR* | 0 | 1 | 0 | 0 | 0 | 0 | 0 | 0 | 0 | 0 | 0 | 0 | 0 | 0 |
| *CS* | 1 | 0 | 0 | 0 | 0 | 0 | 0 | 0 | 0 | 0 | 0 | 0 | 0 | 0 |
| *OFT* | 0 | 0 | 0 | 0.8 | 0.75 | 0 | 0 | 0 | 0 | 0 | 0 | 0 | 0 | 0 |
| *F* | 0 | 0 | 0.75 | 0 | 0.25 | 0 | 0 | 0 | 0 | 0 | 0 | 0 | 0 | 0 |
| *DC* | 0 | 0 | 0.25 | 0.2 | 0 | 0 | 0 | 0 | 0 | 0 | 0 | | 0 | 0 |
| *RC* | 0 | 0 | 0 | 0 | 0 | 0 | 0.29 | 0.13 | 0.201 | 0 | 0 | 0 | 0 | 0 |
| *PS* | 0 | 0 | 0 | 0 | 0 | 0.53 | 0 | 0.62 | 0.6 | 0 | 0 | 0 | 0 | 0 |
| *QD* | 0 | 0 | 0 | 0 | 0 | 0.29 | 0.16 | 0 | 0.11 | 0 | 0 | 0 | 0 | 0 |
| *EP* | 0 | 0 | 0 | 0 | 0 | 0.163 | 0.53 | 0.23 | 0 | 0 | 0 | 0 | 0 | 0 |
| *Co-op* | 0 | 0 | 0 | 0 | 0 | 0 | 0 | 0 | 0 | 0 | 0.75 | 0.83 | 0 | 0 |
| *TC* | 0 | 0 | 0 | 0 | 0 | 0 | 0 | 0 | 0 | 0.8 | 0 | 0.16 | 0 | 0 |
| *R* | 0 | 0 | 0 | 0 | 0 | 0 | 0 | 0 | 0 | 0.2 | 0.25 | 0 | 0 | 0 |
| *P* | 0 | 0 | 0 | 0 | 0 | 0 | 0 | 0 | 0 | 0 | 0 | 0 | 0 | 1 |
| *SE* | 0 | 0 | 0 | 0 | 0 | 0 | 0 | 0 | 0 | 0 | 0 | 0 | 1 | 0 |

Table 12.18 shows the matrix of interdependencies (called the super matrix M) among the sub-criteria with respect to their main criteria. This super matrix M is made to converge to obtain a long-term, stable set of impacts. For convergence, M must be made column stochastic, which is done by raising M to the power of $2^{k+1}$, where $k$ is an arbitrarily large number. In the example, $k = 59$. Table 12.19 shows the converged super matrix.

Table 12.20 shows the relative ratings of the candidate collection centers, S1, S2, and S3, with respect to the sub-criteria. These ratings are obtained after carrying out pairwise comparisons between the candidate collection centers with respect to the sub-criteria and then obtaining the normalized eigen vector.

To obtain pairwise comparisons, interdependencies, and relative ratings (see Tables 12.13 through 12.20), decision makers could be invited to participate in relevant survey questionnaires, individual interviews, focus groups, and on-site observations.

Table 12.21 shows the desirability index calculated for each candidate collection center using Equation (12.2).

The overall performance index for each of the three collection centers is calculated by multiplying the desirability index (see Table 12.21) of each collection center for

**Table 12.19 Converged Super Matrix**

| *Sub-criteria* | *Stabilized Relative Impact* |
|---|---|
| Delivery reliability | 1 |
| Conformance to specs | 1 |
| Order fulfillment LT | 0.44 |
| Flexibility | 0.38 |
| Design capability | 0.18 |
| Returns/warranty | 0.19 |
| Pricing | 0.38 |
| Qty. discounts | 0.15 |
| Stability and eco. perf | 0.27 |
| Co-op and info. exchange | 0.44 |
| Tech. capability | 0.38 |
| Reputation | 0.18 |
| Proximity | 1 |
| Safety and env. | 1 |

each main criterion by the impact of that criterion (see Table 12.21) and summing up over all the criteria. Table 12.22 shows the overall performance indices (efficiencies) for the three collection centers.

## 12.4.2 Application of Goal Programming

This section presents the application of Goal Programming to determine the quantities of end-of-use products to be transported from the candidate collection centers to a remanufacturing facility of interest, while satisfying two important goals of the remanufacturing facility: maximize the total value of purchase and minimize the total cost of purchase. Section 12.4.2.1 gives the nomenclature used in the methodology, and Section 12.4.2.2 presents the problem formulation and a numerical example.

### 12.4.2.1 Nomenclature

$c_i$ = Unit purchasing cost of end-of-use product at collection center $i$
$d_j$ = Demand for end-of-use product $j$
$g$ = Goal index
$i$ = Collection center index, $i = 1, 2, \ldots, s$

**Table 12.20 Relative Ratings of Collection Centers with Respect to Sub-criteria**

| *Sub-criteria/Alternate Collection Centers* | *S1* | *S2* | *S3* |
|---|---|---|---|
| *DR* | 0.33 | 0.141 | 0.524 |
| *CS* | 0.345 | 0.543 | 0.11 |
| *OFT* | 0.274 | 0.068 | 0.657 |
| *F* | 0.109 | 0.309 | 0.581 |
| *DC* | 0.09 | 0.25 | 0.652 |
| *RC* | 0.681 | 0.216 | 0.102 |
| *PS* | 0.309 | 0.581 | 0.109 |
| *QD* | 0.376 | 0.151 | 0.471 |
| *EP* | 0.137 | 0.623 | 0.239 |
| *Co-op* | 0.33 | 0.075 | 0.59 |
| *TC* | 0.137 | 0.239 | 0.623 |
| *R* | 0.67 | 0.23 | 0.12 |
| *P* | 0.292 | 0.092 | 0.615 |
| *SE* | 0.137 | 0.239 | 0.623 |

$k_i$ = Capacity of collection center $i$
$p_i$ = Probability of breakage of end-of-use products purchased from collection center $i$
$p_{max}$ = Maximum allowable probability of breakage
$Q_i$ = Decision variable representing the quantity to be purchased from collection center $i$

**Table 12.21 Desirability Indices**

| *Criteria/Collection Centers* | *S1* | *S2* | *S3* |
|---|---|---|---|
| Reliability | 0.3429 | 0.4432 | 0.2138 |
| Responsiveness | 0.0874 | 0.0528 | 0.2488 |
| Financial issues | 0.0783 | 0.148551 | 0.054 |
| Cultural and strategic issues | 0.1271 | 0.0438 | 0.2299 |
| Others | 0.176 | 0.2027 | 0.6211 |

**Table 12.22 Overall Performance Indices**

| *Collection Center* | *Performance Index* |
|---|---|
| S1 | 0.2318 |
| S2 | 0.2322 |
| S3 | 0.5359 |

$s$ = Number of candidate collection centers
$w_i$ = Performance index of collection center $i$ obtained by carrying out ANP

### 12.4.2.2 Problem Formulation

The following two goals of the remanufacturing facility are considered:

1. Maximize the total value of purchase (TVP)
2. Minimize the total cost of purchase (TCP)

While the first goal involves minimizing the underachievement of the target, the second goal involves minimizing the overachievement of the target. It is at the discretion of the decision maker to add any other goals that are relevant to the situation.

$$\text{Goal 1:—Maximize TVP:} \quad \sum_{i=1}^{s} w_i * Q_i \tag{12.8}$$

$$\text{Goal 2:—Maximize TCP:} \quad \sum c_i * Q_i \tag{12.9}$$

$$\text{Capacity constraint:} \quad Q_i \leq k_i \tag{12.10}$$

$$\text{Demand constraint:} \quad \sum_i Q_i = d_j \tag{12.11}$$

$$\text{Quality constraint:} \quad d_j * p_{\max} \geq \sum_{i=1}^{s} Q_i * p_i \tag{12.12}$$

$$\text{Non-negativity constraint:} \quad Q_i \geq 0 \tag{12.13}$$

**Table 12.23 Data for Goal Programming Model**

| *Collection Center* | *S1* | *S2* | *S3* |
|---|---|---|---|
| Capacity | 300 | 650 | 750 |
| Unit purchasing cost | 1.2 | 0.9 | 1.0 |
| Breakage probability | 0.03 | 0.015 | 0.01 |
| Net demand for the product = 1000 | | | |
| Maximum acceptable breakage probability = 0.025 | | | |

**Table 12.24 Results**

| *Collection Center* | *ANP Rating* | *Quantity Ordered* |
|---|---|---|
| S1 | 0.2318 | 0 |
| S2 | 0.2322 | 650 |
| S3 | 0.5359 | 350 |
| Total value of purchase (TVP) = 338.54 | | |
| Total cost of purchase (TCP) = 935 | | |

The three candidate collection centers from Section 12.4.1 are considered in this numerical example too. Table 12.23 shows the data used for the Goal Programming problem (note that only one product type is considered), and Table 12.24 shows the results obtained by solving the problem using LINGO (v4).

From the application of ANP (see Section 12.4.1), S3 is the highest ranked collection center. When no other system constraints are in place, 750 units might be ordered from S3 before considering other collection centers. But, from the results obtained from the application of Goal Programming, notice that 650 units are ordered from S2 and the remaining 350 units are ordered from S3. This may be attributed to the fact that the unit purchasing cost at S3 is higher than that at S2 (this was not considered in the application of ANP where collection centers were evaluated with respect to qualitative criteria).

The total cost of purchase is 935 and the total value of purchase is 338.54 (aspiration level = 250).

## 12.5 Conclusions

This chapter addressed the following two crucial problems and proposed a quantitative decision-making model for each of them: (1) Which end-of-use products must be selected to reprocess? and (2) Which suppliers must the selected end-of-use products be purchased from? For the first problem, we proposed a Linear Physical Programming model, and for the second problem, we proposed a model that combines Analytic Network Process and Goal Programming.

## References

1. Agarwal, A., Shankar, R. and Tiwari, M.K., Modeling the metrics of lean, agile and leagile supply chain: An ANP based approach, *European Journal of Operational Research*, 173(1), 211–225, 2006.
2. Choi, T.Y., and Hartley, J.L., An exploration of supplier selection practices across the supply chain, *Journal of Operations Management*, 14(4), 333–343, 1996.

3. Gungor, A., Evaluation of connection types in design for disassembly (DFD) using analytic network process, *Computers and Industrial Engineering*, 50(1–2), 35–54, 2006.
4. Ignizio, J.P., *Goal Programming and Extensions*, MA, Lexington Books, D. C. Heath and Company, 1976.
5. Ignizio, J. P., *Linear Programming in Single and Multi Objective Systems*, Prentice Hall, Englewood, Cliffs, NJ, 1982.
6. Kannan, V. R. and Tan, K. C., "Supplier selection and assessment: Their impact on business performance", *The Journal of Supply Chain Management*, 38(4), 11–21, 2002.
7. Lee, E. K., Ha, S. and Kim, S. K., Supplier selection and management system considering relationships in SCM, *IEEE Transactions on Engineering Management*, 48(3), 307–318, 2001.
8. Messac, A., Gupta, S. M. and Akbulut, B., Linear physical programming: A new approach to multiple objective optimization, *Transactions on Operational Research*, 8(2), 39–59, 1996.
9. Muralidharan, C., Anantharaman, N. and Deshmukh, S. G., A multi-criteria group decision making model for supplier rating, *The Journal of Supply Chain Management*, 38(4), 22–33, Fall 2002.
10. Knemeyer, A. M., Ponzurick, G. T. and Logar, M. C., A qualitative examination of factors affecting reverse logistics systems for end-of-life computers, *International Journal of Physical Distribution and Logistics Management*, 32(6), 455–479, 2002.
11. Pochampally, K. K. and Gupta, S. M., Approaches for strategic planning of a reverse supply chain, *Office Automation: Journal of the Japan Society for the Study of Office Automation*, 27(4), 75–82, 2007.
12. Ravi, V., Shankar, R. and Tiwari, M. K., Analyzing alternatives in reverse logistics for end-of-life computers: ANP and balanced scorecard approach, *Computers and Industrial Engineering*, 48(2), 327–356, 2005.
13. Saaty, T.L., *Decision Making with Dependence and Feedback: The Analytic Network Process*, Pittsburgh, PA: RWS Publications, 1996.
14. Saaty, T.L., *The Analytic Hierarchy Process*, New York: McGraw-Hill, 1980.
15. Sarkis, J. and Rasheed, A., Greening the manufacturing function, *Business Horizons*, 38(5), 17–27, 1995.
16. Weber, C.A., Current, J.R., and Benton, W.C., Vendor selection criteria and methods, *European Journal of Operational Research*, 50(1), 2–18, 1991.

# Chapter 13

# A Bi-Objective Supply Chain Scheduling

Tadeusz Sawik

## Contents

## 13.1 Introduction

One of the main issues of supply chain management in the mass customization environment, in which supply chains are driven by customer orders, is coordination of manufacturing and supply of raw materials, and production and distribution of finished products; see, for example, Thomas and Griffin (1996), Erenguc et al. (1999), Kolisch (2000), Chen and Vairaktarakis (2005), Chen and Pundoor (2006), Chen and Hall (2007), and Kreipl and Dickersbach (2008).

In a high-tech industry, a typical customer-driven supply chain may consist of a number of part manufacturers at several locations and a single producer, where parts are supplied by the manufacturers and assembled into finished products, next distributed to customers. In such a supply chain, productivity may vary from plant to plant, and transportation time and cost are not negligible. Due to a limited capacity of both the part manufacturers and the producer, the manufacturing and supply schedules for each supplier of parts and the assembly schedule for finished products should be coordinated in an efficient manner to achieve a high customer service level at low cost.

The main decision issues to consider in such a customer-driven supply chain are briefly described below.

*Order acceptance/rejection.* Decision whether or not to accept a customer request for quotation as a firm order. A request for quotation typically consists of the required quantities of ordered products and sometimes the requested delivery dates. A response to the customer request contains the quantity to be fulfilled, the date of delivery, and the price based on revenue management principles, which may involve penalties associated with deviations from the customer-requested quantities and dates.

*Due date setting.* Quoting due date for each accepted order, when the customer has not requested one or if the latter is the subject of negotiation. The quoted due date is either identical with that requested by the customer or a later one if the requested due date cannot be met in view of the actual workload and available capacity. Typically, the due date setting decisions aim at reaching a high service level (maximum number of orders to be fulfilled by the customer-requested due dates) and can also be based on revenue management principles. The quoted due dates should be considered deadlines that cannot be violated when the orders are executed.

*Order deadline scheduling.* Scheduling of accepted orders by the deadlines to achieve some additional objective, such as minimum cost of holding the inventory of orders completed before the committed delivery dates.

*Material supply scheduling.* Scheduling of material supplies coordinated with order scheduling to meet demand for materials at a low supply cost. Supply scheduling should also be coordinated with manufacturing scheduling for material providers to minimize the total cost of holding material inventory in the supply chain.

The literature on order acceptance and due date setting is limited. An exact method for selecting a subset of orders that maximizes revenues for the static problem in which all order arrivals are known in advance is presented by Slotnick and Morton (1996), and Lewis and Slotnick (2002) developed a dynamic programming approach for the multi-period case. A mixed integer program for a quantity and due date quoting available to promise is presented by Chen et al. (2001). Hegedus and Hopp (2001) consider order delay costs that measure the positive difference between the quoted due date and the requested due date of an order. Sawik (2009b) proposed a bi-objective integer program and lexicographic approach for due-date setting over a rolling planning horizon in make-to-order manufacturing. In addition, a simple critical load index was introduced to quickly identify the system bottleneck and the overloaded periods. The order acceptance strategies based on scheduling methods are presented by Wester et al. (1992) and Akkan (1997). In Wester et al. (1992), the decision of whether or not to accept a new order depends on how much order tardiness it will introduce to the system. Akkan (1997) suggested the accepting of a new order if it can be included in the schedule such that it is completed by its due date, and without changing the schedule for already accepted orders. Ebben et al. (2005) developed a workload-based acceptance strategy in a job shop environment. Another approach is order acceptance based on revenue management principles; see, for example, Bertrand and van Ooijen (2000), Barut and Sridharan (2005), and Geunes et al. (2006).

On the other hand, the literature on deterministic scheduling with due dates objective is abundant, although the research on deadline scheduling is mostly restricted to the single machine case; see, for example, Pinedo (2005).

The above described decision problems can be solved in different ways. In practice the following two main approaches can be applied:

*Integrated (simultaneous) approach.* The coordinated schedules for customer orders, material manufacturing, and material supplies are determined simultaneously, to achieve a high customer service level (i.e., a minimum number of tardy orders) or a high revenue (e.g., a minimum lost revenue due to tardiness or full rejection of orders) at a low cost, in particular, at a minimum cost of holding total supply chain inventory of materials and finished products. The quoted due dates are derived from the completion times of the scheduled orders, whereas the unscheduled orders are considered rejected.

*Hierarchical (sequential) approach.* First, the order acceptance/due date setting decisions are made to select a maximal subset of orders that can be completed by the requested due dates, and for the remaining orders, delayed due dates are determined to minimize the lost revenue owing to tardiness or rejection of orders. The due dates determined must satisfy capacity constraints and are considered deadlines at the order scheduling level. Next, order deadline scheduling is performed to meet all committed due dates and to achieve some additional objective, for example, to minimize cost of holding finished product inventory of orders completed before the deadlines. Finally, scheduling of material supplies (and material manufacturing) coordinated with order scheduling is accomplished to meet demand for materials at a low cost of material supplies and inventory holding.

While the integrated approach contributes to the complexity of the underlying mathematical models and the decision-making procedures, it is capable of reaching a coordinated solution and a global optimum for the entire supply chain scheduling problem. In contrast, the complexity of the hierarchical approach can be significantly reduced; however, it may lead to infeasible solutions obtained at lower decision levels (e.g., at order deadline scheduling) if available capacity is overestimated at higher decision levels (e.g., at due date setting). Therefore, additional coordinating constraints should be incorporated at different decision levels to avoid the infeasibility issue. Furthermore, the hierarchical approach does not guarantee the overall optimal solution to be found.

Most existing supply–production–distribution models study strategic or tactical levels of production–distribution decisions (see Chen 2004) and very few have addressed integrated decisions of supply-production stages of a supply chain, at the detailed scheduling level. In contrast, a review and discussion of the literature on supply chain management by Thomas and Griffin (1996) is concluded with the need for research that addresses supply chain issues at an operational level and that uses deterministic rather than stochastic models. Similar conclusions can also be found in a recent article on supply chain scheduling by Chen and Hall (2007).

Simplified models for integrated scheduling of production and distribution operations were studied by Chen and Vairaktarakis (2005) and Chen and Pundoor (2006), where transportation time and cost are considered. The authors analyzed the computational complexity of various cases of the problem and developed heuristics for NP-hard cases. Hall and Potts (2003) considered a different set of models that treats both delivery lead time and transportation cost, but assumes that delivery is done instantaneously without any transportation time, and Hall et al. (2001) investigated another model with a fixed set of possible delivery dates. The issues of conflict and cooperation between the suppliers and the producer were studied by Chen and Hall (2007), wherein classical scheduling objectives were considered: minimization of the total completion time and of the maximum lateness. Moon et al. (2008) proposed the evolutionary search approach to find feasible solutions for simplified, integrated process planning and scheduling in a supply chain, and Chauhan et al. (2007) considered a real-time scheduling of flow shops with the no-wait constraint in a supply chain environment.

In the literature on production planning and scheduling, the integer programming models have been widely used; for example, see Pochet and Wolsey (2006). For instance, a mixed-integer programming formulation for the integrated assembly scheduling and fabrication lot sizing in make-to-order production was proposed by Kolisch (2000), with the objective function that minimizes the total inventory holding and fabrication setup cost. A mixed-integer programming formulation for the optimal scheduling of industrial supply chains was presented by Amaro and Barbosa-Póvoa (2008). The lexicographic approaches with a hierarchy of integer programming formulations for a multi-objective, long-term production scheduling in make-to-order manufacturing were proposed by Sawik (2007a, 2007b, 2007c), wherein both the maximization of customer satisfaction and leveling of production and inventory are integrated in the objective functions. A monolithic versus hierarchical approach for multi-objective scheduling in a supply chain was proposed by Sawik (2009a), where the supply chain consists of multiple manufacturers (suppliers) of parts with different manufacturing and transportation capacities. The scheduling of manufacturing and delivery of parts and the assembly of products is combined with the supplier selection for each customer order and the due date setting for some orders.

The supply chain considered in this chapter consists of multiple suppliers (manufacturers) of parts, a single producer of finished products, and a set of customers who generate final demand for the products. Each supplier has identical parallel machines for the manufacturing of parts, and the producer has a flexible flow shop for the assembly of products. In the supply chain, the following static and deterministic scheduling problem is considered. Given a set of orders, the problem objective is to determine a coordinated schedule for the manufacture of parts by each supplier, for the delivery of parts from each supplier to the producer, and for the assignment of orders to planning periods over the horizon at the producer, such that a high revenue or a maximum service level is achieved at a minimum total cost of holding the supply chain inventory of parts and products.

The major contribution of this chapter is that it proposes simple integer programming formulations and compares an integrated and a hierarchical approach to the bi-objective, coordinated scheduling in a customer-driven supply chain. In particular, the due date setting decisions in the hierarchical approach are subject to the due date feasibility constraints ensuring the existence of a feasible deadline schedule for orders and a feasible manufacturing and supply schedule for parts. The approach proposed in this chapter is closely related to that presented in Sawik (2009a). However, the selection of a single supplier for each customer order is not considered anymore, which simplifies the integer programs. In addition to maximum service level, a maximum revenue is used as an alternative objective function. Moreover, in the computational experiments, the weighted-sum program is compared with the Tschebyscheff program to determine subsets of nondominated solutions for the integrated approach.

This chapter is organized as follows. Section 13.2 provides a description of the supply chain scheduling problem. The integrated model for bi-objective supply chain scheduling is proposed in Section 13.3, and the hierarchical approach with a set of integer programming formulations is presented in Section 13.4. Numerical examples modeled after a real-world, customer-driven supply chain and some computational results are provided in Section 13.5. Conclusions are made in Section 13.6.

## 13.2 Problem Description

The supply chain under consideration consists of $m$ manufacturers/suppliers of parts, a single producer where finished products are assembled according to customer orders, and a set of customers who generate final demand for the products. In the supply chain, various types of products are assembled by the producer using different part types supplied by the manufacturers. Let $J$ be the set of customer orders known ahead of time, $K$ the set of part types, $L$ the set of product types, $J_l$ the subset of orders for product type $l \in L$, and $I_k$ the subset of suppliers capable of producing part type $k$. Each product type $l \in L$ requires $b_{kl} \geq 0$ parts of type $k \in K$.

Each order $j \in J$ is described by a triple $(a_j, d_j, o_j)$, where $a_j$ is the order ready date (e.g., the earliest period of material availability), $d_j$ is the customer requested due date (e.g., customer required shipping date), and $o_j$ is the size of order (quantity of ordered products of a specific type).

The planning horizon consists of $h$ planning periods (e.g., working days) of equal length $H$ (e.g., hours or minutes) and let $T = \{1, \ldots, h\}$ be the set of planning periods. It is assumed that each customer order must be fully completed in exactly one planning period (e.g., during one day); however, this assumption can be easily relaxed. Large-sized orders that require more than one planning period for completion can be split into single-period suborders to be allocated among consecutive planning periods (e.g., Sawik 2007b, 2009a).

The suppliers manufacture and deliver parts to the producer. The manufacture of parts by supplier $i \in I$, at the earliest can be started $e_i$ periods in advance of period $t = 1$, when the producer can start the assembly of products. The earliness $e_i$ of manufacturing start time for supplier $i$ represents the maximum manufacturing capacity of supplier $i$ available in advance of the start of assembly schedule. Each supplier $i$ has $M_i$ identical parallel machines, and let $q_{ik}$ be the one machine-period manufacturing rate for part type $k \in K$ at supplier $i \in I_k$.

The manufactured parts are next transported to the producer, at most once per period. Different part types can be shipped together and each delivery shipment is limited by the minimum and maximum capacity, respectively $\underline{V}$ and $\overline{V}$ parts. The transportation time of a shipment from supplier $i$ to the producer is assumed constant and equals $\theta_i$ periods. The parts manufactured by supplier $i$ in period $t$ can be shipped to the producer in the same period and can be used for the assembly of products in period $t + \theta_i$, at the earliest.

The producer has a flexible assembly line that consists of $f$ assembly stages in series, and each stage $g \in G = \{1, \ldots, f\}$ consists of $m_g \geq 1$ identical, parallel machines. Each customer order requires processing in various assembly stages; however, some orders may bypass some stages. Let $p_{gj}$ be the one machine-period production rate at assembly stage $g$ for products in order $j$. See Table 13.1 for the indices and parameters described above.

The problem objective is to determine a coordinated schedule for the manufacture of parts by each supplier and for the delivery of parts from each supplier to the

**Table 13.1 Notation: Indices and Parameters**

| **Indices** | |
|---|---|
| $g$ | = Assembly stage, $g \in G = \{1, \ldots, f\}$ |
| $i$ | = Supplier, $i \in I = \{1, \ldots, m\}$ |
| $j$ | = Customer order, $j \in J = \{1, \ldots, n\}$ |
| $k$ | = Part type $k \in K$ |
| $l$ | = Product type, $l \in L$ |
| $t$ | = Planning period, $t \in T = \{1, \ldots, h\}$ |
| **Input Parameters** | |
| $a_j, d_j, D_j, o_j$ | = Ready date, customer requested due date, committed due date, size of order $j$ |
| $b_{kl}$ | = Unit requirement of product type $l$ for part type $k$ |
| $e_i$ | = Earliness of manufacturing start time for supplier $i$ |
| $I_k$ | = Subset of suppliers capable of producing part type $k$ |
| $J_l$ | = Subset of orders for product type $l$ |
| $M_i$ | = Number of parallel machines at supplier $i$ |
| $m_g$ | = Number of parallel machines in assembly stage $g$ |
| $p_{gi}$ | = One machine-period production rate at assembly stage $g$ for products in order $j$ |
| $q_{ik}$ | = One machine-period manufacturing rate at supplier $i$ for part type $k$ |
| $\underline{V}, \overline{V}$ | = Minimum, maximum shipment capacity, respectively |
| $T^{(i)}$ | = $\{-e_i, \ldots, -1, 1, \ldots, h\}$—enlarged, ordered set of planning periods for supplier $i$ |
| $\delta_{jt}$ | = Per unit lost revenue for order $j$ with committed due date $t > d_j$ |
| $\theta_i$ | = Delivery time from supplier $i$ to producer |
| $\varphi_{1ik}, \varphi_{2ik}, \varphi_{3k}, \varphi_{4j}$ | = Inventory holding cost per unit of part stored by supplier, transported to producer, stored by producer, per unit of finished product stored by producer, respectively |

producer along with the schedule for completing each order by the producer, such that a high revenue or a high customer service level is achieved and the total cost of supply chain inventory holding is minimized.

An integrated approach, where manufacturing, supply, and assembly schedules are determined simultaneously, is compared with a hierarchical approach. In the proposed hierarchical approach, first the maximal subset of orders that can be completed by the customer requested due dates is found, and for the remaining orders, delayed due dates are determined to satisfy capacity constraints. Then the deadline assignment of orders to planning periods over the horizon is found to minimize the holding cost of finished product inventory of the customer orders completed before their due dates. Finally, the manufacturing and delivery schedules of the required parts are determined for all suppliers to minimize the total cost of holding the inventory of parts.

## 13.3 Coordinated Supply Chain Scheduling: Integrated Approach

In this section a bi-objective, mixed-integer programming approach is presented for the integrated scheduling of manufacturing and supply of parts and production of finished products in a customer-driven supply chain.

### *13.3.1 Decision Variables*

Let $T^{(i)}$ be the enlarged, ordered set of planning periods with $e_i$ periods added before period $t = 1$, that is,

$$T^{(i)} = \{-e_i, \ldots - 1\} \cup T = \{-e_i, \ldots - 1, 1, \ldots h\}$$

The various decision variables (Table 13.2) used in model **INT** to coordinate different types of schedules in the supply chain are additionally explained below:

- Manufacturing schedule for each supplier $i \in I$:
  - Manufacturing lot variable $u_{ikt}$ that represents the lot size of each part type manufactured by each supplier in each planning period, that is, the number of parts type $k$ manufactured in period $t \in T^{(i)}$ by supplier $i$
- Delivery schedule for each supplier $i \in I$:
  - Delivery lot variable $v_{ikt}$ that represents the lot size of each part type delivered from each supplier to the producer in each planning period, that is, the number of parts type $k$ delivered from supplier $i$ to the producer in period $t \in T^{(i)}$
  - Delivery timing variable $w_{it} = 1$, if delivery of parts from supplier $i$ to the producer is scheduled for period $t \in T^{(i)}$, otherwise $w_{it} = 0$

**Table 13.2 Notation: Decision Variables**

| **Model INT** |
|---|
| $u_{ikt}$ = Manufacturing lot at supplier $i$ of part type $k$ in period $t$ (manufacturing lot variable) |
| $v_{ikt}$ = Delivery lot from supplier $i$ of part type $k$ in period $t$ (delivery lot variable) |
| $w_{it}$ = 1, if delivery of parts from supplier $i$ is scheduled for period $t$; otherwise $w_{it} = 0$ (delivery timing variable) |
| $z_{jt}$ = 1, if customer order $j$ is assigned to planning period $t$; otherwise $z_{jt} = 0$ (order assignment variable) |
| **Model DDS** |
| $x_j$ = 1, if order $j$ is accepted with its customer requested due date; $x_j = 0$ if order $j$ needs delaying to be accepted (order acceptance variable) |
| $y_{jt}$ = 1, if delayed order $j$ is assigned the adjusted due date $t$, ($t > d_j$); otherwise $y_{jt} = 0$ (due date setting variable) |

- Order schedule:
  - Order assignment variable $z_{jt} = 1$ if customer order $j$ is assigned for processing to planning period $t \in T$; otherwise $z_{jt} = 0$

## 13.3.2 Objective Functions

Sales departments often apply revenue management principles for order selection and due date setting. The objective is to maximize a revenue function or, alternatively, minimize the total lost revenue due to tardiness of orders.

Most often, customers value short lead times (due dates) over long lead times. Setting delayed due dates results in a reduction in revenue. Let $\delta_{jt}$ be the lost revenue per unit of each product in the delayed order $j$ completed in period $t$, later than customer requested due date $d_j$. The lost revenue increases with an increase in the delay of completion dates with respect to requested due dates, that is,

$$\delta_{jt+1} > \delta_{j,t};\ j \in J,\ t \in T : t > d_j$$

We assume that setting a delayed due date results in a reduction in revenue proportional to the delay; for example, see Bertrand and van Ooijen (2000). Per unit of lost revenue $\delta_{jt}$ increases by some percent for each day of delay $(t - d_j)$ of completion order $j$ with respect to customer-requested date $d_j$; for example,

$$\delta_{jt} = \beta_j \gamma_j (t - d_j); \quad j \in J, \quad t \in T : t \geq d_j$$

where $\beta_j$ is the per unit revenue for a non-delayed order $j$, and $0 < \gamma_j < 1$ is the rate of per unit daily loss of revenue for order $j$.

In addition, a fixed loss $\varepsilon_j (0 < \varepsilon_j < 1)$ of unit revenue may be applied for each delayed product in order $j$, i.e.,

$$\delta_{jt} = \beta_j [\varepsilon_j + \gamma_j (t - d_j)]; \quad j \in J, \quad t \in T : t > d_j$$

The objective of the integrated scheduling is to determine for each supplier a schedule of manufacturing parts and a schedule of suppling the parts to the producer, and to find for the producer a schedule of assembly of the ordered products, so as to maximize revenue or equivalently to minimize the lost revenue due to tardiness of orders, as a primary optimality criterion. Simultaneously, to achieve a low unit cost, the total supply chain inventory holding cost is minimized. Thus, the secondary objective functions are minimization of the total inventory holding cost. The two objective functions, $f_1$, $f_2$, numbered according to their decreasing importance are defined as:

The lost revenue:

$$f_1 = \sum_{j \in J} \sum_{t \in T : t > d_j} \delta_{jt} o_j z_{jt} \tag{13.1}$$

The cost of inventory holding:

$$\begin{aligned} f_2 = & \sum_{k \in K} \sum_{i \in I_k} \sum_{t \in T} \sum_{r \in T^{(i)} : \tau \leq t} \varphi_{1ik} (u_{ik\tau} - v_{ik, next(\tau, T^{(i)}, \theta_i)}) \\ & + \sum_{k \in K} \sum_{i \in I_k} \sum_{t \in T : t \geq next(-e_i, T^{(i)}, \theta_i)} \varphi_{2ik} \theta_i v_{ikt} \\ & + \sum_{k \in K} \sum_{i \in I_k} \sum_{t \in T} \varphi_{3k} \left( \sum_{r \in T^{(i)} : \tau \leq t} v_{ik\tau} - \sum_{l \in L} \sum_{j \in J_l} \sum_{\tau = 1} b_{kl} o_j z_{j\tau} \right) \\ & + \sum_{t \in T} \sum_{\tau = 1}^{t} \sum_{j \in J : d_j > t} \varphi_{4j} o_j z_{j\tau} \end{aligned} \tag{13.2}$$

where $\varphi_{1ik}$, $\varphi_{2ik}$, $\varphi_{3k}$, and $\varphi_{4j}$ are the unit holding costs of parts stored at supplier $i$, of parts being in-transit from supplier $i$ to producer, of parts stored at producer, and of finished products stored at producer, respectively. The in-transit holding cost represents the cost of transportation parts from suppliers to the producer.

### 13.3.3 Bi-Objective Mixed-Integer Program

This subsection presents the mixed-integer program **INT** for integrated supply chain scheduling.

**Model INT:** Bi-objective integrated scheduling of manufacturing, supply, and assembly:

$$\min \ f_1, f_2 \tag{13.3}$$

subject to

1. *Customer order assignment constraints:*
   - Each customer order is assigned to exactly one planning period for processing:

$$\sum_{t \in T: a_j \leq t} z_{jt} = 1; \quad j \in J \tag{13.4}$$

2. *Producer capacity constraints:*
   - In every period $t$, the total number of machine-periods of production at each assembly stage $g$ cannot be greater than the number of machines available at this stage:

$$\sum_{j \in J} o_j z_{jt} / p_{gj} \leq m_g; \quad g \in G, \ t \in T \tag{13.5}$$

3. *Manufacturing capacity constraints for each supplier:*
   - In every period $t$, the total number of machine-periods of production of all part types at each supplier $i$ cannot be greater than the total number of machines available at this supplier:

$$\sum_{k \in K: i \in I_k} u_{ikt} / q_{ik} \leq M_i; \quad i \in I, \ t \in T^{(i)} \tag{13.6}$$

4. *Part manufacturing and delivery constraints for each supplier:*
   - The cumulative delivery of each part type by period $t$ cannot be greater than the cumulative manufacturing of this part type by period $prev(t, T^{(i)}, \theta_i)$, that is, manufactured not later than $\theta_i$ periods before period $t$.
   - The total delivery of each part type is equal to the total manufacturing of this part type.

- Parts can be delivered only in periods scheduled for delivery, and each shipment is limited by its minimum and maximum capacity, respectively $\underline{V}$ and $\overline{V}$.
- No delivery from supplier $i$ can be scheduled before period $next(-e_i, T^{(i)}, \theta_i)$, that is, no delivery can be scheduled earlier than $\theta_i$ periods after the earliest period $-e_i$ of manufacturing:

$$\sum_{\tau \in T^{(i)}:\tau \leq t} v_{ik\tau} \leq \sum_{\tau=-e_i}^{prev(t, T^{(i)}, \theta_i)} u_{ik\tau}; \quad k \in K,\ i \in I_k,\ t \in T^{(i)}: \quad next(-e_i, T^{(i)}, \theta_i) \leq t < h \tag{13.7}$$

$$\sum_{t \in T^{(i)}} v_{ikt} = \sum_{\tau=-e_i}^{prev(h, T^{(i)}, \theta_i)} u_{ikt}; \quad k \in K,\ i \in I_k \tag{13.8}$$

$$v_{ikt} \leq \overline{V} w_{it}; \quad k \in K,\ i \in I_k,\ t \in T^{(i)} \tag{13.9}$$

$$\sum_{k \in K} v_{ikt} \leq \overline{V}; \quad i \in I,\ t \in T^{(i)} \tag{13.10}$$

$$\sum_{k \in K} v_{ikt} \geq \underline{V} w_{it}; \quad i \in I,\ t \in T^{(i)} \tag{13.11}$$

$$v_{ikt} = 0; \quad k \in K,\ i \in I_k,\ t \in T^{(i)}: \ t < next(-e_i, T^{(i)}, \theta_i) \tag{13.12}$$

$$w_{it} = 0; \quad i \in I,\ t \in T^{(i)}: \ t < next(-e_i, T^{(i)}, \theta_i) \tag{13.13}$$

5. *Part demand satisfaction constraints:*
   - The cumulative deliveries of each part type $k$ by period $t$ from all suppliers $i \in I_k$ cannot be less than the cumulative usage of this part type in periods 1 through $t$.
   - The total delivery of each part type $k$ from all suppliers $i \in I_k$ is equal to the total demand for this part type:

$$\sum_{i \in I_k} \sum_{\tau \in T^{(i)}:\tau \leq t} v_{ik\tau} \geq \sum_{l \in L} \sum_{j \in J_l} \sum_{\tau=1} b_{klo_j} z_{j\tau}; \quad k \in K,\ t \in T: t < h \tag{13.14}$$

$$\sum_{i \in I_k} \sum_{t \in T^{(i)}} v_{ikt} = \sum_{l \in L} \sum_{j \in J_l} b_{klo_j}; \quad k \in K \tag{13.15}$$

6. *Coordinating constraints:*
   – The cumulative requirement for each part type $k$ of the orders assigned to periods 1 through $t$ [right-hand side of Constraint (13.16)] cannot be greater than the total production of this part type at each supplier $i \in I_k$ by period $prev(t, T^{(i)}, \theta_i)$, that is, manufactured at $i$ not later than $\theta_i$ periods before period $t$ [left-hand side of Constraint (13.16)]:

$$\sum_{i \in I_k:\, next(-e_i, T^{(i)}, \theta_i) \leq t} \; \sum_{\tau=-e_i}^{prev(t, T^{(i)}, \theta_i)} u_{ik\tau} \geq \sum_{l \in L} \sum_{j \in J_l} \sum_{\tau=1} b_{kl} o_j z_{j\tau};$$

$$k \in K, t \in T \tag{13.16}$$

7. *Nonnegativity and integrality conditions:*

$$u_{ikt} \geq 0; \quad k \in K, \; i \in I_k, \; t \in T^{(i)} \tag{13.17}$$

$$v_{ikt} \geq 0; \quad k \in K, \; i \in I_k, \; t \in T^{(i)} \tag{13.18}$$

$$w_{it} \in \{0, 1\}; \quad i \in I, \; t \in T^{(i)} \tag{13.19}$$

$$z_{jt} \in \{0, 1\}; \quad j \in J, \; t \in T: \; a_j \leq t \tag{13.20}$$

The coordinating Constraint (13.16) has been introduced to strengthen the mixed-integer program **INT** as it directly links the manufacturing of parts over the planning horizon at each supplier (variable $u_{ikt}$) and the assignment of customer orders to planning periods (variable $z_{jt}$). Thus, Constraint (13.16) directly coordinates the supplier's schedules and the producer's schedule. Constraint (13.16) is implied by Constraints (13.7) and (13.14), and any feasible solution that satisfies Constraint (13.16) also satisfies Constraints (13.7) and (13.14).

The mathematical model presented in this section can be modified or enhanced to consider additional features of the bi-objective supply chain scheduling that can be met in practice. In particular, the primary objective function can be modified. To achieve a high customer service level, the number of orders completed later than the customer-requested dates should be minimized, that is, the maximum service level can be achieved by minimizing the number of tardy orders:

$$f_1^O = \sum_{j \in J} \sum_{t \in T: t > d_j} z_{jt} \tag{13.21}$$

A surrogate objective function that also considers the total lost revenue due to tardiness of orders is minimization of the total number of tardy products:

$$f_1^P = \sum_{j \in J} \sum_{t \in T: t > d_j} o_j z_{jt} \tag{13.22}$$

### 13.3.4 Selected Solution Approaches

This subsection provides a brief description of selected approaches that can be used to determine the non-dominated solution set of the bi-objective mixed-integer program **INT**.

#### 13.3.4.1 Weighted-Sum Program

The non-dominated solution set of the bi-objective program **INT** can be partially determined by the parameterization on $\lambda$ using the following weighted-sum program.

**Model INT$_\lambda$**

$$\min \lambda f_1 + (1-\lambda) f_2 \tag{13.23}$$

subject to Constraints (13.4) through (13.20)
where $0.5 \leq \lambda \leq 1$.

It is well known, however, that the non-dominated solution set of a multi-objective integer program such as **INT**$_\lambda$ cannot be fully determined even if the complete parameterization on $\lambda$ is attempted; for example, see Steuer (1986). To compute unsupported non-dominated solutions, some upper bounds on the objective functions should be added to **INT**$_\lambda$; see, for example, Alves and Climaco (2007).

#### 13.3.4.2 Reference-Point-Based Scalarizing Program

Let $\underline{f} = (\underline{f}_1, \underline{f}_2)$ be a reference point in the criteria space such that $\underline{f}_\iota < f_\iota, \iota = 1, 2$ for all feasible solutions satisfying Constraints (13.4) through (13.20), and denote by $\rho$ a small positive value. The non-dominated solution set of the bi-objective program **INT** can be found by the parameterization on $\lambda$ the following mixed-integer program **INT**$^\lambda$:

**Model INT$^\lambda$**

$$\min\{\delta + \rho(f_1 + f_2)\} \tag{13.24}$$

subject to Constraints (13.4) through (13.20) and

$$\lambda(f_1 - \underline{f}_1) \leq \delta \tag{13.25}$$

$$(1-\lambda)(f_2 - \underline{f}_2) \leq \delta \tag{13.26}$$

$$\delta \geq 0, \tag{13.27}$$

where $0.5 \leq \lambda \leq 1$.

Program **INT**$^\lambda$ is based on the augmented $\lambda$-weighted Tschebyscheff metric.

$$\min\{\lambda|f_1 - \underline{f}_1| + \rho(f_1 + f_2), (1-\lambda)|f_2 - \underline{f}_2| + \rho(f_1 + f_2)\}.$$

## 13.4 Coordinated Supply Chain Scheduling: Hierarchical Approach

In this section the mixed-integer programming formulations are presented for a hierarchical scheduling of customer orders and manufacturing and supplies of parts. The proposed hierarchical approach can be applied within a pull planning framework in customer-driven supply chains. The hierarchical decomposition scheme is based on the lexicographic approach (see Sawik 2009a), that is, it is driven by the relative importance of the objective functions. The hierarchical framework consists of the following three decision-making problems to be solved sequentially (Figure 13.1):

1. Due date setting: **DDS**
2. Order deadline scheduling: **ODS**
3. Part manufacturing and delivery scheduling: **MDS**

The top-level problem **DDS** accounts for the cumulative capacity constraints of the producer, and both the top- and the middle-level problems **DDS** and **ODS** contain the capacity constraints of the suppliers.

The objective of the top-level problem **DDS** is to determine a subset of orders that can be completed by the customer-requested due dates and to update (delay) due dates for the remaining orders such that the capacity constraints are met and

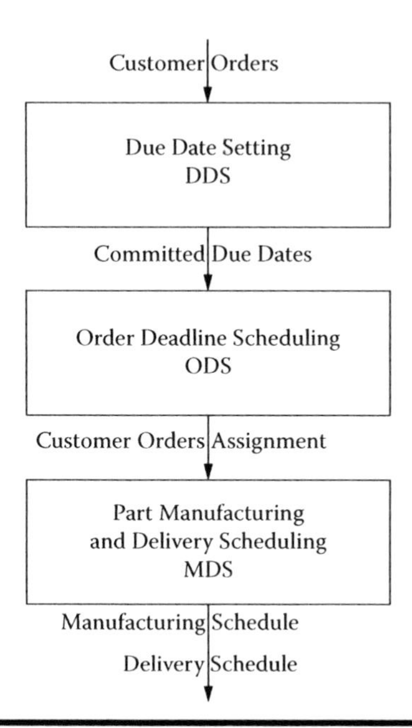


**Figure 13.1 Hierarchical scheduling of a customer-driven supply chain.**

the lost revenue due to the orders delay is minimized. At this level, the necessary conditions (see Sawik 2009a) for feasibility of the customer-requested due dates are checked. If the feasibility conditions [Constraints (13.30) and (13.31) of **DDS**] are satisfied for all requested due dates, then no tardy order occurs and the total revenue is maximal. Otherwise, to reach feasibility later due dates must be committed for some customer orders to minimize the lost revenue for the orders with committed due dates later than the requested due dates. All committed due dates are next interpreted as deadlines that cannot be violated at the order scheduling level.

Given the committed due date for each customer order, the objective of the middle-level problem **ODS** is to determine a feasible, non-delayed assignment of orders to planning periods over the planning horizon to minimize the finished product inventory holding cost of customer orders completed before their due dates.

Finally, given the cumulative requirements for all part types over the planning horizon, the objective of the base-level problem **MDS** is to determine the integrated manufacturing and delivery schedule of parts for each supplier such that the total cost of holding the inventory of parts in the supply chain is minimized.

The mixed-integer programs **DDS**, **ODS**, and **MDS** are presented below (for notation used and definitions of the decision variables, see Tables 13.1 and 13.2).

### 13.4.1 Due Date Setting

In addition to the manufacturing lot variable $u_{ikt}$, the mixed-integer program **DDS** contains the following two binary decision variables:

- Order acceptance variable: $x_j = 1$, if order $j$ is accepted with its customer-requested due date; $x_j = 0$ if order $j$ needs delaying to be accepted.
- Due date setting variable: $y_{jt} = 1$, if order $j$ unacceptable with requested due date (with $x_j = 0$) is assigned a later due date $t$, ($t > d_j$); otherwise $y_{jt} = 0$.

**Model DDS:** *Due date setting*

Minimize the lost revenue

$$f_1 = \sum_{j \in J} \sum_{t \in T: t > d_j} \delta_{jt} o_j y_{jt}, \tag{13.28}$$

subject to

1. *Order acceptance or due date setting constraints:*
   - Each customer order is either accepted with its requested due date or is assigned a later due date to reach production and delivery schedule

feasibility:

$$x_j + \sum_{t \in T: t > d_j} y_{jt} = 1; \quad j \in J \tag{13.29}$$

2. *Producer cumulative capacity constraints:*
   - For any period $t \leq d$, the cumulative demand on capacity in stage $g$ of all the customer orders accepted with requested due dates $d_j$ (or adjusted due dates $\tau$) not greater than $d$ and ready dates $a_j$ (or requested due dates $d_j$, respectively) not less than $t$ must not exceed the cumulative capacity available in this stage in periods $t$ through $d$:

$$\sum_{j \in J:\, t \leq a_j \leq d_j \leq d} o_j x_j / p_{gj} + \sum_{j \in J} \sum_{\tau \in T: t \leq d_j < \tau \leq d} o_j y_{j\tau} / p_{gj} \leq m_g(d - t + 1);$$

$$d, t \in T, \; g \in G: \; t \leq d \tag{13.30}$$

3. *Supplier cumulative capacity constraints*:
   - Constrant (13.6)
   - The cumulative demand on part type $k$ required in periods 1 through $t$ cannot be greater than the total cumulative production of this part type at each supplier $i \in I_k$ by period $prev(t, T^{(i)}, \theta_i)$, that is, manufactured at $i$ not later than $\theta_i$ periods before period $t$:

$$\sum_{l \in L} \sum_{j \in J_l: d_j \leq t} b_{kl} o_j x_j + \sum_{l \in L} \sum_{j \in J_l} \sum_{\tau \in T: d_j < \tau \leq t} b_{kl} o_j y_{j\tau}$$

$$\leq \sum_{i \in I_k:\, next(-e_i, T^{(i)}, \theta_i) \leq t} \; \sum_{\tau = -e_i}^{prev(t, T^{(i)}, \theta_i)} u_{ik\tau}; \quad k \in K, \; t \in T \tag{13.31}$$

4. *Nonnegativity and integrality conditions*: (13.17) and

$$x_j \in \{0, 1\}; \quad j \in J \tag{13.32}$$

$$y_{jt} \in \{0, 1\}; \quad j \in J, \; t \in T: \; t > d_j \tag{13.33}$$

The Objective Function (13.28) represents the lost revenue of orders for which the committed due dates are later than due dates requested by the customers. Alternatively, the objective function $f_1$ (13.28) can be replaced with $f_1^O$ (21) or $f_1^P$

(22), defined as

$$f_1^O = \sum_{j \in J} (1 - x_j) \tag{13.34}$$

$$f_1^P = \sum_{j \in J} o_j(1 - x_j) \tag{13.35}$$

The Capacity Constraints (13.30) and (13.31) ensure that each order $j \in J$ can be completed on or before its requested due date $d_j$ (if $x_j = 1$) or on its delayed due date t > $d_j$ (if $x_j = 0$ and $y_{jt} = 1$). If the capacity constraints hold for all customer-requested due dates, then the objective function $f_1$ (13.28) takes on zero value, because $y_{jt} = 0 \; \forall j \in J, t \in T: \; t > d_j$.

The solution to the mixed-integer program **DDS** determines the subset $\{j \in J: x_j = 1\}$ of customer orders accepted with the customer-requested due dates $d_j$ and the subset of the remaining orders $\{j \in J: x_j = 0\}$ with the postponed due dates such that the lost revenue due to the delay of orders is minimized. Denote by $D_j$ the requested or updated due date for each order $j \in J$, that is,

$$D_j = \begin{cases} d_j & \mathit{if} \;\; x_j = 1 \\ \sum_{t \in T: t > d_j} t y_{jt} & \mathit{if} \;\; x_j = 0 \end{cases}$$

The updated due dates $D_j$ are interpreted as the deadlines at the order scheduling level.

### 13.4.2 Order Deadline Scheduling

Given the updated due dates $D_j$, $j \in J$, the next decision step is to determine a non-delayed assembly schedule for the producer, that is, the assignment of customer orders to planning periods by the committed due dates (deadlines) over the horizon to minimize the finished products inventory holding cost.

The order assignment variable $z_{jt}$ introduced in the integrated model **INT** is the basic variable used below.

**Model ODS:** *Deadline scheduling of customer orders*

Minimize the holding cost of finished product inventory

$$f_2' = \sum_{t \in T} \sum_{\tau=1}^{t} \sum_{j \in J: d_j > t} \varphi_{4j} o_j z_{j\tau} \tag{13.36}$$

subject to

1. *Customer order non-delayed assignment constraints:*

– Each customer order is assigned to exactly one planning period not later than its due date:

$$\sum_{t \in T:\, a_j \le t \le D_j} z_{jt} = 1; \quad j \in J \tag{13.37}$$

2. *Producer capacity constraints:*
   – In every period, the demand on capacity at each assembly stage cannot be greater than the maximum available capacity in this period:

$$\sum_{j \in J} o_j z_{jt} / p_{gj} \le m_g; \quad g \in G,\ t \in T \tag{13.38}$$

3. *Supplier cumulative capacity constraints*:
   – Constraint (13.6).
   – The cumulative demand on part type $k$ required for customer orders assigned to periods 1 through $t$ cannot be greater than the total cumulative production of this part type at each supplier $i \in I_k$ by period $prev(t, T^{(i)}, \theta_i)$, that is, manufactured at $i$ not later than $\theta_i$ periods before period $t$:

$$\sum_{l \in L} \sum_{j \in J_l} \sum_{\tau \in T: a_j \le \tau \le D_j, \tau \le t} b_{kl} o_j z_{j\tau} \le \sum_{i \in I_k:\, next(-e_i, T^{(i)}, \theta_i) \le t}$$

$$\times \sum_{\tau = -e_i}^{prev(t, T^{(i)}, \theta_i)} u_{ik\tau}; \quad k \in K,\ t \in T \tag{13.39}$$

4. *Nonnegativity and integrality conditions* (13.17) and

$$z_{jt} \in \{0, 1\}; \quad j \in J,\ t \in T:\ a_j \le t \le D_j \tag{13.40}$$

### 13.4.3 Part Manufacturing and Delivery Scheduling

The solution to **ODS** determines the optimal assignment of customer orders to planning periods $\{z_{jt}, j \in J, t \in T\}$ and thereby the optimal master schedule for the finished products. Denote by $O_{kt}$ the optimal cumulative usage in periods 1 through $t$ of each part type $k$, that is, the optimal cumulative requirement for each part type $k$:

$$O_{kt} = \sum_{l \in L} \sum_{j \in J_l} \sum_{\tau = 1}^{t} b_{kl} o_j z_{j\tau}; \quad k \in K,\ t \in T \tag{13.41}$$

Now, the base-level problem can be formulated. Given the cumulative requirements $O_{kt}, t \in T$ for all part types $k \in K$ by each period $t$, determine the manufacturing and delivery schedule of parts for each supplier.

**Model MDS:** *Part manufacturing and delivery scheduling*
Minimize the cost of holding the supply chain inventory of parts

$$f_2'' = \sum_{k\in K}\sum_{i\in I_k}\sum_{t\in T}\sum_{\tau\in T^{(i)}:\tau\le t} \varphi_{1ik}(u_{ik\tau} - v_{ik,next(\tau,T^{(i)},\theta_i)})$$

$$+\sum_{k\in K}\sum_{i\in I_k}\sum_{t\in T:t\ge next(-e_i,T^{(i)},\theta_i)} \varphi_{2ik}\theta_i v_{ikt}$$

$$+\sum_{k\in K}\sum_{t\in T}\varphi_{3k}\left(\sum_{i\in I_k}\sum_{\tau\in T^{(i)}:\tau\le t} v_{ik\tau} - O_{kt}\right) \tag{13.42}$$

subject to

1. *Manufacturing capacity constraints*: (13.6).
2. *Part manufacturing and delivery constraints*: (13.7) through (13.13).
3. *Part demand satisfaction constraints*:
   - The cumulative deliveries of each part type $k$ by period $t$ cannot be less than the cumulative demand for this part type by this period.
   - The total delivery of each part type $k$ is equal to the total demand of this part type:

$$\sum_{i\in I_k}\sum_{\tau\in T^{(i)}:\tau\le t} v_{ik\tau} \ge O_{kt}; \quad k\in K,\ t\in T: t<h \tag{13.43}$$

$$\sum_{i\in I_k}\sum_{t\in T^{(i)}} v_{ikt} = O_{kh}; \quad k\in K \tag{13.44}$$

4. *Coordinating constraints:*
   - The total production of each part type $k$ at each supplier $i \in I_k$ by period $prev(t, T^{(i)}, \theta_i)$, that is, manufactured at $i$ not later than $\theta_i$ periods before period $t$ cannot be less than the cumulative requirement for this part type by period $t$:

$$\sum_{i\in I_k:\ next(-e_i,T^{(i)},\theta_i)\le t}\ \sum_{\tau=-e_i}^{prev(t,T^{(i)},\theta_i)} u_{ik\tau} \ge O_{kt}; \quad k\in K,\ t\in T \tag{13.45}$$

5. *Nonnegativity and integrality conditions*: Constraints (13.17) through (13.19).

Coordinating Constraint (13.45) is implied by (13.7) and (13.43), and it directly links the manufacturing of parts over the planning horizon with the producer cumulative requirement for the part types [cf. Constraint (13.16)].

In the Objective Function (13.42), $f_2''$ is the inventory holding cost of manufactured parts stored by the suppliers, being in transportation to the producer or stored by the producer.

Recall that in the hierarchical approach the total supply chain inventory holding cost is split among two decision levels. Thus, the final value of the objective function $f_2$ [Objective Function (13.2)] can be calculated as the sum of its partial solution values, determined for the middle-level problem **ODS** and the base-level problem **MDS**:

$$f_2 = f_2' + f_2'' \tag{13.46}$$

## 13.5 Computational Examples

This section presents some computational examples to compare the integrated and the hierarchical approaches and to illustrate possible applications of the proposed mixed-integer programming models. The examples are modeled after a real-world, customer-driven supply chain for high-tech products; see Figures 13.2 and 13.3.

A brief description of the planning horizon, manufacturers/suppliers, producer, part types, product types, and customer orders is given below:

1. Planning horizon: $h = 14$ periods (days), each of length $H = 2 \times 9$ hours.
2. Customers: $n = 521$ customer orders ranging from 5 to 8,309 products, with various requested due dates, each to be completed in a single period. The total demand is 258,414 products.
3. Manufacturers/suppliers (Figure 13.2):
   - $m = 3$ suppliers:
     - Supplier 1 with $M_1 = 12$ machines and transportation time $\theta_1 = 1$ period.

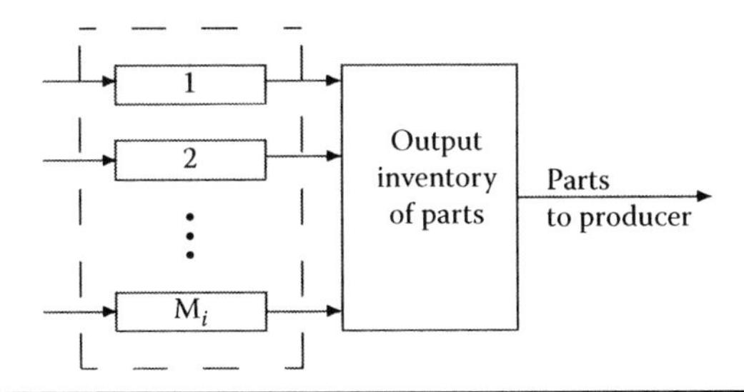


**Figure 13.2 A manufacturer/supplier of parts.**

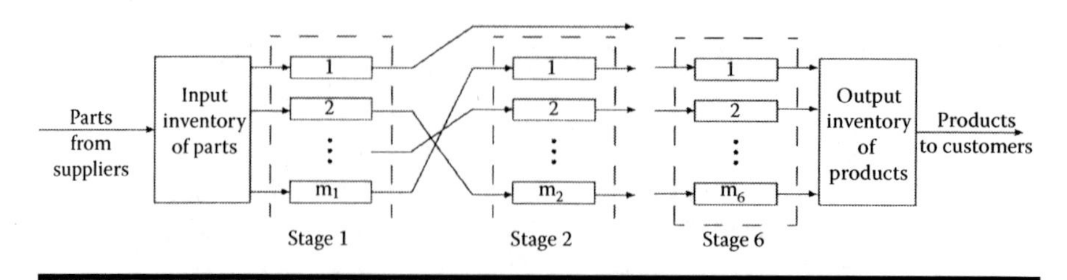


**Figure 13.3 The producer.**

- Supplier 2 with $M_2 = 11$ machines and transportation time $\theta_2 = 2$ periods.
- Supplier 3 with $M_3 = 10$ machines and transportation time $\theta_3 = 3$ periods.

- The earliness of the manufacturing start in advance of the assembly start is $e_i = 2$ days for each supplier $i = 1, 2, 3$.
- The minimum and the maximum allowed shipment from each supplier: $\underline{V} = 5{,}000$ and $\overline{V} = 50{,}000$ parts.
- Parts:
  - 10 product-specific part types.
  - Manufacturing rates (in parts per machine-period): $648 \leq q_{ik} \leq 1{,}080$, $i = 1, 2, 3, k = 1, \ldots, 10$.
- The unit inventory holding costs are $\varphi_{1ik} = 0.001$,$\varphi_{2ik} = 0.0015$, $\varphi_{3k} = 0.002$, $\varphi_{4j} = 0.004\ \forall i, k, j$.

4. Producer (Figure 13.3)
   - $f = 6$ assembly stages in series with parallel machines: $m_g = 10$ parallel machines in each stage $g = 1, 2$; $m_g = 20$ parallel machines in each stage $g = 3, 4, 5$; and $m_g = 10$ parallel machines in stage $g = 6$.
   - Products:
     - Ten product types each to be processed in, at most, four assembly stages.
     - Assembly rates (in products per machine-period): $327 \leq p_{gi} \leq 6680$, $g = 1, \ldots, 6$, $j = 1, \ldots, 521$.

In the above example, the set $L$ of product types is identical with the set $K$ of product-specific part types (i.e., $K = L = \{1, \ldots, 10\}$), and each product type $l$ requires one unit of the corresponding product-specific part type $k$ such that $k = l$, (i.e., $b_{kl} = 1, k \in K, l \in L : k = l, b_{kl} = 0, k \in K, l \in L : k \neq l$); for example, one printed wiring board of a specific design is required per one electronic device of the corresponding type. As a result, for each order $j \in J_l$, the required quantity of product-specific part type $k = l$ equals the quantity $o_j$ of the ordered products $l$.

In the computational experiments, the revenue parameters are identical for all orders and take on the following values:

$$\beta_j = 1, \quad \varepsilon_j = 0.2, \quad \gamma_j = 0.02; \quad j \in J$$

That is, for each product in a non-delayed order, the revenue $\beta_j$ is 100 percent, and for each product in a delayed order, the fixed loss of revenue $\varepsilon_j$ is 20 percent, and the daily loss of revenue $\gamma_j$ is 2 percent.

The characteristics of weighted-sum program $\mathbf{INT}_\lambda$ and Tschebyscheff program $\mathbf{INT}^\lambda$ for the integrated approach with the primary objective of maximizing the service level or the revenue are summarized in Tables 13.3 through 13.6. Results for the hierarchical approach are shown in Table 13.7. The size of each program is represented by the total number of variables, Var.; number of binary variables, Bin.; number of constraints, Cons.; and number of nonzero elements in the constraint matrix, Nonz. The counts presented in these tables are taken from the models after presolving. The two rightmost columns present the solution values and CPU time in seconds (Table 13.7) required to find the optimal solution and to prove its optimality or GAP% if optimality is not proven within a specified CPU time limit (Tables 13.3 through 13.6). The computational experiments were performed on a PC Pentium IV, 1.8GHz, RAM 1GB using AMPL/CPLEX 11 with a traditional branch and bound search and strong branching option.

The solution results for the integrated approach and the characteristics of the mixed-integer programs $\mathbf{INT}_\lambda$ and $\mathbf{INT}^\lambda$ for selected $\lambda$ and for different primary objective functions are summarized, respectively, in Tables 13.3, 13.5 and 13.4, 13.6.

**Table 13.3 Computational Results for the Weighted-Sum Program $\mathbf{INT}_\lambda$: Maximum Revenue**

| $\lambda$ | *Var.* | *Bin.* | *Cons.* | *Nonz.* | $f_1(f_1^O, f_1^P), f_2$ [a] | *GAP%* [a] |
|---|---|---|---|---|---|---|
| 0.99 | 8236 | 7236 | 1944 | 180960 | 199(3,825), 747 | 2.38 |
| 0.9 | 8236 | 7236 | 1944 | 180960 | 199(3,825), 747 | 2.13 |
| 0.8 | 8236 | 7236 | 1944 | 180960 | 199(3,825), 747 | 1.93 |
| 0.7 | 8236 | 7236 | 1944 | 180960 | 200(2,830), 746 | 2.30 |
| 0.6 | 8236 | 7236 | 1944 | 180960 | 201(2,835), 745 | 2.00 |
| 0.5 | 8236 | 7236 | 1944 | 180960 | 202(2,840), 744 | 1.36 |

[a] GAP% for 3600 CPU seconds on a PC Pentium IV, 1.8 GHz, RAM 1 GB, CPLEX 11. $f_1(f_1^O, f_1^P)$ : lost revenue (number of tardy orders, number of tardy products), respectively.

**Table 13.4 Computational Results for the Tschebyscheff Program $\mathbf{INT}^{\lambda}$: Maximum Revenue**

| $\lambda$ | *Var.* | *Bin.* | *Cons.* | *Nonz.* | $f_1(f_1^O, f_1^P)$, $f_2^{a}$ | *GAP%*[a] |
|---|---|---|---|---|---|---|
| 0.99 | 8237 | 7336 | 1916 | 175637 | 199(2,825), 747 | 5.35 |
| 0.9 | 8237 | 7336 | 1916 | 175637 | 200(2,830), 746 | 9.26 |
| 0.8 | 8237 | 7336 | 1916 | 175637 | 201(2,835), 745 | 8.72 |
| 0.7 | 8237 | 7336 | 1916 | 175637 | 202(2,840), 744 | 1.19 |
| 0.6 | 8237 | 7336 | 1916 | 175637 | 235(5,950), 743 | 1.16 |
| 0.5 | 8237 | 7336 | 1916 | 175637 | 252(4,1040), 742 | 1.40 |

[a] GAP% for 3600 CPU seconds on a PC Pentium IV, 1.8 GHz, RAM 1 GB, CPLEX 11. $f_1(f_1^O, f_1^P)$; lost revenue (number of tardy orders, number of tardy products), respectively. $(\underline{f}_1, \underline{f}_2) = (190, 640)$.

For the integrated approach, no proven optimal solution was determined in a one-hour limit of CPU time, and the GAP for the best solution found was ranging from 0.01 percent for the weighted-sum program $\mathbf{INT}_{\lambda}$ and maximum service level as a primary objective to 9.26 percent for the Tschebyscheff program $\mathbf{INT}^{\lambda}$ and maximum revenue. In the tables, the solution value of each objective function is presented along with the corresponding associated values of the complementary objective functions (in parentheses).

**Table 13.5 Computational Results for the Weighted-Sum Program $\mathbf{INT}_{\lambda}$: Maximum Service Level**

| $\lambda$ | *Var.* | *Bin.* | *Cons.* | *Nonz.* | $f_1^O(f_1^P, f_1)$, $f_2^{a}$ | *GAP%*[a] |
|---|---|---|---|---|---|---|
| 0.99 | 8236 | 7336 | 1944 | 180960 | 1(1325,530), 762 | 0.14 |
| 0.9 | 8236 | 7336 | 1944 | 180960 | 3(11304,4178), 708 | 0.01 |
| 0.8 | 8236 | 7336 | 1944 | 180960 | 9(17194, 5743), 672 | 0.30 |
| 0.7 | 8236 | 7336 | 1944 | 180960 | 14(22954,7045), 658 | 0.23 |
| 0.6 | 8236 | 7336 | 1944 | 180960 | 17(24659,7468), 651 | 0.05 |
| 0.5 | 8236 | 7336 | 1944 | 180960 | 19(28009,8474), 649 | 0.17 |

[a] GAP% for 3600 CPU seconds on a PC Pentium IV, 1.8 GHz, RAM 1 GB, CPLEX 11. $f_1^O(f_1^P, f_1)$: number of tardy orders (number of tardy products, lost revenue), respectively.

**Table 13.6 Computational Results for the Tschebyscheff Program INT$^{\lambda}$: Maximum Service Level**

| $\lambda$ | *Var.* | *Bin.* | *Cons.* | *Nonz.* | $f_1^O(f_1^P, f_1), f_2^a$ | *GAP%*[a] |
|---|---|---|---|---|---|---|
| 0.99 | 8237 | 7336 | 1916 | 175637 | 2(2990,882), 728 | 6.98 |
| 0.9 | 8237 | 7336 | 1916 | 175637 | 3(11304,4178), 708 | 0.44 |
| 0.8 | 8237 | 7336 | 1916 | 175637 | 4(11914,4325), 702 | 0.50 |
| 0.7 | 8237 | 7336 | 1916 | 175637 | 5(19199,6215), 698 | 8.42 |
| 0.6 | 8237 | 7336 | 1916 | 175637 | 6(13229,6391), 696 | 6.25 |
| 0.5 | 8237 | 7336 | 1916 | 175637 | 7(20579,6512), 690 | 9.10 |

[a] GAP% for 3600 CPU seconds on a PC Pentium IV, 1.8 GHz, RAM 1 GB, CPLEX 11. $f_1^O(f_1^P, f_1)$: number of tardy orders (number of tardy products, lost revenue), respectively. $(\underline{f}_1^O, \underline{f}_2) = (0, 640)$,

The solution results for the hierarchical approach and the characteristics of the mixed-integer programs **DDS**, **ODS**, and **MDS** are summarized in Table 13.7. While the integrated approach was capable of finding a subset of non-dominated solutions for the coordinated scheduling problem, the hierarchical approach was able to find the best solution with respect to the primary objective function. On the other

**Table 13.7 Computational Results for the Hierarchical Approach**

| *Model* | *Var.* | *Bin.* | *Cons.* | *Nonz.* | *Solution Values* | *CPU*[a] |
|---|---|---|---|---|---|---|
| **Maximum Revenue** | | | | | | |
| DDS | 4363 | 3883 | 1119 | 397563 | $f_1 = 193(f_1^O = 4,\ f_1^P = 800)$ | 2.14 |
| ODS | 4351 | 3871 | 777 | 58776 | $f_2' = 150$ | 12.05 |
| MDS | 942 | 42 | 1278 | 16022 | $f_2'' = 594$ | 1.53 |
| | | | | | $f_2 = f_2' + f_2'' = 744$ | |
| **Maximum Service Level** | | | | | | |
| DDS | 4363 | 3883 | 1119 | 397563 | $f_1^O = 1(f_1 = 964,\ f_1^P = 2190)$ | 1.51 |
| ODS | 4382 | 3902 | 782 | 59329 | $f_2' = 230$ | 9.80 |
| MDS | 942 | 42 | 1278 | 16022 | $f_2'' = 593$ | 1.58 |
| | | | | | $f_2 = f_2' + f_2'' = 823$ | |

[a] CPU seconds for proving optimality on a PC Pentium IV, 1.8 GHz, RAM 1 GB/CPLEX 11. $f_1$, $f_1^O$, $f_1^P$: lost revenue, number of tardy orders, number of tardy products, respectively.

hand, the hierarchical approach may lead to a greater total supply chain inventory holding cost $f_2$ because parts and products inventories are considered separately at two decision levels. Unlike for the integrated model, the computational effort required for the hierarchical approach is very small and proven optimal solutions have been reached in a short CPU time for each problem level. For both approaches, the coordinating constraints [(13.16) and (13.45)] have reduced CPU time up to 10 percent.

Examples of the aggregated production schedule ($\Sigma_{j\in J} o_j z_{jt}$, $t \in T$) for the producer and the aggregated manufacturing schedule ($\Sigma_{k\in K} u_{ikt}$, $t \in T^{(i)}$), and the delivery schedule ($\Sigma_{k\in K} v_{ikt}$, $t \in T^{(i)}$) for each supplier $i \in I$ are shown in Figures 13.4 and 13.5 for the integrated approach and for different primary objective functions.

The results indicate that the shorter the transportation time from a supplier, the greater the total demand for parts of customer orders assigned to that supplier, and the delivery of parts are more frequent and resemble more the just-in-time supplies. The selection of maximum service level as a primary objective function leads to leveled aggregate production of finished products and leveled aggregate manufacturing and delivery of parts.

Figure 13.6 presents total supply chain inventory of parts and products for the integrated approach and for different primary objective functions: maximum revenue (minimum lost revenue, $f_1$) and maximum service level (minimum number of tardy orders, $f_1^O$). The figure compares results obtained for the various weights $\lambda$ in the weighted-sum program and in the Tschebyscheff program. The two solution approaches produce similar results, and the results indicate that the primary objective of maximum revenue leads to slightly higher levels of the total supply chain inventory, virtually independent on weight $\lambda$.

For the hierarchical approach with different primary objective functions, Figure 13.7 shows the supply chain inventory of parts, that is, the total inventory of manufactured parts waiting for shipment, being in transit from suppliers to the producer or delivered to the producer and waiting for processing. In addition, the producer output inventory of finished products waiting for delivery to the customers is presented. The total inventory of parts is varying similarly for both primary objective functions, while the output inventory of finished products is slightly higher for the maximum service level. The results obtained for the hierarchical approach, which is based on a lexicographic optimization, are similar to those found for the integrated approach with a high weight $\lambda$ given to the primary objective function.

Figures 13.6 and 13.7 indicate that the total supply chain inventory of parts and products varies similarly over the planning horizon for both the integrated and the hierarchical approach.

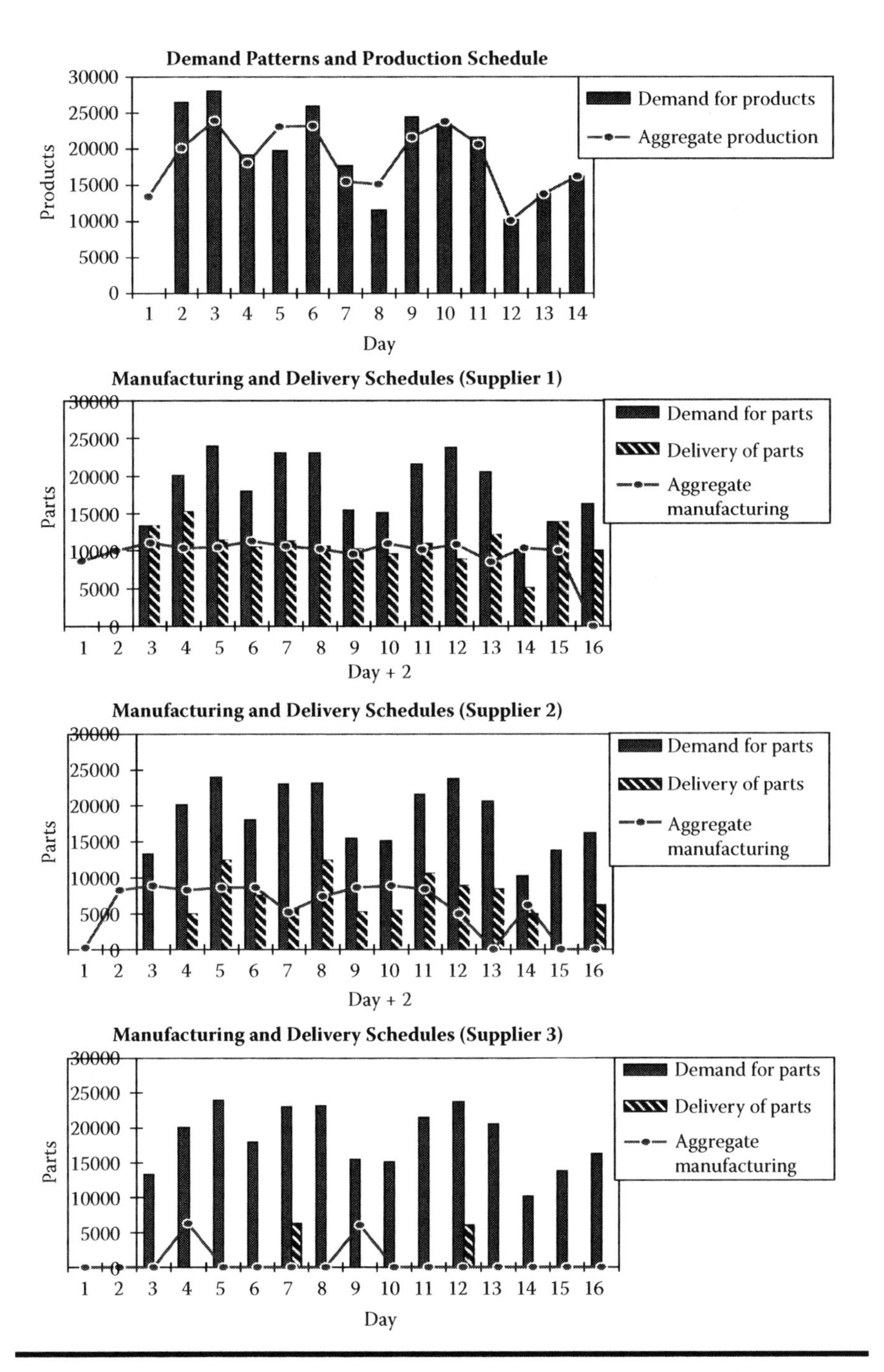


**Figure 13.4 Manufacturing, delivery, and production schedules for the integrated approach (weighted-sum program $INT_{\lambda=0.5}$, primary objective—maximum revenue).**

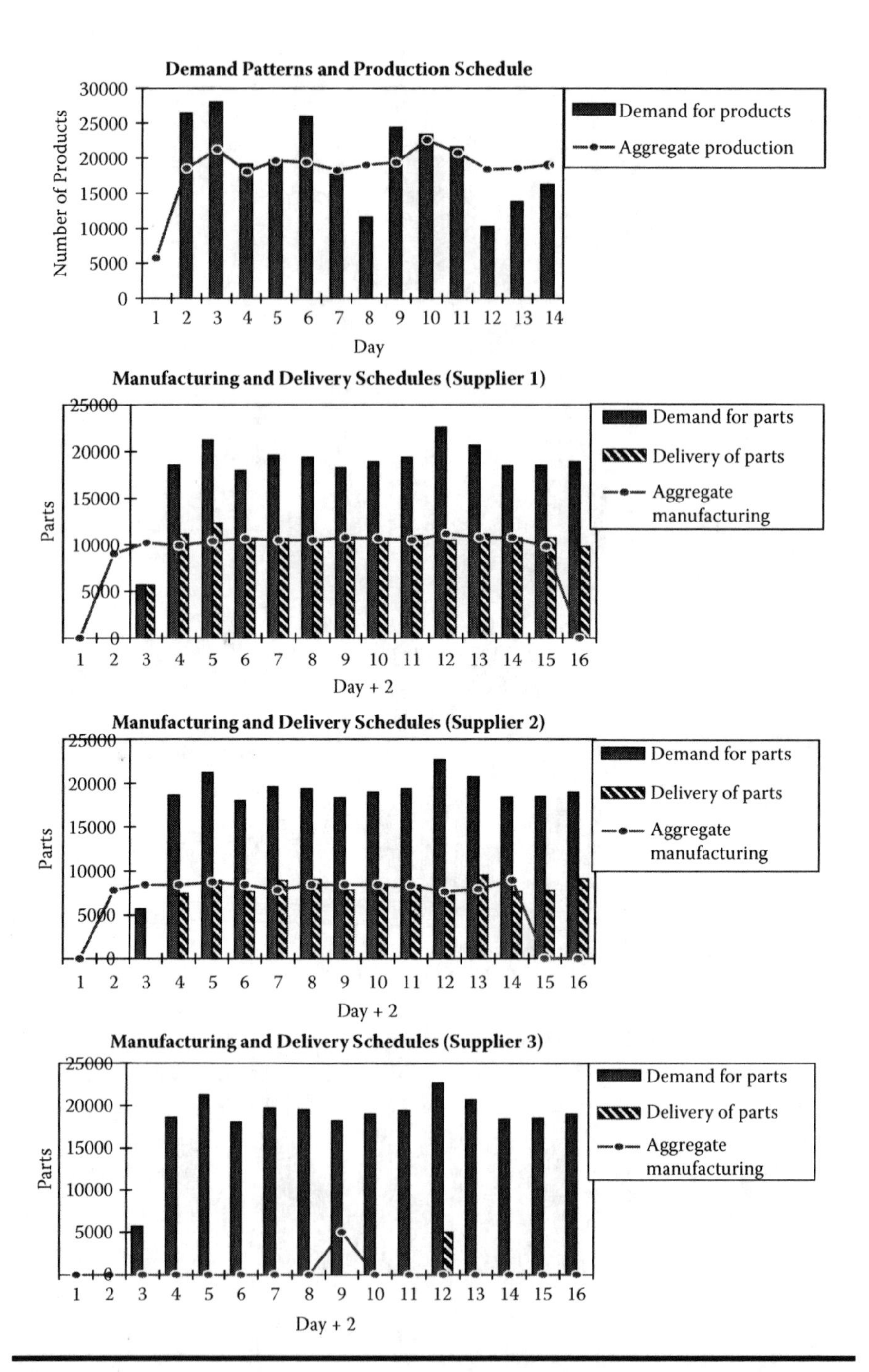


**Figure 13.5 Manufacturing, delivery, and production schedules for the integrated approach (weighted-sum program $INT_{\lambda=0.5}$, primary objective—maximum service level).**

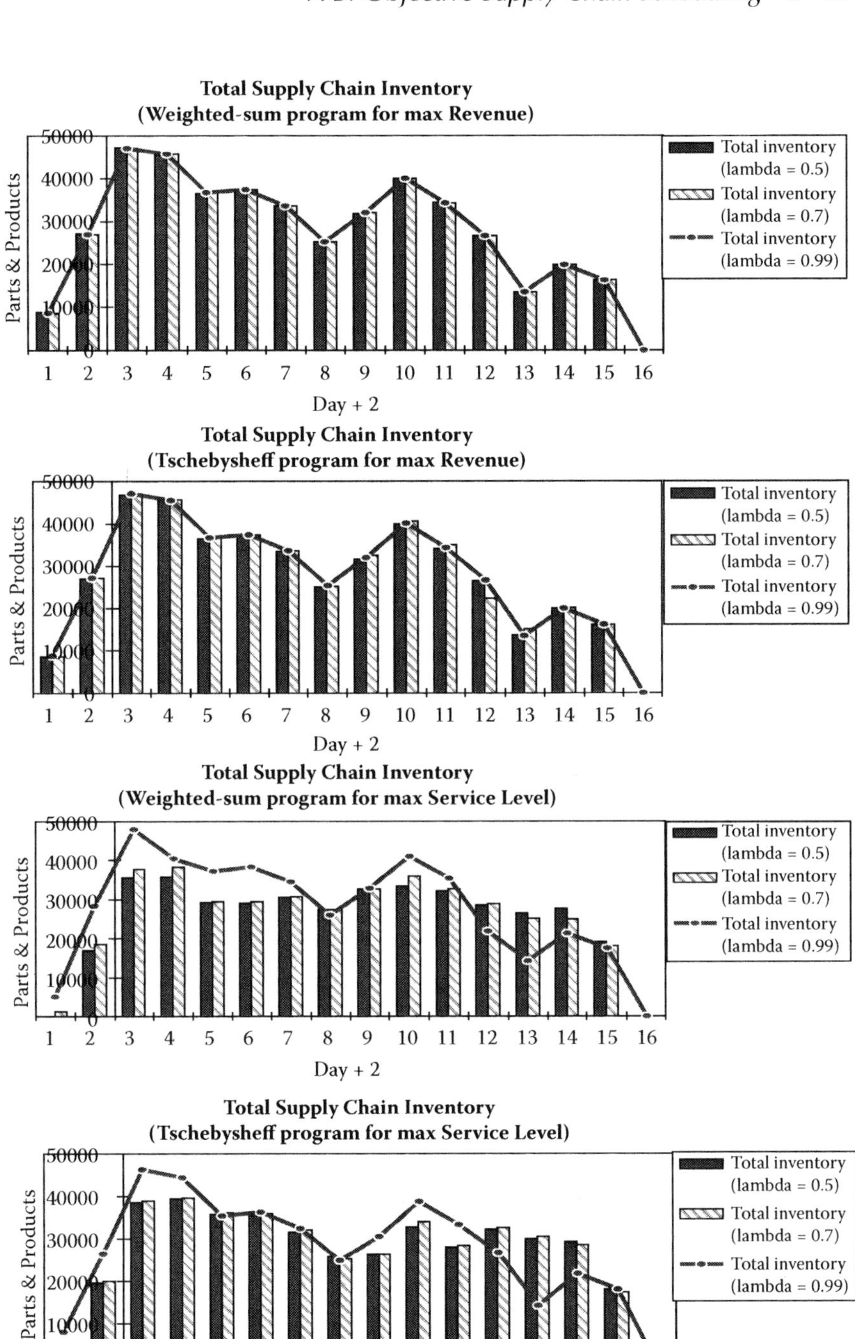


**Figure 13.6 Supply chain inventory for the integrated approach.**

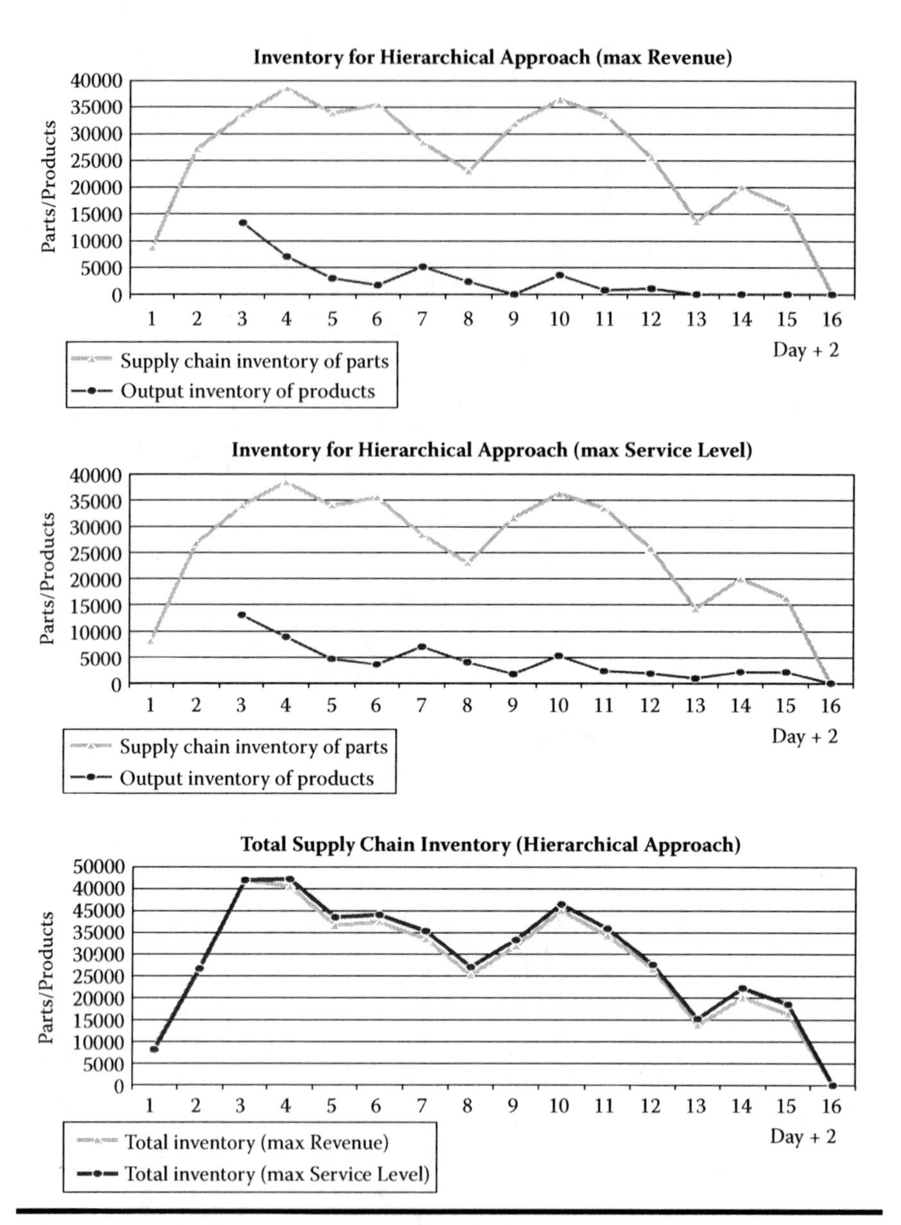


**Figure 13.7 Supply chain inventory for the hierarchical approach.**

## 13.6 Conclusion

In this chapter the integrated and the hierarchical approaches have been proposed and compared for a bi-objective coordinated scheduling in a customer-driven supply chain. The two alternative primary objective functions—that is, maximum revenue or maximum service level—are combined with a secondary objective of minimizing the total supply chain inventory holding cost.

The hierarchical approach coincides with a typical pull planning framework in a make-to-order environment and the corresponding sequential decision making. First, the due dates (deadlines for completing the customer orders) are determined to ensure scheduling feasibility and a high revenue or a high service level; then the orders deadline assignment to planning periods over the horizon is made to minimize product inventory holding cost; and finally, the schedules for manufacturing and delivery of parts from each supplier to the producer are found to minimize the part inventory holding cost. While the hierarchical approach is capable of finding a single proven optimal solution in a very short CPU time, the integrated approach, where all decision problems are solved simultaneously, provides the decision maker with a subset of non-dominated schedules.

For the integrated approach, two basic solution procedures for bi-objective integer programming have been applied and compared: the weighted-sum program and the Tchebycheff program. Both were capable of producing similar subsets of non-dominated schedules; however, the weighted-sum program computationally outperforms the Tchebycheff program (in which the relative importance of the objective functions is weighted in the constraint set) and generates solutions with a smaller GAP.

In the models proposed, various simplifying assumptions have been introduced. For example, the proposed approaches can be enhanced to simultaneously consider both the small (single-period) and the large (multi-period) orders (see Sawik 2007b, 2009a). The two approaches can also be applied for different sets of multiple objectives that can be encountered in customer-driven supply chain operations.

## Acknowledgments

This work was partially supported by a research grant of MNiSzW (N519 576338) and by AGH.

## References

Akkan, C., Finite-capacity scheduling-based planning for revenue-based capacity managament, *European Journal of Operational Research*, 1997, 100, 170–179.

Alves, M.J., and Climaco, J., A review of interactive methods for multiobjective integer and mixed-integer programming, *European Journal of Operational Research*, 2007, 180, 99–115.

Amaro, A.C.S., and Barbosa-Póvoa, A.P.F.D., Supply chain management with optimal scheduling, *Industrial Engineering and Chemical Research*, 2008, 47(1), 116–132.

Barut, S.P., and Sridharan, V., Revenue management in order-driven production systems, *Decision Sciences*, 2005, 36(2), 287–316.

Bertrand, J.W.M., and van Ooijen, H.P.G., Customer order lead times for production based on lead times and tardiness costs, *International Journal of Production Economics*, 2000, 64, 257–265.

Chauhan, S.S., Gordon, V., and Proth, J.-M., Scheduling in supply chain environment, *European Journal of Operational Research*, 2007, 183(3), 961–970.

Chen, Z.-L., Integrated production and distribution operations: taxonomy, models and review, D. Simchi-Levi, S.D. Wu, and Z.-J. Shen, Eds., *Handbook of Quantitative Supply Chain Analysis: Modeling in the E-Business Era*, 2004 (Kluwer Academic Publishers: Boston, MA).

Chen, Z.-L., and Hall, N.G., Supply chain scheduling: conflict and cooperation in assembly system, *Operations Research*, 2007, 55(6), 1072–1089.

Chen, Z.-L., and Pundoor, G., Order assignment and scheduling in a supply chain, *Operations Research*, 2006, 54(3), 555–572.

Chen, Z.-L., and Vairaktarakis, G.L., Integrated scheduling of production and distribution operations, *Management Science*, 2005, 51(4), 614–628.

Chen, C.-Y., Zhao, Z.-Y., and Ball, M.O., Quantity and due date quoting available to promise, *Information Systems Frontiers*, 2001, 3(4), 477–488.

Ebben, M.J.R., Hans, E.W., and Olde Weghuis, F.M., Workload based acceptance in job shop environments, *OR Spectrum*, 2005, 27, 107–122.

Erenguc, S.S., Simpson N.C., and Vakharia, A.J., Integrated production/distribution planning in supply chains: an invited review, *European Journal of Operational Research*, 1999, 115, 219–236.

Geunes, J., Romeijn, H.E., and Taaffe, K., Requirements planning with pricing and order selection flexibility, *Operations Research*, 2006, 54(2), 394–401.

Hall, N.G., Lesaoana, M., and Potts, C.N., Scheduling with fixed delivery dates, *Operations Research*, 2001, 49(1), 134–144.

Hall, N.G., and Potts, C.N., Supply chain scheduling: batching and delivery, *Operations Research*, 2003, 51(4), 566–584.

Hegedus, M.G., and Hopp, W.J., Due date setting with supply constraints in systems using MRP, *Computers & Industrial Engineering*, 2001, 39, 293–305.

Kolisch, R., Integration of assembly and fabrication for make-to-order production, *International Journal of Production Economics*, 2000, 68, 287–306.

Kreipl, S., and Dickersbach, J.T., Scheduling coordination problems in supply chain planning, *Annals of Operations Research*, 2008, 106, 103–122.

Lewis, H.F., and Slotnick, S.A., Multi-period job selection: planning work loads to maximize profit, *Computers & Operations Research*, 2002, 29, 1081–1098.

Moon, C., Lee, Y.H., Jeong, C.S., and Yun, J.S., Integrated process planning and scheduling in a supply chain, *Computers and Industrial Engineering*, 2008, 54(4), 1048–1061.

Pinedo, M., *Planning and Scheduling in Manufacturing and Service Industries*, 2005 (Springer: New York).

Pochet, Y., and Wolsey, L.A., *Production Planning by Mixed Integer Programming*, 2006 (Springer: New York).

Sawik, T., A lexicographic approach to bi-objective scheduling of single-period orders in make-to-order manufacturing, *European Journal of Operational Research*, 2007a, 180(3), 1060–1075.

Sawik, T., Multi-objective master production scheduling in make-to-order manufacturing, *International Journal of Production Research*, 2007b, 45(12), 2629–2653.

Sawik, T., A multi-objective customer orders assignment and resource leveling in make-to-order manufacturing, *International Transactions in Operational Research*, 2007c, 14(6), 491–508.

Sawik, T., Monolithic versus hierarchical approach to integrated scheduling in a supply chain, *International Journal of Production Research*, 2009a, 47(21), 5881–5910.

Sawik, T., Multi-objective due-date setting in a make-to-order environment, *International Journal of Production Research*, 2009b, 47(22), 6205–6231.

Steuer, R.E., *Multiple Criteria Optimization: Theory, Computation and Application*, 1986 (Wiley: New York).

Thomas, D.J., and Griffin, P.M., Coordinated supply chain management, *European Journal of Operational Research*, 1996, 94(1), 1–15.

Wester, F.A.W., Wijngaard, J., and Zijm, W.H.M., Order acceptance strategies in a production-to-order environment with setup times and due dates, *International Journal of Production Research*, 1992, 30(4), 1313–1326.

# Chapter 14

# Applying Data Envelopment Analysis and Multiple Objective Data Envelopment Analysis to Identify Successful Pharmaceutical Companies

Ronald K. Klimberg, George P. Sillup, George Webster, Harold Rahmlow, and Kenneth D. Lawrence

## Contents

## 14.1 Introduction

Since the 1980s, U.S. companies have been embracing the Total Quality Management (TQM) movement. TQM is a philosophy that advocates four basic principles: (1) an intense focus on customer satisfaction, (2) accurate measurement of activities, (3) continuous improvement of products and processes, and (4) empowerment of people (Noori and Radford 1995). In striving to obtain these philosophies and principles, most companies must strike an appropriate balance between economic efficiency and quality. One component of a company's TQM program is a performance measuring process, such as benchmarking. Benchmarking is the process by which a company analyzes the performance and practices of world-class companies and compares their practices and performances to its own. A company can use this knowledge to improve its own operations.

The benchmarking process can also be used to compare similar operations within an organization. The results of this benchmarking process should not be used to punish the less efficient/effective operations/individuals. Instead, it should be used for identifying areas of improvement, for determining how this improvement can be obtained, and ultimately for rewarding those following the "best" practices (Juran 2004). For these reasons, a tool that allows quantitative evaluation of similar operations against a variety of effectiveness and efficiency objectives will be a valuable contribution. A significant trend has been an attempt to incorporate this benchmarking process by measuring and profiling efficient and effective "best" practices.

This is especially true for pharmaceutical corporations where inefficiency can result in a protracted development process and delayed market entry for new medicines that have a finite patent life and average about 15 years to go from compound discovery to market approval (DiMasi et al. 2003, DiMasi and Paquette 2004). A recent estimate places this cost on a pre-tax basis at $802 million although both higher and lower estimates exist (DiMasi et al. 2003). On an after-tax basis, assuming sufficient revenues to capture R&D tax benefits, the estimated cost of developing an average drug is $480 million (Grabowski and Vernon 2000a). Furthermore, only three of ten new pharmaceutical products generate enough after-tax present-value returns in excess of average after-tax R&D costs (Grabowski and Vernon 2000b).

Quantitative forecasting models, even rather sophisticated models, are easier to develop today as result of our improving computer technology. These quantitative forecasting techniques use historical data to predict the future. Most quantitative forecasting techniques can be categorized into either time series approaches or causal models. Time series forecasting techniques are forecasting techniques that only use the time series data itself and no other data to build the forecasting models. These time series approaches isolate and measure the impact of the trend, and seasonal and cyclical time series components. Causal models use a set of predictor/independent variables, possibly including the time series components, believed to influence the forecasted variable. One of the most popular causal model approaches is regression analysis. Regression techniques employ the statistical method of least squares to establish a statistical relationship between the forecasted variable and the set of predictor/independent variables.

Many forecasting situations involve producing forecasts for comparable units. A comparable unit could be an individual, a group of individuals, a department, a company, and so on. Each comparable unit should be performing a similar set of tasks. When applying regression analysis, the established statistical relationship is an average relationship using one set of weights assigned to the predictor/independent variables. However, when used with a set of comparable units, the relative importance of each of the predictor/independent variables will most likely vary from comparable unit to comparable unit. For example, if advertising is an independent variable, one comparable unit might emphasize advertising more (or less) than other comparable units. Either way is neither better nor worse; it is just how that particular comparable unit emphasizes advertising. As a result, the regression model could provide forecast estimates that are too high or too low in some cases.

In this chapter we apply and extend some of our recent work in which we introduced a methodology that incorporates a new variable that captures the unique weighting of each comparable unit into regression forecasting analysis (Klimberg et al. 2004, 2005). This new variable is the relative efficiency of each comparable unit that is generated by a nonparametric technique called data envelopment analysis (DEA). We extend this methodology by introducing two new variables that capture two distinct dimensions of each comparable unit using an extension to DEA called multiple objective data envelopment analysis (MODEA). After we provide a brief introduction to DEA and MODEA, we review modeling approaches specific to the pharmaceutical industry. Subsequently, we present our regression forecasting methodology and the results of applying this methodology to a data set of fifty pharmaceutical companies, followed by conclusions and potential future extensions.

## 14.2 Background

### *14.2.1 Data Envelopment Analysis (DEA)*

One of the major concerns of managers in evaluating the performance of an operation within any type of organization is efficiency. Efficiency measures whether resources, equipment, and/or people are being put to good use. One dimension of the efficiency of an operation is the manner by which that organization selects and uses resources to produce its products. The more product produced for a given amount of resources, the more efficient (i.e., less wasteful) the operation is. Mathematically, efficiency can be defined as the ratio of weighted outputs to weighted inputs:

$$E = \text{Efficiency} = \frac{\text{Weighted sum of outputs}}{\text{Weighted sum of inputs}}$$

Charnes et al. (1978) developed an innovative nonparametric methodology called data envelopment analysis (DEA) to measure the relative efficiency of comparable units. In DEA terminology, these are called decision-making units (DMUs). DEA takes into account multiple inputs and outputs to produce a single aggregate measure of relative efficiency for each DMU. It requires only that the selected inputs and outputs be quantifiable. The technique analyzes these multiple inputs and outputs in their natural physical units without reducing or transforming them into some common measurement, such as dollars. Finally, DEA uniquely evaluates all the DMUs and all their inputs and outputs simultaneously, and conservatively identifies the sets of relatively efficient and relatively inefficient DMUs. Thus, the solution of a DEA model provides a manager with a summary that contains comparable DMUs grouped together and ranked by relative efficiency.

The DEA model of Charnes et al. (1978) finds the relative efficiency of each DMU, $\mathrm{w}_r$, by solving the following linear fractional DEA formulation (DEA1):

$$\text{MAX } \mathrm{w}_r = \frac{\sum_{j=1}^{J} \mathrm{u}_j \mathrm{O}_{\mathrm{jr}}}{\sum_{i=1}^{\mathrm{I}} \mathrm{v}_i \mathrm{I}_{\mathrm{ir}}} \tag{14.1}$$

s.t.

$$\frac{\sum_{j=1}^{J} \mathrm{u}_j \mathrm{O}_{jk}}{\sum_{i=1}^{I} \mathrm{v}_i \mathrm{I}_{ik}} \leq 1 \quad \forall k \tag{14.2}$$

$$\mathrm{u}_j \geq 0 \quad \forall j,$$

$$\mathrm{v}_i \geq 0 \quad \forall i$$

where:

$i = 1, \ldots, I$ Inputs used at DMU
$j = 1, \ldots, J$ Outputs produced at DMU
$k = 1, \ldots, r, \ldots, K$ DMUs

Parameters:

$\mathrm{O}_{jk}$ = Amount of the $j$-th output for the $k$-th DMU
$\mathrm{I}_{ik}$ = Amount of the $i$-th input for the $k$-th DMU

Decision variables:

$u_j$ = Weight assigned to the $j$-th output
$v_i$ = Weight assigned to the $i$-th input

This formulation finds the set of weights, $u_j$ and $v_i$, that maximizes the efficiency, $w_r$, of DMU$r$ [Equation (14.1)]. The constraints in Equation (14.2) require the ratio of the sum of weighted outputs to the sum of weighted inputs (i.e., the efficiency of each DMU, including the $r$-th DMU) to be no larger than 1. A similar DEA formulation is solved sequentially for each DMU. A DMU is considered relatively inefficient ($w_r < 1$) if increasing its outputs without increasing inputs, or decreasing its inputs without decreasing outputs, is possible. The degree of inefficiency for a DMU is measured relative to a set of more efficient DMUs. On the other hand, a DMU identified as being efficient ($w_r = 1$) does not necessarily imply absolute efficiency, that is, a DMU is only relatively efficient as compared to the other DMUs being considered.

Charnes et al. (1978) converted DEA1 into a linear program by

- arbitrarily setting the denominator in the objective function [Equation (14.1)] to 1: $\sum_{i=1}^{I} v_i I_{ir} = 1$, multiplying both sides of the constraints in Equation (14.2) by the sum of the weighted inputs, yielding the linear equivalent constraint set:

$$\sum_{j=1}^{J} u_j O_{jk} \leq \sum_{i=1}^{I} v_i I_{ik} \quad \forall k.$$

Additionally, a special case called weakly efficient causes the DEA model to be modified in practice. A particular DMU may be considered weakly efficient if, in the DEA solution, its DEA efficiency score is 1 and one or more of its weights are equal to zero. To address this problem, Charnes et al. (1978) require each weight to be greater than $\varepsilon$, an infinitesimal value, to ensure that weakly efficient DMUs are not classified as efficient. The modified DEA linear formulation (DEA2) of Charnes et al. therefore becomes:

**Model: Modified Data Envelopment Analysis Model (DEA2) of Charnes et al.**

$$\text{MAX } w_r = \sum_{j=1}^{J} u_j O_{jr} \tag{14.3}$$

st.

$$\sum_{i=1}^{I} v_i I_{ir} = 1 \tag{14.4}$$

$$\sum_{j=1}^{J} u_j O_{jk} - \sum_{i=1}^{I} v_i I_{ik} \leq 0 \quad \forall k \tag{14.5}$$

$$u_j, v_i \geq \varepsilon \quad \forall j, i$$

where $\varepsilon$ is a small infinitesimal value.

Since the publication of the 1978 paper by Charnes et al., there have been thousands of theoretical contributions and practical applications in various fields using DEA (Seiford 1996). DEA has been applied to many diverse areas such as: healthcare, military operations, criminal courts, university departments, banks, electric utilities, mining operations, manufacturing productivity, and railroad property evaluation (Emrouznedjad et al. 2008, Gattoufi and Reisman 2004, Klimberg 1998, Klimberg and Kern 1992, Seiford 1996, Seiford and Thrall 1990, Tavares 2002).

### *14.2.2 Multiple Objective Data Envelopment Analysis (MODEA)*

Over the past several years, organizations—private as well as public—have not only concentrated on increasing efficiency by increasing productivity and lowering costs, but also on improving the quality of the product delivered. As a result, when managers evaluate comparable units, they are concerned with operational efficiency as well as other objectives, such as effectiveness or quality. Most real-world situations are inherently multiobjective situations. Ignoring the multiobjective nature neither addresses the overall performance nor does it allow the decision maker to address the various trade-offs between these objectives. A technique that concentrates on a single objective at a time therefore obscures the nature of the manager's real problem.

The single-objective DEA approach takes into account multiple inputs and outputs to produce a single aggregate measure of performance. To include other dimensions of performance, appropriate input and output variables can be added to the single-objective DEA model. However, by collapsing all these input and output variables to produce one aggregate performance measure or by solving several independent single objectives, DEA models would discount performance in these other dimensions. An approach in which these other dimensions/objectives are individually measured is multiple objective DEA (MODEA). MODEA enables managers to address multiple objectives simultaneously, thereby allowing decision makers to address more fully the trade-offs between objectives (Klimberg 1998, Klimberg and Puddicombe 1999).

MODEA produces objective ratings for each objective within each comparable unit. The MODEA formulation is:

$$\text{MAX} \sum_{p=1}^{P} w_{rp} = \sum_{p=1}^{P} \sum_{j=1}^{J} u_{jp} O_{jrp} \tag{14.6}$$

st.

$$\sum_{i=1}^{I} \mathrm{v_{ip}} I_{irp} = 1 \quad \forall \tag{14.7}$$

$$\sum_{j=1}^{J} \mathrm{u_{jp}} O_{jkp} - \sum_{i=1}^{I} \mathrm{v_{ip}} I_{ikp} \leq 0 \quad \forall \tag{14.8}$$

$$\mathrm{u_{jp}}, \mathrm{v_{ip}} \geq \varepsilon \quad \forall i, j, p$$

where:

$i = 1, \ldots, I$ Inputs used at DMU
$j = 1, \ldots, J$ Outputs produced at DMU
$k = 1, \ldots, r, \ldots, K$ DMUs
$p = 1, \ldots, P$ Objectives

- Parameters:
  $\mathrm{O}_{jkp}$ = Amount of the $j$-th output of the $p$-th objective for the $k$-th DMU
  $\mathrm{I}_{ikp}$ = Amount of the $i$-th input of the $p$-th objective for the $k$-th DMU
- Decision variables:
  $\mathrm{u}_{jp}$ = Weight assigned to the $r$-th output of the $p$-th objective
  $\mathrm{v}_{ip}$ = Weight assigned to the $i$-th input of the $p$-th objective

The above formulation finds the set of weights, $\mathrm{u}_{jp}$ and $\mathrm{v}_{ip}$, that maximize the sum of each efficiency objective, $\mathrm{w}_{rp}$, for all the DMUs, [Equation (14.6)]. Constraints (14.7) and (14.8) are similar to Constraints (14.4) and (14.5), respectively, except now we have a set of these constraints for each objective. As presented, the MODEA formulation is simply $p$ single-objective DEAs.

With more than one objective, we are most likely to have one or more variables that may appear in one or more objectives as either an input or an output. We call such a variable a *common variable*. In most cases, a common variable's contribution in one objective should be somewhat related to its contribution provided to another objective. The percentage contribution of each common variable should be correlated, that is, either relatively close to or following a defined relationship (x times greater than the other) and should not be of significant difference. The percentage contributions of common variables can be simultaneously controlled with MODEA by adding weight restriction constraints to the MODEA model. We restrict the percentage contributions for these variables appearing in one or more objective(s) such that they coincide more with the decision-maker's perceptions and conventions of good operating practices. This examination of the weights and their contributions is of paramount concern when we consider more than one objective.

The MODEA approach enables managers to address multiple objectives simultaneously. It accomplishes this by controlling the weights used in one or more of the objectives by common variables, therefore allowing decision makers to more fully address the trade-offs between objectives. MODEA has been applied to several diverse areas, such as concrete facilities, physicians, and manufacturing plants (Klimberg 1998, Klimberg and Puddicombe 1999, Klimberg et al. 2001).

### 14.2.3 Forecast Modeling in the Pharmaceutical Industry

Forecast modeling has been and is being used to predict the complex processes of the pharmaceutical industry, a welcome shift from less sophisticated methods employed in some global markets (Choo 2000). Much of this modeling work reported in the literature is directed at forecasting the demand and understanding the impact of pricing for new medicines, particularly for first-in-class branded drugs in new markets (Comanor and Schweitzer 2007, Connor et al. 2003, Latta 1998b). Some forecast modeling approaches are adaptations of earlier modeling theories and their applications, such as Latta's comprehensive use of Gompertz or "S" curves, which can be described as an intercept (minimum value), an asymptote (maximum value), a slope (rate of change), and a time horizon indicating how long the growth will occur (Latta 2007). These curves present new drugs as a function of time and, when combined with the idea of Rogers' adopter segments, provide a framework for forecasting innovator drug use over time and draw upon earlier applications of the Bass Model for medical innovation (Sillup 1992).

Two comprehensive sources outline numerous forecast modeling techniques for the pharmaceutical industry as well as suggest methods for the different phases of a drug within its product life cycle, for example, market entry versus product in the mid-term of its patent life (Cook 2006, Jain and Malehorn 2006). Other modeling techniques assess market penetration or risks associated with product introduction. For example, there is a method to combine patient information, physician information, and competitive marketing activities to enhance the validity of physicians' choice response using a prescription choice model in a hierarchical Bayesian framework (Pack 2003). There are also simulations to quantify the risk of incorrectly specifying the relationship between a component of an event, such as price, and the resulting variable, sales (Triantis and Song 2007). While these explanations can be most helpful to those constructing demand forecasts for unit sales and/or revenues, other parts of the literature describe modeling methods for those seeking applications associated with the supply chain.

The challenge of considering the impact on supply chain for pharmaceuticals has been around as long as there have been drug manufacturing processes. It appeared about 30 years ago in the literature when Boschi et al. (1979) discussed a subjective probability approach to forecasting research and development (R&D) to quantify the uncertain future success of research and to forecast the future success expected from the R&D activities. More recently, Latta's work using manufacturing resource planning (MRP) forecasts considered the impact on the supply chain by enabling forecasters to generate assumptions about the market and use available data from different sources to elicit demand forecasts and account for unplanned change (e.g., account seasonality) (Latta 1998a).

These efforts are certainly meritable but their primary focus is about predicting the rate of market penetration of a new or existing pharmaceutical. Generally, views specifically directed to the supply chain are engineering based and assess the ways to enhance the efficiency of component elements of the supply chain for a product. An example is the evaluation of batch processing methods to examine packaging solutions for biological vaccines and highlights that a produced batch of vaccines can only be reworked as long as it has not been bottled (Teunter and Flapper 2006). This is significant at any time in a product's life cycle but especially during clinical studies when patients have no other therapeutic alternative and very few batches are being manufactured. While this type of evaluation is important to the supply chain, other work looks at predicting how the component elements of supply chain activity can influence the market forecast of a product. Jain, through his modeling work and from a published interview with a principal forecaster at Wyeth Pharmaceuticals, identified three major elements critical to forecasting success: products, markets, and supply chain (Jain 2002). Of the three, understanding and forecasting the efficiency of a supply chain and its multiple inputs and outputs will have great influence on product forecast as well as managerial decisions.

### 14.2.4 Regression Forecasting Methodology

Klimberg et al. (2004) proposed a three-step regression forecasting methodology to be applied to a data set containing multiple input and output variables from a set of comparable units. Subsequently, Klimberg et al. (2005) modified the methodology to consider multiple objectives. The comprehensive methodology for one or multiple objectives is now the following four-step process:

- Step 1: Define objective(s). Define the (multiple) objective(s) and their potential inputs and outputs relevant to the particular problem as the ratio of weighted outputs to weighted inputs:

$$\text{E} = \text{Efficiency} = \frac{\text{Weighted sum of outputs}}{\text{Weighted sum of inputs}}$$

  Identify one output, for each objective, as the principal/critical variable. Additionally, if the situation has multiple objectives, from the set of principal variables select one as the primary overall variable for all the objectives. This primary variable is what will be forecasted.
- Step 2: Stepwise regression. Run a stepwise regression, for each objective, using the objective's corresponding principal/critical output variable as the

dependent variable and all the input variables as regression independent variables. If the data set is relatively small and the number of input variables is not significantly large, the stepwise regression may not be necessary. The purpose of the stepwise regression is to decrease the number of input variables to be included to only those variables that are statistically significant and to improve the discriminating power of the DEA analysis.

- Step 3: MODEA/DEA analysis. Using the defined objective(s) from Step 1 and the set of statistically significant input variables from the stepwise regression in Step 2, run the appropriate MODEA/DEA model. An efficiency score will be produced for each objective and for each DMU. These efficiency scores are surrogate measures of the unique emphasis of the variables and of the performance by each DMU. In addition to each efficiency score being comprised of a different set of input and output values, each comparable unit's efficiency score includes a unique set of weights. The MODEA/DEA process attempts to find objectively the set of weights that will maximize a comparable unit's efficiency. Therefore, the MODEA/DEA model has selected the best possible set of weights for each comparable unit. The variation of these weights from comparable unit to comparable unit allows each comparable unit to have its own unique freedom to emphasize the importance of each of these input and output variables in its own way. How well they do this is measured by the efficiency score. So, we will use these efficiency scores as surrogate measures of the unique emphasis of the variables and of performance, and include them in Step 4.
- Step 4: Stepwise regression. Run a stepwise regression using the primary output variable, defined in Step 1, as the dependent variable and all the significant input variables and the MODEA/DEA efficiency scores as regression-independent variables. This regression, including the efficiency scores, should be more statistically accurate than not using them.

## 14.3 Example

We collected AY2003 data from fifty leading pharmaceutical companies, including the following variables:

| | | | |
|---|---|---|---|
| 1 | Total inventory | 5 | Common shares for diluted EPS |
| 2 | Total receivables | 6 | R&D expenses |
| 3 | Total assets | 7 | Stockholders' equity |
| 4 | Employees | 8 | Net sales |

| | | | |
|---|---|---|---|
| 9 | Net income | 14 | Short-term debt |
| 10 | Cash flow | 15 | Long-term debt |
| 11 | Operating cycle | 16 | Market capitalization |
| 12 | Fixed assets | 17 | Common stock |
| 13 | WACC: Weighted average cost of capital | | |

Firm performance is typically measured using ratios that analyze liquidity (the ability to pay current liabilities with current assets), activity (the ability to generate revenues with a set of assets), profitability (the ability to create values for investors), and solvency (the ability to remain a going concern in the long run). Central to this analysis is the DuPont system of financial analysis, which measures ROE by its components of cost efficiency, asset use effectiveness, and financial leverage:

- *Return on sales:* net income after tax divided by sales
- *Total asset turnover:* sales divided by total assets
- *Degree of financial leverage:* total assets divided by stockholders' equity

Table 14.1 lists the summary statistics of applying the DuPont system of financial analysis to the fifty pharmaceutical companies. Return on sales ranges from a low of −8.2 to a high of 0.34. Seventeen companies have return on sales values less than or equal to 0, and twenty-three firms between 0 and 0.2. So, about a third of the companies have negative return on sales and almost half have return on sales between 0 and 0.2. In terms of total asset turnover, the minimum value is slightly above 0, 0.05, with eighteen companies with total asset turnover values less than 0.5 and only seven companies with total asset turnover values greater than or equal to 1. Therefore, most (i.e., 86 percent) of the pharmaceutical companies have total asset turnover values less than 1. Finally, thirty-five companies have degree of financial leverage measures less than or equal to 0.5, and three companies actually have negative

**Table 14.1 Summary Statistics of the DuPont System Measures**

| | *Return on Sales* | *Total Asset Turnover* | *Financial Leverage* |
|---|---|---|---|
| Minimum | −8.20 | 0.06 | −0.44 |
| Maximum | 0.34 | 2.53 | 1.5 |
| Average | −0.75 | 0.66 | 0.34 |
| Median | 0.09 | 0.59 | 0.32 |

**Table 14.2 Weighted Average Cost of Capital and Operating Efficiency of 50 Pharmaceutical Companies**

| | *Operating Efficiency* | | | | |
|---|---|---|---|---|---|
| *WACC* | *CF and NI* $> \$1b$ | *CF and NI* $> \$.1b$ | $0 <$ *CF and NI* $< \$.1b$ | *CF or NI* $< 0$ | *Total* |
| $\leq< 8.5\%$ | 4 | 3 | 2 | 4 | 13 |
| $\geq 8.5 - < 9.0\%$ | 5 | 1 | 2 | 3 | 12 |
| $\geq 9.0 - < 9.9\%$ | 1 | 4 | 2 | 4 | 11 |
| $\geq 10.0\%$ | 3 | 1 | 4 | 6 | 14 |
| Total | 13 | 9 | 10 | 17 | 50 |

values. Overall, the performance of these fifty pharmaceutical companies is rather wide ranging.

A company's assets are financed by either debt or equity. WACC (weighted average cost of capital) is the average of the cost of each of these sources of financing weighted by their respective usage in the given situation. By taking a weighted average, we can see how much interest the company must pay for every dollar it borrows. WACC is expressed as a percentage. It is the appropriate discount rate to use for cash flows similar in risk to the overall firm as well as a measure of the amount of value a company creates for its shareholders. Table 14.2 encapsulates the WACC for the fifty pharmaceutical companies. The WACC values are fairly evenly distributed. Additionally, cash flow (CF) and net income (NI) provide indicators of operating efficiency. Table 14.2 also tabulates the operating efficiency of the fifty pharmaceutical firms. As with the WACC, the operating efficiencies of the fifty pharmaceutical companies, using CF and NI, are basically widespread.

For the single-objective DEA model, we defined efficiency as

$$\text{Efficiency} = \frac{1, 2, 5, 9, 10, 11, 13}{3, 4, 6, 7, 15, 17}$$

The numbers refer to the seventeen variables listed in the first paragraph of this section. For the MODEA model, we defined the following two objectives:

$$\text{Z1: } \textit{Operational} \text{ efficiency} = \frac{9, 10}{3, 4, 6, 13} \qquad \text{Z2: } \textit{Financial} \text{ efficiency} = \frac{5, 7, 13, 16}{14, 15, 17}$$

The results of the DEA and MODEA models are summarized in Table 14.3. The single-objective DEA efficiency scores are significantly spread out, ranging from a low of 13.9 to a high of 99.9 with a little over half of the companies (twenty-seven)

**Table 14.3 Data Envelopment Analysis (DEA) and Multiple Objective Data Envelopment Analysis (MODEA) Summary Statistics**

| | | *MODEA* | |
|---|---|---|---|
| | *Single Objective DEA* | *Z1* | *Z2* |
| Minimum | 13.9 | 0.004 | 37.4 |
| Maximum | 99.9 | 100 | 100 |
| Average | 55.7 | 67.3 | 91.5 |
| Median | 48.4 | 67.7 | 100 |
| # 100% | 0 | 5 | 31 |

having efficiencies less than 50 percent. These DEA results are quite similar to the results discussed above. The MODEA model for operational efficiency (Z1) shows similar results, with almost half of the companies (twenty-seven) having an operational efficiency less than 50 percent and with only five firms being 100 percent operationally efficient. On the other hand, in terms of financial efficiency, even though the range of efficiencies is wide, thirty-six companies have financial efficiencies greater than 90 percent and, furthermore, over half the firms were found to be 100 percent financially efficient. The MODEA model separated the financial efficiency from the operational efficiency, with these input and output variables, and seems to conclude that most of these fifty pharmaceutical companies are financially efficient but vary considerably in their operational efficiency.

In Figure 14.1, the region where most of the return on sales values occur, <0.2, the operation efficiencies are widely dispersed. A similar dispersed pattern is observed for most of the total asset turnover values in Figure 14.2.

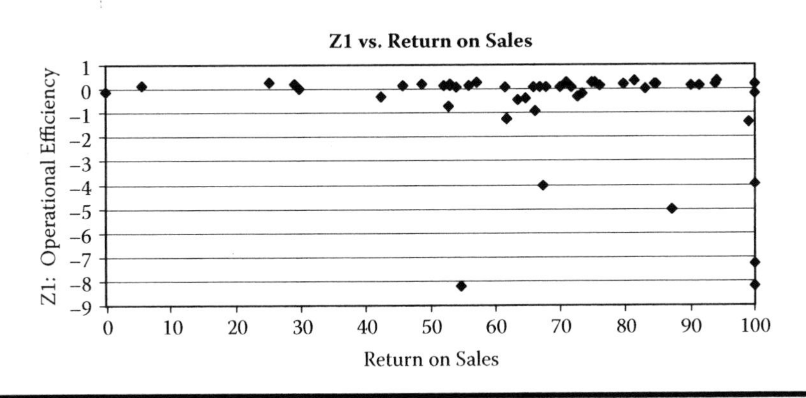


**Figure 14.1 Graph of operational efficiency, Z1, versus return on sales.**

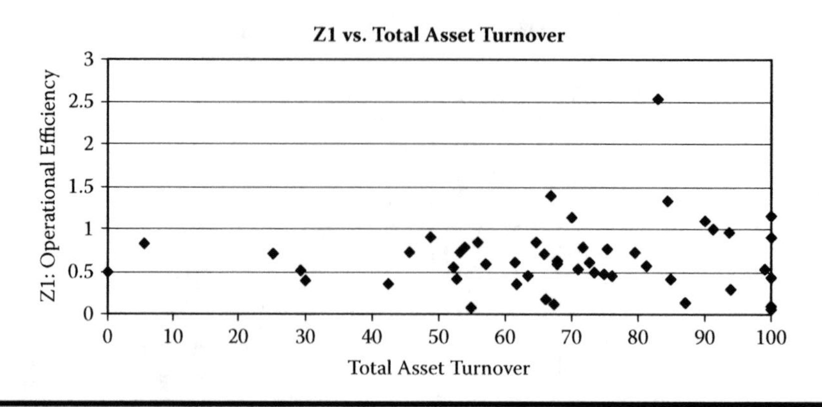


**Figure 14.2 Graph of operational efficiency, Z1, versus total asset turnover.**

Likewise in Figure 14.3, most of the financial leverage values are less than 0.5 and have financial efficiencies greater than 80 percent.

Table 14.4 summarizes the results of the regression models from Step 4 of our methodology. The primary variable to predict was identified as net income. The first stepwise regression, using all the input variables, found the variables employees, long-term debt, assets, and short-term debt significant and had an $R^2$ of 93.0 and a standard error of 589. The stepwise regression model did not change when the single-objective DEA efficiency scores were added as an independent variable. Because the single-objective DEA efficiency scores were not added, a simple regression was executed. The DEA efficiency score has a very weak relationship, $R^2$ of 1.2 to net income. Adding the MODEA, two efficiency scores improved the $R^2$slightly but decreased the standard error by about 15 percent.

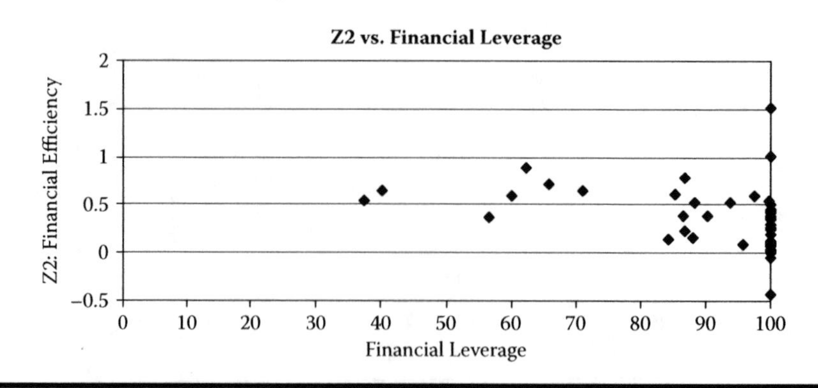


**Figure 14.3 Graph of financial efficiency, Z2, versus financial leverage.**

**Table 14.4 Graph of Operational Efficiency, Z2, and Financial Leverage**

| | $R^2$ | *Standard Error* |
|---|---|---|
| Stepwise—Input variables only:<br>Employees, long-term debt, assets, short-term debt | 93.9 | 589 |
| Stepwise—Input variables and DEA efficiency scores:<br>Employees, long-term debt, assets, short-term debt | 93.9 | 589 |
| Regression—DEA efficiency score only | 1.2 | 2288 |
| Stepwise—Input variables and MODEA efficiency scores:<br>Employees, long-term debt, WACC, common stock, Z1, Z2 | 95.8 | 496 |
| Regression—MODEA efficiency scores only | 13.3 | 2165 |

## 14.4 Discussion

Overall, the performance of these fifty pharmaceutical companies is rather wide ranging, as determined by the DuPont system of financial analysis, which measures ROE by its components of cost efficiency, asset use effectiveness, and financial leverage. This wide range of performance is also observed for operating efficiencies when measured by WACC, CF, and NI. These findings are not inconsistent with prior retrospective statistical analyses of pharmaceutical companies using industry-level data that depicts a pretty close link among profitability, CF, and R&D investment (Nelson and Winter 1982, Vernon 2003, 2005). Additional work, primarily the profitability of pharmaceuticals and R&D investment, also connects the R&D investment decision and expected results on a per-project basis. An example is Scherer's more empirical adaptation of Grabowski's earlier work, which is comprehensively detailed in Abbot and Vernon's 2007 article (Scherer 2001, Abbott and Vernon 2007).

Results using DEA and MODEA suggest some poignant observations. The single-objective DEA efficiency scores are significantly spread out (from 13.9 low to 99.9 high) and indicate that a little over half of the companies (twenty-seven of fifty) have efficiencies of less than 50 percent. The MODEA model for operational efficiency shows similar results for twenty-seven of the fifty companies. They have an operational efficiency less than 50 percent with only five being 100-percent operationally efficient. However, in terms of financial efficiency, thirty-six companies have financial efficiencies greater than 90 percent, although the range of efficiencies is wide. Furthermore, over half the firms are found to be 100 percent financially efficient with NI as the primary predictive variable. Because the MODEA model separated the financial efficiency from the operational efficiency, we were able to determine which of the companies varied considerably in their operational efficiency despite being financially efficient.

These results get more interesting when the R&D process of discovering, developing, and introducing a new drug to the market is considered. For a compound to obtain approval for sale from regulatory agencies, it must successfully pass through several stages of research and development, to include discovery research, pre-clinical testing in animals, and clinical studies with human subjects, of which there are three phases, I, II, and III (Smith 2009, Evens and Covinsky 2009). Consequently, a common practice is to capitalize expenditures through Phase III, which culminates in approval by regulatory agencies such as the FDA (Abbott and Vernon 2007). This capitalized value, which incorporates expenditures on both successful and failed R&D projects, represents the true economic cost of bringing a new drug to market.

However, operational decisions must be made in both Phases I and II, assuming the drug continues to generate good safety and efficacy data, such as manufacturing sufficient quantities of the medicine to treat patients enrolled in clinical studies. An excellent way to evaluate these operational decisions is Abbott and Vernon's (2007) work to rewrite equations in terms of these developmental stages rather than the pharmaceutical company's fiscal year. Their model of a Phase I Go/No Go Decision focuses on one of the most critical developmental decisions to begin clinical studies and, consequently, incur huge expenditures. Phase I Go/No Go also determines the first actual data on the average costs, times, and technical success rates of particular R&D investment projects, primarily based on expected net present value (NPV) (Abbott and Vernon 2007, Smith 2009).

This approach accounts for temporal variations in pharmaceutical development and establishes the basis to identify successful projects and to evaluate and improve the operating efficiency of those successful projects using MODEA. Assuming NPV is a key predictive value along with NI, MODEA calculated in terms of the clinical phases could prove to be a viable approach to determine operating efficiency. While this can have general applicability, unavailability of the information sets used by pharmaceutical companies may make results pertinent to a specific company's decision. In the pharmaceutical industry, the vast majority of R&D projects fail for reasons related to safety, efficacy, or commercial viability. For those that continue successfully, MODEA offers a method to evaluate operating efficiency and thereby improve financial performance.

## 14.5 Conclusions and Implications for Future Research

Acknowledging the challenges faced by a pharmaceutical forecaster, there is a need to strike a delicate balance between over-engineering the forecast with complex equations that few stakeholders understand and an overly simplistic approach that relies too heavily on anecdotal information and opinion (Cook 2006). In this chapter we presented an easy-to-understand but comprehensive alternative based on a regression forecasting methodology for forecasting comparable units, which included surrogate measures of the unique weighting of the variables and of performance. These surrogate measures were generated by applying DEA and MODEA. The results of applying this regression forecasting methodology included adding financial and operational efficiency scores for a data set of fifty pharmaceutical companies that improved the models.

Additionally, it demonstrated that this may provide a promising, rich approach to forecasting comparable units, particularly for the supply chain of the pharmaceutical development process. Here, DEA and MODEA can introduce greater specificity for assessing forecasts. Of numerous efforts that are currently being employed within the pharmaceutical industry, such as monitoring monthly intermittent demand forecasts for a major international pharmaceutical company using a commercially available statistical forecasting system, DEA and MODEA have the potential to improve both operating and financial efficiency (Syntetos et al. 2009).

Going forward, we plan to perform further testing with other data sets, in particular with data over time. While there are potential applications within other industries, the pharmaceutical industry remains a fertile environment because the costs of drug development appear to be increasing while the potential market for new medicines seems to be shrinking. Thus, techniques with the ability to adjust to development phases and improve efficiency of critical decisions, such as Phase I Go/No Go decisions, would be a welcome addition to the industry's supply chain management.

## References

Abbott, T.A., and Vernon, J.A. (2007), The cost of US pharmaceutical price regulation: a financial simulation model of R&D decisions, *Managerial and Decision Economics*, 28(4/5), 293–306.

Boschi, R.A., Balthasar, H.U., and Menke, M.M. (1979), Quantifying and forecasting exploratory research success, *Research Management*, 225), 14.

Charnes A., Cooper, W.W., and Rhodes, E. (1978), Measuring efficiency of decision making units, *European Journal of Operational Research*, 2, 429–444.

Choo, L. (2000), Forecasting practices in the pharmaceutical industry in Singapore, *The Journal of Business Forecasting Methods & Systems*, 19(2), 18.

Comanor, W.S., and Schweitzer, S.O. (2007), Determinants of drug prices and expenditures, *Managerial and Decision Economics*, 28, 357–370.

Connor, P., Alldus, C., Ciapparelli, C., and Kirby, L. (2003), Long-term pharmaceutical forecasting: IMS Health's experience, *Journal of Business Forecasting Methods & Systems*, 22(1), 10.

Cook, A.G. (2006), *Forecasting for the Pharmaceutical Industry: Models for New Product and In-Market Forecasting and How to Use Them*, Farnham Surrey, U.K.: Gower Publishing.

DiMasi, J.A., Hansen, R.W., and Grabowski, H.G. (2003), The price of innovation: new estimates of drug development costs, *Journal of Health Economics*, 22, 151–185.

DiMasi, J.A., and Paquette, C. (2004), The economics of follow-on drug research and development: trends in entry rates and the timing of development, *Pharmacoeconomics*, 22(Suppl. 2), 1–14.

Emrouznedjad, A., Parker, B.R., and Tavares, G. (2008), Evaluation of research in efficiency and productivity: a survey and analysis of the first 30 years of scholarly literature in DEA, *Socio-Economic Planning Sciences*, 44, 151–157.

Evens, R.P. and Covinsky, J. (2009), R&D planning and goverance, in Evens, R.R. (Ed.), *Drug and Biological Development from Molecule to Product and Beyond*, pp. 31–64, New York, NY: Springer Science + Business Media, LLC.

Gattoufi, S.O.M., and Reisman, A. (2004), A taxonomy for data envelopment analysis, *Socio-Economic Planning Sciences*, 38(2–3), 141–158.

Grabowski, H.G., and Vernon, J.M. (2000a), The determinants of pharmaceutical research and development expenditures, *Journal of Evolutionary Economics*, 10, 201–215.

Grabowski, H.G., and Vernon, J.M. (2000b), The distribution of sales revenues from pharmaceutical innovation, *Pharmacoeconomics*, 18(Suppl. 1), 21–32.

Jain, C.L. (2002), Forecasting process at Wyeth Ayerst Global Pharmaceuticals, *The Journal of Business Forecasting Methods & Systems*, 20(4), 3–6.

Jain, C.L., and Malehorn, J. (2006), *Benchmarking Forecasting Practices: A Guide to Improving Forecasting Performance*, Great Neck, NY: Graceway Publishing.

Juran, J.M. (2004), *Architect of Quality: The Autobiography of Dr. Joseph M. Juran*, New York, NY: McGraw-Hill.

Klimberg, R.K. (1998), Model-based health decision support systems: data envelopment analysis (DEA) models for health systems performance evaluation and benchmarking, in Tan, J. (Ed.), *Health Decision Support Systems*, Gaithersburg, MD: Aspen Publications.

Klimberg, R.K., and Kern, D. (1992), Understanding data envelopment analysis (DEA), Boston University School of Management Working Paper, pp. 92–44.

Klimberg, R.K., Lawrence, S.M., and Lawrence, K.D. (2004), Forecasting sales of comparable units with data envelopment analysis (DEA), *31st Conference of the Northeast Business & Economics Association*, New York, September.

Klimberg, R.K., Lawrence, S.M., and Lawrence, K.D. (2005, March/April), An application of multiple objective data envelopment analysis to forecasting sales, *NEDSI Meeting*, Philadelphia, PA.

Klimberg, R.K., and Puddicombe, M. (1999), A multiple objective approach to data envelopment analysis, *Advances in Mathematical Programming and Financial Planning*, 5, 201–232, JAI Press.

Klimberg, R.K., Van Bennekom, F.C., and Lawrence, K.D. (2001), Beyond the balanced scorecard, *Advances in Mathematical Programming and Financial Planning*, 6, 19–33, Elsevier Science.

Latta, M. (1998a), Manufacturing resource planning for ethical pharmaceuticals using market models, *The Journal of Business Forecasting Methods & Systems*, Fall, 17(3), 12–17.

Latta, M. (1998b), Using market models to forecast demand for ethical pharmaceuticals, *The Journal of Business Forecasting Methods & Systems*, 17(1), 3–4 and 6–8.

Latta, M. (2007), How to forecast the demand of a new drug in the pharmaceutical industry, *Journal of Business Forecasting*, 26(3), 21–28.

Nelson, R., and Winter, S.G. (1982), *An Evolutionary Theory of Economic Change*, Cambridge, MA: Harvard University Press.

Noori, H., and Radford, R. (1995). *Production and Operations Management: Total Quality and Responsiveness*, New York, NY: McGraw-Hill.

Pack, J. (2003), A Simulated Pre-Launch Market Evaluation Model for New Pharmaceutical Products, Proquest Dissertations and Theses, Section 0054, Part 0338, 143 pages; Ph.D. dissertation, New York: Columbia University, Publication Number AAT 3088398.

Scherer, F.M. (2001), The link between gross profitability and pharmaceutical R&D spending, *Health Affairs*, 20, 216–220.

Seiford, L.M. (1996), Data envelopment analysis: the evaluation of the state of the art (1978–1995), *The Journal of Productivity Analysis*, 9, 99–137, 1996.
Seiford, L.M., and Thrall, R.M. (1990), Recent developments in DEA: the mathematical programming approach to frontier analysis, *Journal of Econometrics*, 46, 7–38.
Sillup, G.P. (1992), Forecasting the adoption of new medical technology using the Bass model, *Journal of Health Care Marketing*, 12(4), 42–51.
Smith, L.J. (2009), Types of clinical studies, in Evens, R.P. (Ed.), *Drug and Biological Development from Molecule to Product and Beyond*, pp. 107–121, New York, NY: Springer Science + Business Media, LLC.
Syntetos, A.A., Konstantinos, N., Boylan, J.E., Fildes, R., and Goodwin, P. (2009), The effects of integrating management judgment into intermittent demand forecasts, *International Journal of Production Economics*, 118(1), 72–81.
Tavares, G. (2002), A bibliography of data envelopment analysis (1978–2001), RUTCOR, Rutgers University; also available at http: //rutcor.rutgers.edu/pub/rrr/reports2002.1_2002.pdf
Teunter, R.H., and Flapper, S.D. (2006), A comparison of bottling alternatives in the pharmaceutical industry, *Journal of Operations Management*, 24(3), 215–234.
Triantis, J.E., and Song, H. (2007), Pharmaceutical forecasting model simulation guidelines, *Journal of Business Forecasting*, 26(2), 31–37.
Vernon, J.A. (2003), Simulating the impact of price regulation on pharmaceutical innovation, *Pharmaceutical Development and Regulation*, 1(1), 55–56.
Vernon, J.A. (2005), Examining the link between price regulation and pharmaceutical R&D investment, *Health Economics*, 14(1), 1–17.

# Index

## L



## M

## N

## O

## T



## U



## V



## W